Nuclear Physics with Effective Field Theory II

PROCEEDINGS FROM THE INSTITUTE FOR NUCLEAR THEORY

Series Editors: Wick C. Haxton *(Univ. of Washington)*
Ernest M. Henley *(Univ. of Washington)*

Published

Vol. 1: Nucleon Resonances and Nucleon Structure
ed. G. A. Miller

Vol. 2: Solar Modeling
eds. A. B. Balantekin and J. N. Bahcall

Vol. 3: Phenomenology and Lattice QCD
eds. G. Kilcup and S. Sharpe

Vol. 4: N* Physics
eds. T.-S. H. Lee and W. Roberts

Vol. 5: Tunneling in Complex Systems
ed. S. Tomsovic

Vol. 6: Nuclear Physics with Effective Field Theory
eds. M. J. Savage, R. Seki and U. van Kolck

Vol. 7: Quarkonium Production in High-Energy Nuclear Collisions
eds. B. Jacak and X.-N. Wang

Vol. 8: Quark Confinement and the Hadron Spectrum
eds. A. Radyushkin and C. Carlson

Vol. 9: Nuclear Physics with Effective Field Theory II
eds. P. F. Bedaque, M. J. Savage, R. Seki and U. van Kolck

Vol. 10: Exclusive and Semi-Exclusive Processes at High Momentum Transfer
eds. C. Carlson and A. Radyushkin

Vol. 11: Chiral Dynamics: Theory and Experiment III
eds. A. M. Bernstein, J. L. Goity and U.-G. Meißner

Vol. 12: The Phenomenology of Large N_C QCD
ed. R. F. Lebed

Vol. 13: The r-Process: The Astrophysical Origin of the Heavy Elements and Related Rare Isotope Accelerator Physics
eds. Y.-Z. Qian, E. Rehm, H. Schatz and F.-K. Thielemann

Vol. 14: Open Issues in Core Collapse Supernova Theory
eds. A. Mezzacappa and G. M. Fuller

Vol. 15: Opportunities with Exotic Beams
eds. T. Duguet, H. Esbensen, K. M. Nollett and C. D. Roberts

Proceedings from the Institute for Nuclear Theory — Vol. 9

An INT Workshop

Nuclear Physics with Effective Field Theory II

University of Washington, 25–26 February 1999

Editors

Paulo F. Bedaque
University of Washington

Martin J. Savage
University of Washington

Ryoichi Seki
California State University and
California Institute of Technology

Ubirajara van Kolck
California Institute of Technology

World Scientific
Singapore • New Jersey • London • Hong Kong

Published by

World Scientific Publishing Co. Pte. Ltd.
5 Toh Tuck Link, Singapore 596224
USA office: 27 Warren Street, Suite 401-402, Hackensack, NJ 07601
UK office: 57 Shelton Street, Covent Garden, London WC2H 9HE

British Library Cataloguing-in-Publication Data
A catalogue record for this book is available from the British Library.

Cover description: "Red Square" on the University of Washington campus near sunset. At the center is the "Broken Obelisk," a steel sculpture that was a gift to the University of Washington from the Virginia Wright Fund. It is one of the few sculptures by Barnett Newman, who was known primarily for his abstract color field paintings. The work is 26 feet high and weighs two tons. Red Square is paved with red bricks and sits atop the Central Plaza parking garage.

Proceedings from the Institute for Nuclear Theory — Vol. 9
NUCLEAR PHYSICS WITH EFFECTIVE FIELD THEORY II
Proceedings of the Workshop

ISBN-13 978-981-02-4181-0
ISBN-10 981-02-4181-X

This book is dedicated to the memory of

Andrea V. Seki

whose tenacious spirit inspired all who knew her.

SERIES PREFACE

The National Institute for Nuclear Theory Series

The National Institute for Nuclear Theory (INT) was established by the US Department of Energy in March, 1990. The goals of the INT include:

1. Creating a productive research environment where visiting scientists can focus their energies and exchange ideas on key issues facing the field of nuclear physics, including those crucial to the success of existing and future experimental facilities;
2. Encouraging interdisciplinary research at the intersections of nuclear physics with related subfields, including particle physics, astrophysics, atomic physics, and condensed matter;
3. Furthering the development and advancement of physicists with recent Ph.D.s;
4. Contributing to scientific education through graduate student research, INT summer schools, undergraduate summer research programs, and graduate student participation in INT workshops and programs;
5. Strengthening international cooperation in physics research through exchanges and other interactions.

While the INT strives to achieve these goals in a variety of ways, its most important efforts are the three-month programs, workshops, and schools it sponsors. These typically attract 300 visitors to the INT each year. In order to make selected INT workshops and summer schools available to a wider audience, the INT and World Scientific established the series of books to which this volume belongs.

In February, 1999, the INT sponsored a workshop on Nuclear Physics with Effective Field Theory in order to review recent progress in the fast-moving area. This was the second such workshop, the first having been held one year earlier. Organized by Paulo Bedaque, Martin Savage, Ryoichi Seki, and Bira van Kolck, the workshop was focused on recent progress and remaining challenges in two-nucleon and three-nucleon systems, as well as the prospects for extending calculations to many-body systems. This ninth volume of the INT series summarizes the talks and discussions of this workshop.

Earlier series volumes include the proceedings of the 1991 and 1993 Uehling summer schools on Nucleon Resonances and Nucleon Structure and on Phenomenology and Lattice QCD; the 1994 INT workshop on Solar Modeling; the 1997 Jefferson Lab/INT workshop on Nucleon Resonance Physics; the tutorials of the spring 1997 INT program on Tunneling in Complex Systems; the 1998 Caltech/INT workshop on Nuclear Physics with Effective Field Theory; the proceedings of the 1998 RHIC Winter Workshop on Quarkonium Production in Relativistic Nuclear Collisions; and the proceedings of Confinement III. We will continue to publish those proceedings of INT workshops and schools which we judge to be of broad interest to the physics community.

Wick C. Haxton and Ernest Henley
Seattle, Washington, December 1998

Preface

This is the second workshop on *Nuclear Physics with Effective Field Theory*. The first workshop, held at Caltech in February 1998, was instrumental in establishing the viability of EFT methods in nuclear physics. It inspired a significant amount of work and understanding in the following twelve months. Progress was made mostly in the study of the role of pions in the two-nucleon system, and in the renormalization of three-body systems, opening the road to a description of many-nucleon systems. Several questions remained to be answered, and in fact still remain; most importantly, perhaps, is the question of convergence. This second meeting focused on the recent progress and identifying the main issues that should be addressed in the near future.

Even with the substantial increase in the number of physicists attending this workshop, the openess, intensity and focus of discussions were impressive. It was particularly encouraging to see that more than half of the participants were postdocs or students, which attests to the vitality of this field.

The workshop was held at the National Institute for Nuclear Theory on February 25 and 26, 1999. We thank Wick Haxton, Director of the National Institute for Nuclear Theory, for providing the funding, and Stephen Ellis, chairman of the University of Washington Physics Department, for his hospitality and for his entertaining remarks that opened the workshop. Maria Francom, Shizue Shikuma, Nancy Tate and Linda Vilett provided invaluable organizational help and took care of the behind-the-scenes work involved in hosting such a meeting. Without their assistance, the workshop would not have materialized. We are also indebted to Maria Francom for the help with these proceedings.

Paulo F. Bedaque, Martin J. Savage, Ryoichi Seki, and Ubirajara van Kolck

Table of Contents

Session 4: *F. Ravndal*, Chair; *J. McGovern*, Assistant

Session 5: *M. Birse*, Chair; *M. N. Butler*, Assistant

Session 6: *R. Seki*, Chair; *H. Grießhammer*, Assistant

OVERVIEW

U. van KOLCK
W. K. Kellogg Radiation Laboratory,
California Institute of Technology,
Pasadena CA 91125, USA
E-mail: vankolck@krl.caltech.edu

I briefly review the main developments since the first workshop on the nuclear applications of Effective Field Theories, held in 1998. A summary of some of the goals P. Bedaque, M. Savage, R. Seki and I had in mind for this second workshop is also presented.

1 Pre-history

The use of Effective Field Theories (EFTs) in nuclear physics has grown out of attempts to link the successful phenomenology of non-relativistic potentials to the underlying theory of strong interactions, QCD. To a large extent, the reason why such a phenomenological description looks so different from the field theory of quark and gluons arises from the fact that QCD has intrinsic mass scales. Different aspects of the QCD dynamics play different roles depending on how the typical momentum Q of a given process compares with these mass scales. EFT is a useful technique because it was designed to explore the existence of scales in order to separate short- and long-range physics.

QCD has an intrinsic scale $M_{QCD} \sim 1$ GeV, associated to the masses of most hadrons (m_ρ, m_N, ...) and to the scale of chiral symmetry breaking ($4\pi f_\pi$, where $f_\pi \sim 92$ MeV is the pion decay constant). While perturbative QCD applies to $Q \gg M_{QCD}$, much of traditional nuclear physics concerns processes involving momenta at a lower scale, $M_{nuc} \sim 100$ MeV, corresponding to pion physics (m_π, f_π) and the structure of light nuclei (inverse root-mean-square radii $\langle r^2\rangle^{-1/2}$, ...). Because nucleons at this momentum scale are non-relativistic, a third scale $Q^2/M_{QCD} \sim 10$ MeV enters via energies (kinetic energies M_{nuc}^2/m_N, binding energies per nucleon B/A, ...).

EFTs are built in terms of the degrees of freedom that are relevant at low energies, so that scattering amplitudes have the correct analytic structure in a region around $Q = 0$. In the case of nuclear physics, the relevant degrees of freedom are nucleons, pions, and possibly delta isobars. An important restriction on allowed interactions comes from symmetries that are inferred from experiment. Although the QCD dynamics is complicated at low energies, we know that it has an approximate, chiral $SU(2) \times SU(2)$ symmetry. The hadronic spectrum also suggests that the vacuum breaks this symmetry down to $SU(2)$ of isospin. The EFT of QCD at $Q \sim M_{nuc}$ is the collection of all possible interactions among the relevant degrees of freedom, with the given symmetries. This is still an infinite collection of interactions that can be combined in an infinite number of ways without violating unitarity, causality, and cluster decomposition. The EFT is predictive only if a power counting can be established which allows an ordering of possible contributions. The minimal assumption about the strength of interactions is that, after M_{nuc} has been accounted for explicitly, all parameters are given in terms of M_{QCD} according to their mass dimension. This is in accord with phenomenological models [1], where there is no evidence that hadronic interactions have small couplings. It implies that the only possible expansion parameter is Q/M_{QCD}.

This EFT already has a long history of applications to mesonic and one-baryon processes, going by the name of Chiral Perturbation Theory (ChPT). In this context it is easy to develop a power counting in Q/M_{QCD} based on the fact that, for processes where all external momenta are $O(Q)$, all energies in intermediate states are also $O(Q)$. This power counting shows that scattering amplitudes, except at special kinematic points such as the delta pole, are perturbative. Ref. [2] offers a good review of this sector of the EFT.

The situation is a lot more complicated and thus interesting when two or more baryons are present. Weinberg [3] has identified the origin of this complication: because baryons have masses $O(M_{QCD})$, intermediate states can arise in which particles differ in energy from the initial state only by an amount of $O(Q^2/M_{QCD})$. This $O(M_{QCD}/Q)$ infrared enhancement can cause a breakdown of perturbation theory for amplitudes involving more than one nucleon, leading to the formation of bound states (nuclei).

Once the source of enhancement is understood, the challenge is to isolate parts of multi-nucleon amplitudes that still obey some sort of expansion. Weinberg [3] proposed that the standard ChPT power counting should continue to hold for "irreducible" diagrams, which are those that cannot be made disconnected by cutting only lines with large propagators. In the case of a system of A nucleons, irreducible diagrams are just A-nucleon irreducible. Weinberg defined the nuclear potential as the sum of A-nucleon-irreducible diagrams. When soft external probes —photons and/or pions with momenta of $O(Q)$— are present, irreducible diagrams comprise those where the external particles are attached to. Their sum is referred to as the kernel. The full amplitude is then constructed by sewing the ChPT expansions of irreducible sums together with small-energy intermediate states. Whatever the power counting for the full amplitude is, this procedure should correctly account for its first few orders, even though it might include some less important, higher-order contributions as well.

This approach has been applied to several processes [3,4,5,6], and it has met with considerable success. For example, the nuclear potential was constructed to $O((Q/M_{QCD})^3)$ using time-ordered perturbation theory in Ref. [4]. (The equivalent covariant-perturbation-theory version of this potential was considered in Ref. [7].) Some of the diagrams for the two-nucleon (NN) potential are depicted in Fig. 1: *(i)* in leading order there is static one-pion exchange (OPE) and two momentum-independent contact terms; *(ii)* due to parity and time-reversal invariance, there are no first-order corrections; *(iii)* second-order corrections include energy dependence and dressing in OPE, static two-pion exchange (TPE) without and with deltas, and several momentum-dependent contact interactions; *(iv)* third-order corrections consists of further TPE without and with a delta; and *(v)* higher-order terms —not included— contain several two-loop diagrams and a host of new contact terms. The spin-isospin structure of the potential exhausts that of previous phenomenological parametrizations, and the fixed radial profile incorporates novel two-pion-exchange terms. One-loop diagrams in the potential require regularization, and the infinite number of loops generated by iteration of the potential require further, non-perturbative regularization. For numerical convenience, an overall smooth momentum cut-off $\exp(-Q^2/\Lambda^2)$ was used. (A regulator includes some arbitrarily-high orders in Q/Λ, but as long as $\Lambda \sim M_{QCD}$ they are of the same size as terms in Q/M_{QCD} already neglected.) The Schrödinger equation was solved, and the phase shifts and deuteron properties calculated in terms of the parameters in the Lagrangian. Fits to values from the Nijmegen phase-shift analysis (PSA) [8] were performed for cut-off values $\Lambda = 500, 780, 1000$ MeV. Good agreement and little cut-off dependence were found up to laboratory energies of about 100 MeV. The agreement is best in the higher-l partial waves; these are perturbative up to relatively high energies due to inverse factors of the angular momentum l, and, apart from pion exchange, depend only on contact terms with many derivatives that do not appear at low orders. The lower-l partial waves, on the other hand, are non-perturbative already at low momenta, and are sensitive to the leading contact terms. A sample of the results [5] for lower partial-wave phase shifts can be found in Fig. 2 and for deuteron properties in Table 1. (Note that the theoretical electromagnetic moments in this table refer only to the one-nucleon component of the current; a full calculation including two-nucleon currents that appear to the same order has not been attempted.) A similar analysis stressing the relative importance of different orders (but neglecting TPE) can be found in Refs. [9,10]. Aspects of TPE, delta effects, and higher partial waves have been elaborated upon in Refs. [7,11]. More recently, the long-range chiral TPE has been incorporated in a Nijmegen phase-shift re-analysis of pp data [12], where it successfully substitutes for heavy-meson

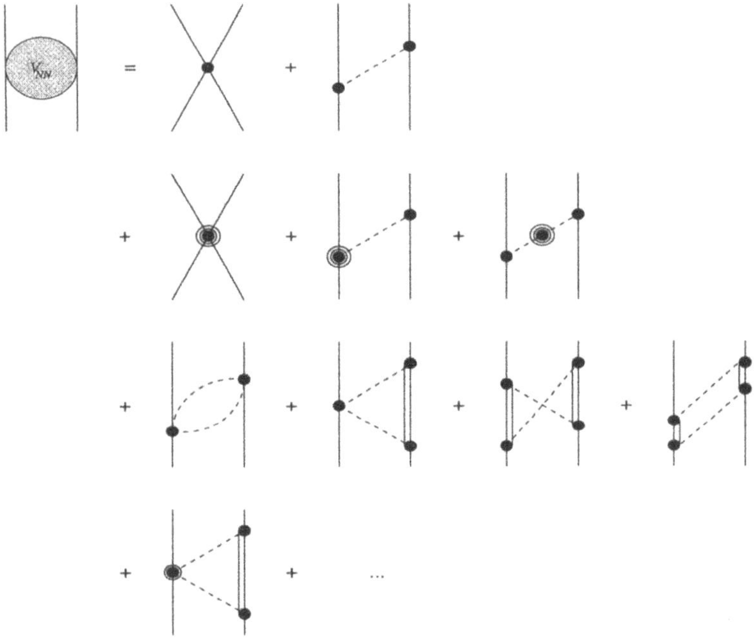

Figure 1: Some time-ordered diagrams contributing to the two-nucleon potential V_{NN} in Chiral Perturbation Theory. (Double) solid lines represent nucleons (and/or deltas), dashed lines pions, a heavy dot an interaction in the lowest-order Lagrangian, a dot within a circle an interaction in the first-order Lagrangian, and a dot within two circles an interaction in the second-order Lagrangian. Diagrams on the right-hand side are ordered according to the power ν of Q, first line corresponding to $\nu = \nu_{min}$, second and third lines to $\nu = \nu_{min} + 2$, fourth line to $\nu = \nu_{min} + 3$, and "..." denoting $\nu \geq \nu_{min} + 4$. All orderings with at least one pion or delta in intermediate states are implied. Not shown are diagrams contributing only to renormalization of parameters.

exchange.

Note that this approach also explains a long-standing puzzle of the strong interactions: why nuclear binding energies are so much smaller than M_{QCD}. Neglecting dimensionless factors, pion exchange sets the scale of the leading-order interactions at $\sim 1/f_\pi^2$, while intermediate states with only nucleons contribute $\sim m_N Q/4\pi$. Schematically, the nuclear amplitude is thus

$$T = \frac{1}{f_\pi^2}\left(1 + \frac{m_N Q}{4\pi f_\pi^2} + \ldots\right). \tag{1}$$

Since $m_N \sim 4\pi f_\pi$, the amplitude can develop a pole at a momentum $Q \sim f_\pi$, corresponding to a binding energy $B \sim f_\pi^2/m_N = O(M_{nuc}^2/M_{QCD}) \ll$

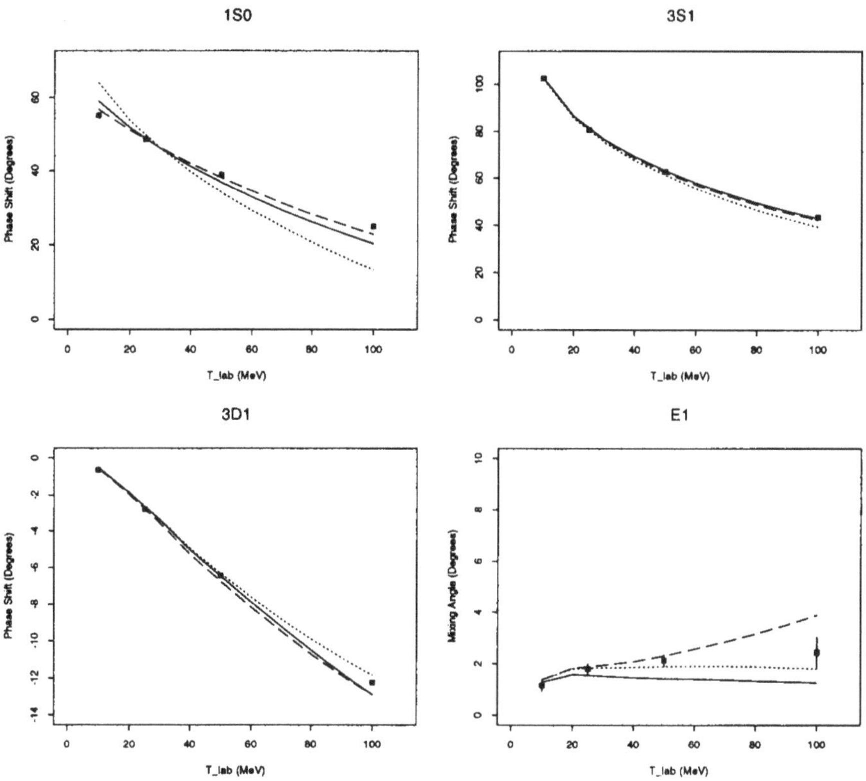

Figure 2: $^1S_0$, $^3S_1$, and $^3D_1$ NN phase shifts and ε_1 mixing angle in degrees as functions of the laboratory energy in MeV: chiral expansion up to $\nu = 3$ for cut-offs of 500 (dotted), 780 (dashed), and 1000 MeV (solid line); and Nijmegen PSA (squares).

M_{QCD}. At this level, scattering at low energies can be approximated by an effective range expansion with scattering length $a \sim 1/M_{nuc}$, effective range $r_e \sim 1/M_{nuc}$, *etc.*

A review of nuclear applications of this EFT can be found in Ref. [13], and in some of the contributions to these proceedings.

2 The first workshop

Despite the successes, a number of issues have been raised in the past few years. On one hand, there have been concerns regarding the role of regularization and renormalization in this non-perturbative context. For example,

Table 1: Effective chiral Lagrangian fits at $\nu = 3$ for various cut-offs Λ and experimental values for the deuteron binding energy (B), magnetic dipole moment (μ_d), electric quadrupole moment (Q_E), asymptotic D/S ratio (η), and d-state probability (P_D).

Deuteron quantities	Λ (MeV) 500	780	1000	Experiment
B (MeV)	2.15	2.24	2.18	2.224579(9)
μ_d (μ_N)	0.863	0.863	0.866	0.857406(1)
Q_E (fm^2)	0.246	0.249	0.237	0.2859(3)
η	0.0229	0.0244	0.0230	0.0271(4)
P_D (%)	2.98	2.86	2.40	

there have been claims that dimensional regularization does not work and that no systematic cut-off EFT can be constructed for nuclear systems [14]. On the other hand, the role of fine-tuning was discussed. As it turns out, the scattering lengths in the two S-wave NN channels are substantially larger than $1/M_{nuc}$, reflecting typical momenta in the deuteron and in the singlet virtual state somewhat smaller than f_π. This can only be achieved by some fine-tuning of short-range interactions. If we write the coefficients of leading, non-derivative contact interactions as $C_0 = 4\pi/m_N\aleph$, the new scale $\aleph$ is on average about 30 MeV instead of f_π. It was pointed out that Weinberg's power counting does not incorporate this fine-tuning, and claimed that it is not consistent when scattering lengths are large [15]. Suggestions were made that fine-tuning could be incorporated through a dibaryon field [16] or through a new power counting where pions are treated perturbatively [17].

A good test-ground for the discussion of these issues is afforded by the very-low-energy region of $Q \sim \aleph$, where only the lowest partial waves are significant. At such low momenta the pion can be integrated out, its contributions being accounted for via contact interactions. This "pionless" theory is immensely simpler than the "pionful" theory and can be treated mostly by analytical means. Since the infinite sum of reducible NN diagrams reduces to a geometric series, all that is required is the evaluation of a single loop, and renormalization can be carried out easily with several regularization schemes. It is conceivable that for some range of momenta effects of the addition of an explicit pion are perturbative, because of the pion's derivative interactions, if the pion mass is small enough. Conversely, above some momentum scale M_{NN} pion effects will pile up and a perturbative treatment will become insufficient.

The first workshop [18], sponsored by the Kellogg Radiation Lab and the Institute for Nuclear Theory, and held at Caltech in February of 1998, was organized in order to put to rest many of the doubts about the effectiveness of EFTs in nuclear physics. In the overview of that workshop [19], I listed some of the issues we wanted to tackle:

(i) Are all regularizations equivalent? Is there a problem with dimensional regularization plus minimal subtraction? Can a cut-off be removed at the end of the calculation?

(ii) Is there fine-tuning in the NN S waves? If so, how is it incorporated in an EFT?

(iii) Is there a consistent power counting for the full NN amplitude? If so, what is the expansion parameter? What is the radius of convergence?

(iv) How does one account for corrections to the leading order: in the potential or otherwise? Is there a finite number of counterterms to a given order?

(v) Is a dibaryon field necessary and/or useful?

(vi) Can we make predictions for other processes? How important are many-body effects?

(vii) What is the exact power counting in the region where pions are perturbative? At which momenta do pion effects become non-perturbative?

(viii) Does an EFT with non-perturbative pions make sense? Is the corresponding power counting Weinberg's?

(ix) Can we acquire insight into the power counting by studying other non-relativistic theories with "long-range" interactions that support bound states, such as non-relativistic QCD?

All of the questions concerning the pionless theory have been answered. It has been shown [20,17,21] that consistent power countings exist for the full amplitude, but their forms depend on lack or existence of fine-tuning. In the NN case fine-tuning is essential for the appearance of non-perturbative effects at $Q \sim \aleph$, where an expansion in Q/M_{nuc} can be made. Counterterms scale with inverse powers of $\aleph$. The leading order consists of all iterations of non-derivative contact interactions, and subleading orders are built from insertions of contact interactions with more derivatives. To any given order the number of counterterms is finite, all regularizations are equivalent, and the amplitude is equivalent to the effective range expansion. Some higher-order terms can be resummed through the iteration of a potential within regularization schemes that contain a mass scale $\Lambda \sim M_{nuc}$. The (bare) potential in this case is just the sum of contact interactions. A dibaryon field also resums certain higher-order terms, which is useful under certain circumstances. Predictions were made for processes other than NN scattering, an example being the quartet channel in nucleon-deuteron (Nd) scattering [22]. The issue of the importance of many-body forces was hinted at. Answers to the questions regarding the pionful theory were not as fully developed. A power counting for perturbative pions was established [17], but the radius of convergence remained obscure. A proof that one could consistently go beyond this range by using non-perturbative pions remained elusive.

The positive outcome of the first workshop is illustrated by a quote from the summary [23]:

> Many of the results discussed at the conference are very recent (there was a flurry of papers on nucl-th in the week prior to the meeting), but it appears that it should now be possible to apply the effective Lagrangian technique, which has been so successful in the pion and single-nucleon sectors, also to the problem of NN bound states. There still appear to be some very interesting unresolved questions in the three-body problem that were discussed by Bedaque.

3 Recent developments

Most of the work carried out in the year since has explored the roads open by the Caltech meeting: the scope of the perturbative-pion theory in the two-nucleon system, and the role of three-nucleon interactions in the three-nucleon system.

In the two-nucleon sector, most of the effort has been placed in extending the pionless EFT to higher energies. For $Q < m_\pi$, explicit inclusion of the pion is not necessary but not forbidden either. Since $m_\pi \sim M_{nuc}$, effects of the finite range of the interaction cannot destroy the power counting, and keeping them explicitly only amounts to including some small, higher-order terms. They will show up as specific contributions to effective range parameters. As we increase Q, these range effects ought to become more important relative to shorter-range contributions. For $Q \gtrsim m_\pi$ a full dynamical pion is required.

A crucial problem is how to calculate pion-range effects. Because of approximate chiral symmetry, pion interactions bring factors of Q or m_π. Factors of Q provide suppression at small energies, and in the chiral limit ($m_\pi \rightarrow 0$) it might happen that pions are amenable to perturbation theory up to a scale $M_{NN} > M_{nuc}$. In fact, a naive dimensional estimate, $M_{NN} \sim 4\pi M_{nuc}^2/M_{QCD} \sim M_{nuc}$, is implied in Weinberg's power counting, where each additional pion exchange brings a $Q^2/M_{nuc}^2(Q^2 + M_{nuc}^2)$ from the vertices and pion propagator, and a $M_{QCD}Q/4\pi$ from the two-nucleon propagator and integral. When actual values are inserted, one finds [17] $M_{NN} \sim 300$ MeV. Likewise, OPE should be small compared to the leading contact interaction $C_0 \sim 4\pi/M_{QCD}\aleph$ as long as $\aleph < M_{NN}$. Assuming that indeed $\aleph < M_{NN}$ and $m_\pi < M_{NN}$, a simple power counting follows from taking $Q \sim \aleph \sim m_\pi < M_{NN}$ and counting powers of the common light scale Q. This direct extension of the pionless power counting was proposed by Kaplan, Savage, and Wise [17] (see also Ref. [24]).

A number of recent papers have tried to improve the estimate of M_{NN}. The most straightforward way to do this is to calculate the NN amplitude in the first few orders in the Q/M_{NN} expansion, and to explicitly demonstrate convergence. Ref. [17] has analyzed NN scattering to subleading order. In leading order only the contact interaction in the first line of Fig. 1 appears, and it is iterated to all orders; in subleading order there are the contact interaction in the second line of Fig. 1 and static (or "potential") OPE. The results of a fit [17] with three-parameters per isospin channel to the Nijmegen PSA [8] is shown in Fig. 3. These results are very encouraging. However, their interpretation is complicated by the existence of fitting parameters. There have been claims that they indicate only that $M_{NN} \sim m_\pi$ [25,10], but these claims have been attributed to deficient fitting procedures [26]. Ref. [27] examined predictions for the effective range parameters and concluded that they are poor. In any case, a more definitive argument depends on a sub-subleading calculation where iterated OPE first appears. (It is quite possible that at M_{nuc} OPE is numerically less important than contact interactions, but that at the same time iterated OPE is as or more important than OPE.)

A more indirect test of the perturbative-pion idea is the consistent calculation of a number of other processes involving two nucleons and external probes. This has been pursued in several papers [28]. Results are again good, but it is puzzling that they do not seem to be any better than those of the pionless theory, or even model-dependent extensions of the effective range expansion. This might be indication that the range of the perturbative-pion theory is no larger than m_π. However, the pionless theory seems to work farther in momentum than it was supposed to. Perhaps the perturbative-pion theory will provide an explanation for this unexpected fact, if one can understand the apparently accidental agreement between the two theories.

The new power counting shifts pion effects to higher orders if compared to the original counting. Iterating pions to all orders at $Q < M_{NN}$ (as done in Weinberg's program) only sums up a bunch of high-order, presumably less important terms. In this region the perturbative-pion theory is a convenient rearrangement of the non-perturbative-pion theory. As Q approaches M_{NN},

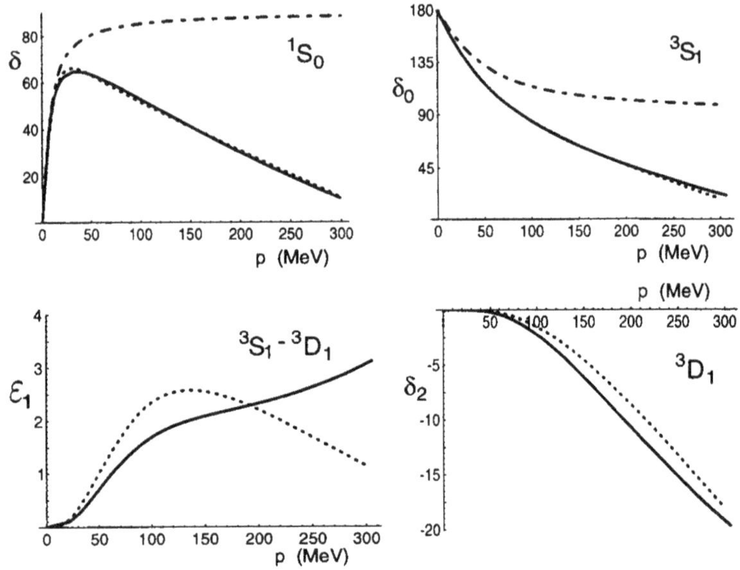

Figure 3: $^1S_0$, $^3S_1$, and $^3D_1$ NN phase shifts and ε_1 mixing angle in degrees as functions of center-of-mass momentum in MeV: perturbative-pion EFT in leading (dot-dashed) and subleading (dashed line) orders; and Nijmegen PSA (solid line). Figure courtesy of M. Savage.

multiple-pion exchange becomes comparable to OPE and/or some pion exchange becomes comparable to contact interactions. Pions then have to be iterated to all orders but it has not yet been proved that this can be done in a model-independent way —or, in other words, that Weinberg's power counting is consistent in this regime. Again looking at the NN system, perturbative arguments exist against consistency [17], but non-perturbative albeit numerical evidence exists in favor [5,9,10]. Successes in reactions with external probes are also encouraging, for good predictions can be made even at energies where the perturbative-pion theory should fail (near the pion production threshold).

It is in the many-nucleon sector that the EFT of perturbative pions holds most promise. We aim at an eventual description of heavy nuclei that is consistent with QCD and hadronic physics and retains as much predictive power as possible. However, computational demands grow rapidly with the number of nucleons. The problem simplifies considerably if non-perturbative aspects are limited to contact interactions. In both pionless and perturbative-pion EFTs, only non-derivative contact interactions need to be included to all orders. These EFTs differ only at subleading order, and this difference can be accounted for in perturbation theory.

The main concern is to determine the relative orders of many-nucleon forces. Since contact interactions containing six or more nucleon fields are not constrained by two-nucleon data, the number of new leading-order parameters will dictate the degree of usefulness of the EFT.

The three-nucleon system presents the first test of the EFT approach beyond the dilute deuteron. By the time of the Caltech workshop, Bedaque, Hammer, and van Kolck [22] had shown that the three-nucleon system in the spin $J = 3/2$ channel can be very successfully *predicted* in terms of NN parameters up to a high order, as the comparison displayed in Fig. 4 between EFT and both a PSA [29] and a threshold measurement [30] shows. In this channel nucleons are affected by the Pauli principle, which balances their pairwise attraction. It is not *a priori* obvious that the three-body system is as simple in the ultraviolet in the absence of the Pauli principle. This is clearly a problem of non-perturbative renormalization that goes well beyond the simple geometric series of the two-body system. We have determined the order a three-body force first appears and the number of parameters it contains [31]. Lessons should be generalizable to the more-nucleon case, where some explorations of the connections with more standard approaches have already begun [32].

Most of the developments summarized in this section are elaborated upon in the rest of these proceedings. A more complete review of the status of nuclear EFTs just before the workshop can be found in Ref. [13].

4 Aims of this workshop

This second workshop was organized with the intention of evaluating progress since the first workshop and defining problems to be dealt with next. It was planned along the same lines of the Caltech meeting: two intense days of talks and discussions among active researchers in the field. The larger number of participants reflected the growth of interest on the subject, but attendance was limited by the chosen format. A considerable fraction of the (actually, most active) participants were postdocs and graduate students.

Some of the issues the organizers hoped would be discussed in relation to the two-body sector were:

(i) What can simple potentials models teach us about convergence and fitting procedure in the perturbative-pion EFT, or about the comparison between perturbative-pion and non-perturbative-pion EFTs?

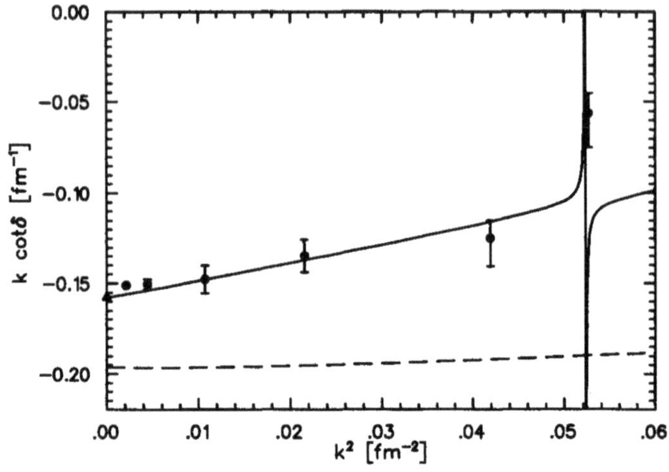

Figure 4: k cotδ in fm^{-1}, where δ is the Nd phase shift in the $J = 3/2$ channel, as function of the square of the relative momentum k in fm^{-2}: pionless EFT in leading (dashed) and sub-subleading order (solid line); a threshold measurement (triangle); and a PSA (circles).

(ii) What is the range of validity of the perturbative-pion EFT in NN scattering?

(iii) Can we obtain good phenomenology of other two-nucleon processes in the perturbative-pion EFT? How does it compare to the pionless theory?

(iv) Are processes involving external probes well described also in the non-perturbative-pion EFT?

With respect to the more-body sector:

(v) How does renormalization work in a non-perturbative context with more than two bodies? Are three-nucleon forces, for example, contaminated by the fine-tuning scale $\aleph$? At which order do they first appear?

(vi) Does the EFT retain predictive power in relevant nuclear channels, such as the triton?

(vii) Are light nuclei within the perturbative-pion EFT? For example, is the triton obtained in leading order realistic?

The first day was mostly dedicated to the perturbative-pion theory in the two-body system, that is, to points *(i)*-*(iii)*. David Kaplan and Jim Steele discussed convergence of both versions of the pionful theory in the context of models adjusted to $^1S_0$ phase shifts. Mark Wise reviewed the ideas behind the perturbative-pion theory, while Martin Savage described many of its applications, in comparison to the pionless theory. Gautam Rupak and Tom Mehen discussed the many non-trivial aspects involved in the calculation and fitting to sub-subleading order, including the role of non-static (or "radiation") pions. Tom Cohen argued against the usefulness of a perturbative treatment of pions by comparing predicted against observed effective range parameters. Ulf Meißner discussed isospin violation. The morning of the second day also bore on recent applications of the non-perturbative-pion EFT to processes involving external probes of the two-nucleon system, item *(iv)*. Silas Beane introduced the general strategy and its application to pion- and photon-deuteron elastic scattering. Bira van Kolck described coherent pion photoproduction on the deuteron and pion production in NN collisions. Mannque Rho focused on pion effects in several two-nucleon reactions. Points *(v)*-*(vi)* were considered in the last afternoon. Paulo Bedaque showed the novel, non-perturbative renormalization that takes place in three-body systems, while Hans Hammer presented results for the three-nucleon system, including scattering in the two S-wave nucleon-deuteron channels and binding of triton. Matthias Lutz discussed his attempts to describe nuclear matter, and Wick Haxton explained how the shell model can be recast as an effective theory. Peter Lepage delivered the summary talk.

Talks were 20 or 30 minutes long followed by 10 minutes of discussion. Two hour-long discussion sessions were held at the end of each day for evaluation of past and future progress. Sessions were chaired by Martin Savage, Mike Luke, Barry Holstein, Dick Furnstahl, Finn Ravndal, Mike Birse, and Ryoichi Seki; discussions were recorded by Dan Phillips, Sean Fleming, Roxanne Springer, Judith McGovern, Malcolm Butler, and Harald Grießhammer.

These proceedings document the talks and discussions, and, I hope, convey the spirit of those days.

Acknowledgments

I thank Paulo Bedaque, Martin Savage and Ryoichi Seki for a pleasurable collaboration in the organization of the meeting, Wick Haxton and the INT for the financial support of the workshop, and especially Maria Francom for the flawless local running of the event. I am grateful to Seki also for suggestions on this manuscript. My research is supported in part by the NSF grant PHY

94-20470.

References

1. J.L. Friar, *Few-Body Syst.* **22**, 161 (1997); in Ref. [18].
2. V. Bernard, N. Kaiser, and U.-G. Meißner, *Int. J. Mod. Phys.* E **4**, 193 (1995).
3. S. Weinberg, *Phys. Lett.* B **251**, 288 (1990); *Nucl. Phys.* B **363**, 3 (1991); *Phys. Lett.* B **295**, 114 (1992).
4. C. Ordóñez and U. van Kolck, *Phys. Lett.* B **291**, 459 (1992); U. van Kolck, U. of Texas Ph.D. dissertation (1993); U. van Kolck, *Phys. Rev.* C **49**, 2932 (1994); J.L. Friar, D. Hüber, and U. van Kolck, *Phys. Rev.* C , to appear (1999).
5. C. Ordóñez, L. Ray, and U. van Kolck, *Phys. Rev. Lett.* **72**, 1982 (1994); *Phys. Rev.* C **53**, 2086 (1996).
6. M. Rho, *Phys. Rev. Lett.* **66**, 1275 (1991); T.-S. Park, D.-P. Min, and M. Rho, *Phys. Rep.* **233**, 341 (1993); S.R. Beane, C.Y. Lee, and U. van Kolck, *Phys. Rev.* C **52**, 2914 (1995); R.M. Rockmore, *Phys. Lett.* B **356**, 153 (1995); T.-S. Park, D.-P. Min, and M. Rho, *Phys. Rev. Lett.* **74**, 4153 (1995); *Nucl. Phys.* A **596**, 515 (1996); S.R. Beane, V. Bernard, T.-S.H. Lee, U.-G. Meißner, and U. van Kolck, *Nucl. Phys.* A **618**, 381 (1997); P. Wilhelm, *Phys. Rev.* C **56**, 1215 (1997); S.R. Beane, V. Bernard, T.-S.H. Lee, and U.-G. Meißner, *Phys. Rev.* C **57**, 424 (1998); A. Misra and D.S. Koltun, `nucl-th/9810075`.
7. J.L. Friar and S.A. Coon, *Phys. Rev.* C **49**, 1272 (1994); J.L. Friar, `nucl-th/9901082`; N. Kaiser, R. Brockmann, and W. Weise, *Nucl. Phys.* A **625**, 758 (1997); N. Kaiser, S. Gerstendörfer, and W. Weise, *Nucl. Phys.* A **637**, 395 (1998); E. Epelbaoum, W. Glöckle, and U.-G. Meißner, *Nucl. Phys.* A **637**, 107 (1998); E. Epelbaoum, in Ref. [18].
8. V.G.J. Stoks, R.A.M. Klomp, M.C.M. Rentmeester, and J.J. de Swart, *Phys. Rev.* C **48**, 792 (1993).
9. G.P. Lepage, `nucl-th/9706029`.
10. J.V. Steele and R.J. Furnstahl, *Nucl. Phys.* A **645**, 439 (1999).
11. L.S. Celenza, A. Pantziris, and C.M. Shakin, *Phys. Rev.* C **46**, 2213 (1992); C.A. da Rocha and M.R. Robilotta, *Phys. Rev.* C **49**, 1818 (1994); *Phys. Rev.* C **52**, 531 (1995); *Nucl. Phys.* A **615**, 391 (1997); J.-L. Ballot, M.R. Robilotta, and C.A. da Rocha, *Int. J. Mod. Phys.* E **6**, 83 (1997); *Phys. Rev.* C **57**, 1574 (1998); M.J. Savage, *Phys. Rev.* C **55**, 2185 (1997).

12. M.C.M. Rentmeester, R.G.E. Timmermans, J.L. Friar, and J.J. de Swart. nucl-th/9901054.
13. U. van Kolck, *Prog. Part. Nucl. Phys.* **43**, 409 (1999).
14. T.D. Cohen, *Phys. Rev.* C **55**, 67 (1997); D.R. Phillips and T.D. Cohen, *Phys. Lett.* B **390**, 7 (1997); K.A. Scaldeferri, D.R. Phillips, C.-W. Kao, and T.D. Cohen, *Phys. Rev.* C **56**, 679 (1997); S.R. Beane, T.D. Cohen, and D.R. Phillips, *Nucl. Phys.* A **632**, 445 (1998); D.R. Phillips, in Ref. [18]; K.G. Richardson, M.C. Birse, and J.A. McGovern, hep-ph/9708435; M.C. Birse, in Ref. [18].
15. D.B. Kaplan, M.J. Savage, and M.B. Wise, *Nucl. Phys.* B **478**, 629 (1996).
16. D.B. Kaplan, *Nucl. Phys.* B **494**, 471 (1997).
17. D.B. Kaplan, M.J. Savage, and M.B. Wise, *Phys. Lett.* B **424**, 390 (1998); *Nucl. Phys.* A **534**, 329 (1998); D.B. Kaplan, in Ref. [18]; M.J. Savage, in Ref. [18].
18. *Nuclear Physics with Effective Field Theory*, ed. R. Seki, U. van Kolck, and M.J. Savage (World Scientific, 1998).
19. U. van Kolck, in Ref. [18].
20. U. van Kolck, in *Proceedings of the Workshop on Chiral Dynamics 1997, Theory and Experiment*, ed. A. Bernstein, D. Drechsel, and T. Walcher (Springer-Verlag, 1998), hep-ph/9711222; *Nucl. Phys.* A **645**, 273 (1999); in Ref. [18].
21. J. Gegelia, nucl-th/9802038; nucl-th/9805008 v2.
22. P.F. Bedaque and U. van Kolck, *Phys. Lett.* B **428**, 221 (1998); P.F. Bedaque, H.-W. Hammer, and U. van Kolck, *Phys. Rev.* C **58**, R641 (1998); P.F. Bedaque, in Ref. [18].
23. A. Manohar, in Ref. [18].
24. M. Lutz, in hep-ph/9606301.
25. J. Gegelia, nucl-th/9806028 v2.
26. T. Mehen and I.W. Stewart, *Phys. Lett.* B **445**, 378 (1999); *Phys. Rev.* C **59**, 2365 (1999).
27. T.D. Cohen and J.M. Hansen, *Phys. Lett.* B **440**, 233 (1998); T.D. Cohen and J.M. Hansen, *Phys. Rev.* C **59**, 13 (1999).
28. J.-W. Chen, H.W. Grießhammer, M.J. Savage, and R.P. Springer, *Nucl. Phys.* A **644**, 221 (1998); *Nucl. Phys.* A **644**, 245 (1998); M.J. Savage and R.P. Springer, *Nucl. Phys.* A **644**, 235 (1998); D.B. Kaplan, M.J. Savage, and M.B. Wise, *Phys. Rev.* C **59**, 617 (1999); D.B. Kaplan, M.J. Savage, R.P. Springer, and M.B. Wise, *Phys. Lett.* B **449**, 1 (1999); J.-W. Chen, nucl-th/9810021; M.J. Savage, K.A. Scaldeferri, and M.B. Wise, nucl-th/9811029; M. Binger, nucl-th/9901012.

29. W.T.H van Oers and J.D. Seagrave, *Phys. Lett.* B **24**, 562 (1967); A.C. Phillips and G. Barton, *Phys. Lett.* B **28**, 378 (1969).
30. W. Dilg, L. Koester, and W. Nistler, *Phys. Lett.* B **36**, 208 (1971).
31. P.F. Bedaque, H.-W. Hammer, and U. van Kolck, *Phys. Rev. Lett.* **82**, 463 (1999); *Nucl. Phys.* A **646**, 444 (1999).
32. G. Gelmini and B. Ritzi, *Phys. Lett.* B **357**, 431 (1995); J.L. Friar, D.G. Madland, and B.W. Lynn, *Phys. Rev.* C **53**, 3085 (1996); J.J. Rusnak and R.J. Furnstahl, *Nucl. Phys.* A **627**, 495 (1997); R.J. Furnstahl and J.C. Hackworth, *Phys. Rev.* C **56**, 2875 (1997); B.D. Serot and J.D. Walecka, *Int. J. Mod. Phys.* E **6**, 515 (1997); M. Lutz, *Nucl. Phys.* A **642**, 171c (1998); T. Papenbrock and G.F. Bertsch, *Phys. Rev.* C **59**, 2052 (1999).

SESSION 1

Session Chair: *M. Luke (University of Toronto)*
Notes taken by: *D. R. Phillips (University of Washington)*

EFFECTIVE FIELD THEORY EXPANSIONS FOR NONRELATIVISTIC SCATTERING

D. B. KAPLAN
Institute for Nuclear Theory, University of Washington, Seattle, WA 98195-1550, USA
E-mail: dbkaplan@phys.washington.edu

I analyze a toy model to give insights into the workings of an effective field theory (EFT) for nonrelativistic scattering. By comparing an exactly solvable Schrödinger problem with its EFT formulation, I show how the the contact terms of the EFT and the scattering parameters of the short range interaction are related. I then try to clarify the difference between the KSW and Weinberg expansions and discuss the validity of treating pions perturbatively.

1 Introduction

Effective field theory for nucleon interactions [1,2] with pions included perturbatively does not seem to converge very well to data, in particular above momenta $p \sim 300$ MeV. If nothing can be done to remedy the situation it will be disappointing for two reasons: first of all, I doubt that effective field theory techniques will significantly help one perform few-body bound state calculations unless one can show that only contact interactions need to be summed up nonperturbatively; second of all, the validity of the effective field theory approach should extend somewhat beyond $p \sim 300$ MeV if one is to successfully analyze nuclear matter. There have been various reasons suggested for why the theory with perturbative pions has problems: that one pion exchange cannot be included perturbatively; that a resonance in the correlated two pion channel (the "σ") cannot be ignored; that the effective range of the short range physics is too large; that unfortunate cancellations make NNLO calculations exceptionally unreliable; and so on.

In an effort to sort out these various issues, Jim Steele and I have taken a step backward from real NN interactions in order to analyze a toy model for which one can solve both the full theory and the effective theory exactly [4]. The toy model we consider is one for nonrelativistic scattering of two "nucleons" with arbitrary short distance interactions, and a long range interaction which is a delta function shell, $\propto \delta(r - r_0)$. As expected, after solving the effective theory exactly and relating it to the exact solution of the full theory, one finds

- the effective theory can reproduce scattering very well, up to momentum scales characterizing the short range interactions;

- the contact interaction coefficients of the effective theory are directly related to the scattering parameters of the short distance physics, and are independent of the long range physics.

However, the true value of the toy model is to test the chiral expansion of the effective theory, comparing the Weinberg approach [3,5] (expansion of the potential) with the KSW approach [6] (expansion of the amplitude). We find that the former is more accurate, but that the KSW expansion converges quite quickly to the exact answer, provided that one follows a certain algorithm for fixing the contact interaction coefficients. Furthermore, one can see that the KSW expansion always works well as one approaches the chiral limit.

Subsequent to giving this talk on work in progress, we applied the same algorithm to more realistic potentials, with large values for the effective range, and found problems with the KSW expansion in these cases[4], which I describe at the end of this talk write-up.

2 Short-range Interactions

I assume that interactions between two particles I will call "nucleons" can be described exactly by a potential $V = V_S + V_L$ consisting of a short- and long-range part. I simplify things even further in this section by taking the long-range potential V_L to be zero. Within the context of field theory, it is most natural to express the scattering problem in terms of the Feynman scattering amplitude. Focusing exclusively on the $^1S_0$ partial wave below, the amplitude is given by

$$\mathcal{A} \equiv \frac{4\pi}{M}\left(\frac{e^{2i\tilde{\delta}}-1}{2ip}\right) = \frac{4\pi/M}{p\cot\tilde{\delta} - ip} , \tag{1}$$

where $p = \sqrt{ME_{cm}}$ is the center of mass momentum of the two nucleons and $\tilde{\delta}$ is the phase shift for the short-range potential V_S alone.

A successful EFT should describe the scattering amplitude in terms of an expansion in powers of momentum over the characteristic short-distance scale Λ_S. It was already evident fifty years ago that such an expansion was possible with the discovery of the effective range expansion[7], which expressed the quantity $p\cot\tilde{\delta}$ as

$$p\cot\tilde{\delta} = -\frac{1}{\tilde{a}} + \frac{1}{2}\tilde{r}_0 p^2 + \sum_{n=2}^{\infty} \tilde{v}_n p^{2n} , \tag{2}$$

where $\tilde{a}$ is the scattering length, $\tilde{r}_0$ is the effective range, and the $\tilde{v}_n$ are higher terms in the Taylor expansion. All the effective range parameters are expected to scale with the appropriate power of Λ_S with the exception of $\widetilde{a}$, which can take on any value. I will focus on the case where the scattering length is abnormally large $|\widetilde{a}| \gg 1/\Lambda_S$ (as occurs in actual NN scattering), which signals the presence of a state that is weakly bound or almost bound.

2.1 The Exact Solution of the Effective Theory

The effective field theory description for this system is valid for momenta $p \lesssim \Lambda_S$ and consists solely of contact interactions in a derivative expansion between the nucleons. The fundamental NN vertex at tree-level is[a]

$$\mathcal{A}^{\rm EFT}_{\rm tree} = -\frac{4\pi}{M}\,\widehat{C}(p^2) = -\frac{4\pi}{M}\left(\widehat{C}_0 + \widehat{C}_2 p^2 + \widehat{C}_4 p^4 + \ldots\right) . \tag{3}$$

By using dimensional regularization and the PDS subtraction scheme, it is possible to compute the exact scattering amplitude for the EFT in terms of $\widehat{C}$, resulting in [6]

$$\mathcal{A}^{\rm EFT} = -\frac{4\pi/M}{\widehat{C}^{-1}(p^2;\mu) + \mu + ip} , \tag{4}$$

where μ is the renormalization scale. Comparing with Eq. (1) one sees that the EFT formally agrees with the exact result, provided that the $\widehat{C}$ couplings are chosen to satisfy

$$\frac{1}{\widehat{C}(p^2;\mu)} + \mu = -p\cot\widetilde{\delta} . \tag{5}$$

Expanding both sides of Eq. (5) in powers of p^2 and employing the effective range expansion Eq. (2), solutions for the running couplings $\widehat{C}_{2n}(\mu)$ can be obtained. The first few are

$$\begin{aligned}
\widehat{C}_0(\mu) &= -\frac{1}{\mu - 1/\widetilde{a}} , \\
\widehat{C}_2(\mu) &= \frac{1}{2}\widetilde{r}_0\,[\widehat{C}_0(\mu)]^2 , \\
\widehat{C}_4(\mu) &= \frac{[\widehat{C}_2(\mu)]^2}{\widehat{C}_0(\mu)} + \widetilde{v}_2\,[\widehat{C}_0(\mu)]^2 .
\end{aligned} \tag{6}$$

An important feature of these solutions is that the EFT exhibits a nontrivial infrared fixed point at $(\mu\widehat{C}_0) = -1$, $(\mu^{2n+1}\widehat{C}_{2n}) = 0$ [8]. It is obvious this must

[a] The $\widehat{C}$ couplings used throughout are a factor of $M/4\pi$ times the conventionally normalized C's found in the literature.

occur because the naively irrelevant operator $\widehat{C}_0$ has been forced to affect low-momentum scattering. Note also that the parameters $\widehat{C}_{2n}/(\widehat{C}_0)^2$ correspond directly to the effective range parameters $\{\frac{1}{2}\widetilde{r}_0, \widetilde{v}_n\}$.

Although in the present example the PDS scheme allows for a formal summation of all the diagrams of the EFT, more realistic interactions require a systematic expansion which approximates the exact EFT result Eq. (4). I now explore two different expansion schemes.

2.2 The Weinberg Expansion of the Effective Theory

The Weinberg approach[3,5,9,10] consists of expanding the kernel of the Lippmann-Schwinger equation, and then iterating it to all orders. The coefficients of the dimensionful contact interactions are taken to scale with Λ_S according to their naive dimension, namely

$$\widehat{C}_{2n}p^{2n} \sim (1/\Lambda_S)(p/\Lambda_S)^{2n} . \tag{7}$$

The kernel is then expanded in powers of p/Λ_S, where p is an external momentum so that counting powers of p gives a way of organizing corrections to the potential. At N^{th} order in the expansion, $V_N^{\rm EFT} \equiv \widehat{C}^{[N]}(p^2;\mu) = \sum_{n=0}^{N} \widehat{C}_{2n}p^{2n}$.

Although traditionally the Weinberg approach is analyzed with a momentum cutoff, there are advantages in using dimensional regularization with the PDS subtraction scheme [6]. At N^{th} order in the expansion, the amplitude is given by Eq. (4), with the exact interaction $\widehat{C}(p^2;\mu)$ replaced by $\widehat{C}^{[N]}(p^2;\mu)$. The N couplings $\widehat{C}_{2n}$ can be fixed by requiring that the theory correctly reproduce the first N terms in the effective range expansion. The prediction for the phase shift $\widetilde{\delta}$ at leading order (LO), next-to-leading order (NLO), and next-to-next-to-leading order (NNLO) are then given by

$$\begin{aligned}
(p\cot\widetilde{\delta})_{\rm LO} &= -\frac{1}{\widetilde{a}} ; \\
(p\cot\widetilde{\delta})_{\rm NLO} &= (p\cot\widetilde{\delta})_{\rm LO} + \frac{\frac{1}{2}\widetilde{r}_0 p^2}{1+\frac{1}{2}\eta\,\widetilde{r}_0 p^2} ; \\
(p\cot\widetilde{\delta})_{\rm NNLO} &= (p\cot\widetilde{\delta})_{\rm NLO} + \frac{p^4\left(\widetilde{v}_2+\frac{1}{4}\eta\,\widetilde{r}_0^2\right)}{\left(1+\frac{1}{2}\eta\,\widetilde{r}_0 p^2\right)\left(1+\frac{1}{2}\eta\,\widetilde{r}_0 p^2+\eta\, p^4\left[\widetilde{v}_2+\frac{1}{4}\eta\,\widetilde{r}_0^2\right]\right)} ,
\end{aligned} \tag{8}$$

where $\eta \equiv (1/\widetilde{a}-\mu)^{-1}$. The explicit μ dependence in the physical quantities is a result of neglecting counterterms necessary for renormalization at any finite order in the expansion.

For small μ, these expressions fail above $p \sim (\widetilde{a}\widetilde{r}_0)^{-1/2}$ which is much below the expected EFT breakdown of $p \sim \Lambda_S$ (e.g., see the $\mu = 0$ treatment in Ref. [11], or the equivalent momentum subtraction used in Ref. [3]). This failure is no surprise, since for small μ the $\widehat{C}_{2n}$ couplings given in Eq. (6) do not obey the naive scaling assumed in Eq. (7), which was the starting point for Weinberg's expansion. Evidently, the assumed power counting is only obtained for $\mu \gtrsim \Lambda_S$, in which case one sees that the result in Eq. (8) coincides with the effective range expansion up to higher order terms in p^2/Λ_S^2.

2.3 The KSW Expansion of the Effective Theory

The KSW expansion starts from the assumption $\widetilde{a} \gg 1/\Lambda_S$ and that the rest of the effective range parameters $\{\frac{1}{2}\widetilde{r}_0, \widetilde{v}_n\}$ scale with appropriate powers of Λ_S, according to their dimension. In this case, the amplitude Eq. (1) may be expanded as

$$\mathcal{A} = -\frac{4\pi/M}{(1/\widetilde{a}+ip)}\left[1+\frac{\widetilde{r}_0/2}{(1/\widetilde{a}+ip)}p^2+\frac{(\widetilde{r}_0/2)^2}{(1/\widetilde{a}+ip)^2}p^4+\frac{\widetilde{v}_2}{(1/\widetilde{a}+ip)}p^4+\ldots\right] \quad (9)$$

provided one is not near the kinematic point corresponding to a pole in $\mathcal{A}$. This expansion can be realized in by applying the following power counting rules outlined in [12], where basically $1/\widetilde{a}$, $\vec{p}$ and m_π are counted as a small parameter Q. Summing all Feynman diagrams to a given order in Q and making use of Eq. (6) reproduces the expansion Eq. (9). Since this is a consistent expansion of a physical quantity, results are μ-independent at each order. The most peculiar feature of the expansion is the scaling of $\widetilde{a}$ as $\mathcal{O}(Q^{-1})$. This allows for an expansion in powers of p/Λ_S while keeping powers of $p\widetilde{a}$ to all orders, but it will cause subtleties when the scale m_π is introduced.

3 Adding Toy Pions

I now consider the more interesting case of an interaction with both long- and short-range structure. The short-range interaction will remain unspecified for generality, giving the phase shift $\widetilde{\delta}$ when acting alone. For the long-distance part of the potential V_L, I choose the specific form of a delta-shell at a radius given by the pion Compton wavelength and a strength depending on m_π in a manner consistent with chiral symmetry

$$V_L(r) = -\frac{g_\pi m_\pi}{M}\,\delta\left(r-\frac{1}{m_\pi}\right) . \quad (10)$$

This caricature of the real one-pion exchange (OPE) potential allows us to analytically explore issues involving EFT expansions. The Yukawa part of the

$^1S_0$ channel OPE interaction in momentum space is

$$\tilde{V}_{\rm OPE} = -\frac{4\pi}{M\Lambda_{\rm NN}}\frac{m_\pi^2}{q^2+m_\pi^2}\ , \qquad \Lambda_{\rm NN} \equiv \frac{16\pi f_\pi^2}{g_A^2 M} \simeq 300\ {\rm MeV}, \tag{11}$$

where $m_\pi = 140$ MeV is the pion mass, $f_\pi = 93$ MeV is the pion decay constant, and $g_A = 1.29$ is the axial coupling. The Fourier transform of V_L is

$$\tilde{V}_L = -\frac{4\pi g_\pi}{M m_\pi}\frac{\sin q/m_\pi}{q/m_\pi}\ . \tag{12}$$

Comparison of the two expressions suggests that for V_L to have comparable effects to $V_{\rm OPE}$, one should take $g_\pi = m_\pi/\Lambda_\pi$ with $\Lambda_\pi \sim \Lambda_{\rm NN}$. In fact, the properties of V_L and the OPE Yukawa potential are most analogous if one takes

$$\Lambda_\pi = 500\ {\rm MeV}\ . \tag{13}$$

By analogous, I mean that in both cases, the strength of the interaction is approximately 30% of the critical value needed to give a bound state (which occurs at $g_\pi = 1$ in the toy model), and the strength of the phase shifts resulting from the two potentials are roughly comparable; see Fig. 1.

With $\Lambda_S > m_\pi$, it is straightforward to solve the Schrödinger equation

$$\left[-\frac{\nabla^2}{M} + V_S + V_L\right]\psi = E\psi \tag{14}$$

and determine the exact s-wave Feynman scattering amplitude $\mathcal{A}$ to be

$$\mathcal{A} = \mathcal{A}_L + \frac{4\pi/M}{f(\xi)^2\left(p\cot\tilde{\delta} - ip\right) - g_\pi m_\pi e^{2i\xi} f(\xi)}\ , \tag{15}$$

with $\xi = p/m_\pi$ and

$$f(\xi) \equiv 1 - g_\pi \frac{e^{i\xi}\sin\xi}{\xi}\ . \tag{16}$$

As in the previous section, $\tilde{\delta}$ is the phase shift due to V_S alone. The quantity $\mathcal{A}_L$ is the amplitude one finds for V_L alone, when the short-range potential V_S is turned off,

$$\mathcal{A}_L = \frac{4\pi g_\pi}{M m_\pi}\frac{\sin^2\xi}{f(\xi)\,\xi^2}\ . \tag{17}$$

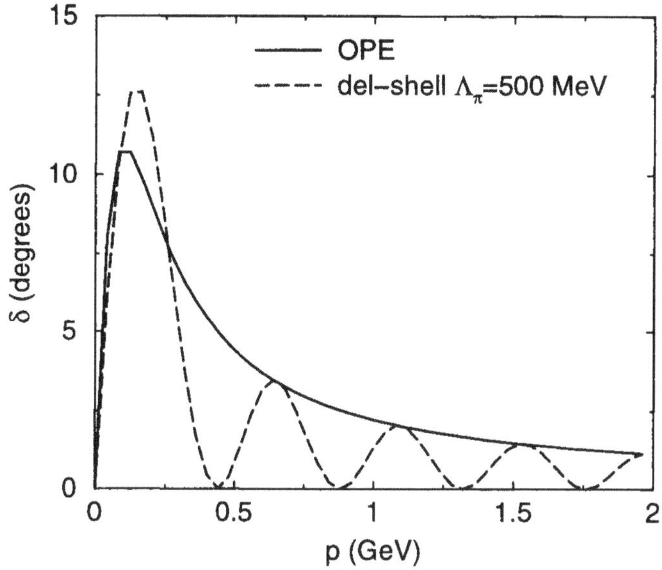

Figure 1: By taking $\Lambda_\pi = 500$ MeV, the envelope of the OPE and delta-shell potential coincide.

Equivalently, one finds the exact phase shift δ to be given by

$$\cot\delta = \frac{(\xi - g_\pi \sin\xi\cos\xi)\cot\widetilde{\delta} - g_\pi\cos^2\xi}{g_\pi\sin^2\xi\ \cot\widetilde{\delta} + \xi + g_\pi\sin\xi\cos\xi} \ . \tag{18}$$

From this one can determine the exact effective range parameters defined by

$$p\cot\delta = -\frac{1}{a} + \frac{1}{2}r_0 p^2 + \sum_{n=2}^{\infty} v_n p^{2n} + \ldots \tag{19}$$

in terms of the short-distance effective range parameters Eq. (2). For example, with $g_\pi \equiv m_\pi/\Lambda_\pi$,

$$a = \frac{1-\widetilde{a}(\Lambda_\pi + m_\pi)}{m_\pi(1-\widetilde{a}m_\pi) - \Lambda_\pi} \tag{20}$$

$$r_0 = \frac{2m_\pi\left(1-\widetilde{a}m_\pi\right)^2 + 2\Lambda_\pi\left(1-3\widetilde{a}m_\pi\right)\left(1-\widetilde{a}m_\pi\right) + 3\widetilde{a}^2\Lambda_\pi^2 m_\pi^2\widetilde{r}_0}{3m_\pi^2\left[1-\widetilde{a}\left(\Lambda_\pi + m_\pi\right)\right]^2} \ . \tag{21}$$

I wish to reproduce the full amplitude $\mathcal{A}$ in Eq. (15) in a systematic expansion of an EFT, replacing the potential V_S with contact interactions. I first show that the Feynman amplitude can be exactly reproduced by the effective theory, as was the case when I considered V_S alone. Subsequently, I address the issue of how to compute these amplitudes in perturbation theory, which is of practical interest for realistic NN scattering, where exact analytical solutions to the EFT are impossible to obtain.

3.1 The Exact Solution of the Effective Theory with Toy Pions

Taking an EFT with contact interactions $\widehat{C}(p^2;\mu)$ and the explicit long-range potential V_L given in Eq. (10), it is again possible to sum up all diagrams using dimensional regularization and the PDS subtraction scheme (see [4]) resulting in

$$\mathcal{A}^{\mathrm{EFT}} = \mathcal{A}_L + \frac{4\pi/M}{-f(\xi)^2\left[\widehat{C}^{-1}(p^2;\mu)+\mu+ip\right] - g_\pi m_\pi e^{2i\xi} f(\xi)}\ , \tag{22}$$

with $f(\xi)$ and $\mathcal{A}_L$ as in Eqs. (16, 17). A comparison between the above expression and the exact amplitude Eq. (15) shows that the EFT reproduces the full theory provided the $\widehat{C}$ couplings are chosen to satisfy

$$\frac{1}{\widehat{C}(p^2;\mu)} + \mu = -p\cot\widetilde{\delta}\ . \tag{23}$$

This is exactly the same result that I derived earlier in the theory without long-range interactions, Eq. (5). Thus, the $\widehat{C}_{2n}$ interactions truly represent the short-distance physics alone, and are independent of V_L. (This need not in general be true for the more realistic case of nonzero overlap in space between V_S and V_L.)

3.2 The Weinberg Expansion

In Weinberg's power counting scheme, the scale Λ_π is considered a short-distance scale and so insertions of V_L are of the same order as insertions of $\widehat{C}_0$, as seen from Eq. (12). Thus to N^{th} order in the expansion, the effective potential is given by

$$V_N^{\mathrm{EFT}} = V_L + \sum_{n=0}^{N-1} \widehat{C}_{2n} p^{2n}\ . \tag{24}$$

The Nth order expression for the Feynman amplitude is just given by the exact EFT solution Eq. (22) with $\widehat{C}(p^2;\mu)$ replaced by the truncated series of $\widehat{C}_{2n}p^{2n}$ in Eq. (24). The discussion now is entirely analogous to that of §2.2. Again the amplitude is not renormalized and therefore is μ-dependent. However, as long as $\mu \gtrsim \Lambda_S$ and the N couplings $\widehat{C}_{2n}$ are chosen to reproduce the effective range expansion (up to higher order terms), the exact expression for the amplitude Eq. (15) is obtained with $p \cot \widetilde{\delta}$ replaced by the first N terms in the short-distance effective range expansion.

3.3 The KSW Expansion

The Naive KSW Expansion and Its Failings Near the Pole

Once pions are included in the EFT, the KSW expansion treats m_π as order Q as well, generalizing the rules of §2.3. This is consistent with chiral perturbation theory, which treats p and m_π to be of the same order and small. It follows that V_L in Eq. (12) is $\mathcal{O}(Q^0)$, which is subleading compared to $\widehat{C}_0$, which is $\mathcal{O}(Q^{-1})$. Therefore, pions enter the amplitude perturbatively [6], unlike in the Weinberg approach. It is advantageous to treat the OPE potential perturbatively, since perturbative long-range interactions are analytically tractable, at least up to NNLO. While this is not a major benefit in two-nucleon systems, it is likely to be very valuable in many-body problems due to computational simplifications.

However, there is a subtlety associated with implementing this method, having to do with the large scattering amplitude which arises from the delicate interplay between short-range physics and pion exchange. Making use of the exact solution for $\widehat{C}_0$ from Eq. (6), one finds that the leading contribution to the amplitude at $\mathcal{O}(Q^{-1})$ is

$$\mathcal{A}_{-1}^{\rm EFT} = -\frac{4\pi/M}{1/\widetilde{a} + ip} \ . \tag{25}$$

This has a pole at $p_\star = i/\widetilde{a}$, in contrast to the exact amplitude which has a pole at $p_\star \simeq i/a$. Even if $1/a$ and $1/\widetilde{a}$ differ by an amount small compared to m_π, since the threshold cross section is $4\pi a^2$, this expansion will fail to accurately describe low-momentum scattering if either a or $\widetilde{a}$ are particularly large. For example, in the extreme case where the scattering length a is infinite, then the true amplitude at threshold is also infinite. However, $\widetilde{a}$ would be finite in that case, and so Eq. (25) would require the subleading perturbative corrections to the cross section to be infinite in order to produce the correct result!

The situation is similar to the problem of calculating e^+e^- scattering amplitudes at the Z^0 pole in perturbation theory. Even though the true value for the Z^0 mass and the tree level value only differ by an $\mathcal{O}(\alpha)$ contribution, practical calculations are done by perturbing around the exact Z^0 mass and introducing a mass shift at each order in perturbation theory to cancel the subsequent radiative corrections.

Similarly, in the present example, a small difference between $1/\tilde{a}$ and $-ip_\star$ does not ensure an accurate description of scattering for $p \sim |1/a|$. For large and positive a, where a shallow bound state exists and the pole in $\mathcal{A}$ is on the physical sheet (such as the deuteron pole in $^3S_1$ NN scattering), an expansion around the true location of the pole is necessary. For large and negative a, where the pole is kinematically inaccessible (as in the $^1S_0$ channel), it suffices to reproduce the exact scattering length at leading order. This is the procedure of matching at LO already carried out in the literature [6], but it needed to be understood from the perspective of the short-distance scales given in Eq. (6), because extending the procedure to NLO differs from the conventional approach. The following discussion has much in common with that of Mehen and Stewart [13] and Rupak and Shoresh [14].

The Pole Expansion

The resolution, as in the example of the Z^0 pole, is to reorder the expansion, making use of the fact that the difference between $1/\tilde{a}$ and either $1/a$ or $-ip_\star$ is $\mathcal{O}(Q^2)$. For the case relevant to NN scattering in the $^1S_0$ channel, I write

$$\frac{1}{\tilde{a}} = \frac{1}{a} + \sum_{k=2}^{\infty} \alpha_k \ , \tag{26}$$

where the α_k are $\mathcal{O}(Q^k)$ functions of a, m_π, Λ_π and μ. For example, the long-range potential I am considering gives

$$\frac{1}{\tilde{a}} = \frac{\Lambda_\pi + m_\pi - a\, m_\pi^2}{1 + a\,\Lambda_\pi - a\, m_\pi} \qquad \Longrightarrow \qquad \alpha_n = -\Lambda_\pi \left(\frac{am_\pi - 1}{a\Lambda_\pi}\right)^n \ . \tag{27}$$

Substituting the expression for the short-distance scattering length Eq. (26) into $\widehat{C}_0$ from Eq. (6) gives an expansion in Q:

$$\widehat{C}_0(\mu) \equiv \sum_{k=-1}^{\infty} \widehat{C}_0^{[k]}(\mu) \ , \tag{28}$$

where $C_0^{[k]}$ is of order Q^k. The first three contributions are

$$\widehat{C}_0^{[-1]} = -\frac{1}{\mu - 1/a} \ ,$$

$$\widehat{C}_0^{[0]} = -\alpha_2 \left(\widehat{C}_0^{[-1]}\right)^2 , \tag{29}$$
$$\widehat{C}_0^{[1]} = {\alpha_2}^2 \left(\widehat{C}_0^{[-1]}\right)^3 - \alpha_3 \left(\widehat{C}_0^{[-1]}\right)^2 .$$

This expansion ensures that the EFT yields the exact scattering length a at each order in the KSW expansion, with $\widehat{C}_0^{[k]}$ corrections to cancel the perturbative contributions of the pions arising at each order. Note that $\widehat{C}_0$ as defined in Eq. (28) remains m_π-independent, even though the $\widehat{C}_0^{[k]}$ terms each have a rather complicated m_π dependence.

Once $\widehat{C}_0$ is expanded, the exact solutions for the $\widehat{C}$ couplings in Eq. (6) imply that each $\widehat{C}_{2n}$ coefficient must be similarly expanded. For instance,

$$\widehat{C}_2^{[-2]} = \tfrac{1}{2}\widetilde{r}_0 \left(\widehat{C}_0^{[-1]}\right)^2 , \quad \widehat{C}_2^{[-1]} = \widetilde{r}_0 \left(\widehat{C}_0^{[-1]}\widehat{C}_0^{[0]}\right) \quad \ldots$$
$$\widehat{C}_4^{[-3]} = \frac{\left(\widehat{C}_2^{[-2]}\right)^2}{\widehat{C}_0^{[-1]}} , \quad \widehat{C}_4^{[-2]} = 2\frac{\widehat{C}_2^{[-1]}\widehat{C}_2^{[-2]}}{\widehat{C}_0^{[-1]}} + \alpha_2 \left(\widehat{C}_2^{[-2]}\right)^2 + \widetilde{v}_2 \left(\widehat{C}_0^{[-1]}\right)^2 \quad \ldots . \tag{30}$$

That the leading behavior of each $\widehat{C}_{2n}$ is determined by lower dimensional operators is further evidence of the EFT being tuned to lie near a nontrivial fixed point.

Note that in the KSW expansion, the parameter $\widetilde{r}_0$ first enters at $\mathcal{O}(Q^0)$ through an insertion of $\widehat{C}_2^{[-2]}p^2$, while the next new parameter $\widetilde{v}_2$ does not enter until $\mathcal{O}(Q^2)$ through an insertion of $\widehat{C}_4^{[-2]}p^4$. In general, the short-distance effective range parameters Eq. (2) enter the expansion of $\mathcal{A}$ through the operators $C_{2n}^{[-2]}p^{2n} \sim \mathcal{O}(Q^{2n-2})$ for each $n \geq 1$. Thus in the KSW expansion, one new parameter is encountered for every *two* powers of Q. The NNLO amplitude $\mathcal{A}_1$, for example, involves graphs not included in $\mathcal{A}_0$; nevertheless, it is completely parameterized by the same two numbers a and $\widetilde{r}_0$. The fact that new parameters appear in the expansion of $\mathcal{A}$ with only even powers of Q occurs because the effective range expansion of $p\cot\widetilde{\delta}$ is in even powers of momentum. A valid fitting procedure for the EFT should reflect this behavior, with one free parameter at LO to fix the scattering length a, one free parameter at NLO to fix $\widetilde{r}_0$, and subsequent free parameters appearing at every other order $\mathcal{O}(Q^{2n-2})$ to fix each $\widetilde{v}_{n\geq 2}$.

An Algorithm for Fixing Coupling Constants

The Weinberg expansion has a simple algorithm for fixing the low-energy constants of the EFT: Taylor expand the EFT result for $p\cot\delta$ and match to the effective range expansion at each order. The procedure to fix the low-energy constants in the KSW expansion is more obscure. When the nucleon interactions happen to be fine-tuned, high order perturbative pion contributions can in principle have relatively large effects on low energy observables due to cancellations at lower orders in the expansion [b]. Our simple long-distance potential V_L allows for an analytical determination of all the $C_{2n}^{[k]}$ couplings. However, for realistic situations one needs an algorithm for fixing these couplings. The above discussion makes it possible to surmise a prescription (easily extended to the $^3S_1$ with its bound state)

1. Expand each $\widehat{C}_{2n}$ coupling as

$$\widehat{C}_{2n} = \sum_{k=-(n+1)}^{\infty} \widehat{C}_{2n}^{[k]} ,$$

   where $\widehat{C}_{2n}^{[k]}$ is $\mathcal{O}(Q^k)$ in the expansion, and compute the amplitude to the desired order.

2. Use the threshold amplitude to fix $\widehat{C}_0^{[-1]}$ so that the LO result reproduces the experimental scattering length a, and fix the higher order $\widehat{C}_0^{[k]}$ by requiring the scattering length be unchanged at each higher order.

3. Determine the renormalization group equations in the PDS scheme for the $\widehat{C}_{2n}$ couplings to the order one is working. The beta function for $n \geq 1$ is always of the form

$$\mu \frac{d\widehat{C}_{2n}}{d\mu} = 2\mu \widehat{C}_0 \widehat{C}_{2n} + \mu \frac{dF_{2n}}{d\mu} \tag{31}$$

   where F_{2n} only depends on couplings $\widehat{C}_{2m}$ with $m < n$ and possibly the long-distance physics.[c] Note that for $n > 1$, the F_{2n} are actually the leading contribution in the Q-expansion of Eq. (31). Solving this equation leads to

$$\widehat{C}_{2n} = \widetilde{v}_n \widehat{C}_0{}^2 + F_{2n} , \tag{32}$$

[b] For the toy models I am considering, the Weinberg expansion does not suffer from this problem, as the long-range interactions are included completely at lowest order. However, for real NN scattering the Weinberg method also involves a chiral expansion of the long-range interactions beyond OPE, and so in principle subleading contributions could similarly exhibit large effects on low energy scattering, complicating the determination of the C coefficients.

[c] The toy pions Eq. (10) are well behaved at the origin so never contribute to the F_{2n}'s. Actual pion exchange has a finite overlap with the EFT contact interactions, so $\log\mu$ terms could occur. However, at least to NNLO in the Q-expansion, only F_0 has any such contribution.

where $\widetilde{v}_n$ is an undetermined constant of integration treated as $\mathcal{O}(Q^0)$. This solution should then be expanded in Q to produce each of the $\widehat{C}_{2n}^{[k]}$'s to be used in the calculated amplitude $\mathcal{A}$.

4. Within the Q-expansion, the phase shift is computed consistently from this amplitude $\mathcal{A}$ in terms of the parameters $\widetilde{r}_0$, $\widetilde{v}_n$, which are the short-distance effective range parameters. These parameters are then determined by finding the best fit to the data with an appropriate weighting in p.

In this manner, contact is made between the plethora of terms in the KSW expansion, and the effective range parameters $\{\frac{1}{2}\widetilde{r}_0,\ \widetilde{v}_n\}$ characterizing the short distance physics. At a given order, the KSW and Weinberg expansions have the same number of free parameters to fit. This general algorithm will correctly reproduce the KSW expansion in the presence of pions as explicitly shown in §4.

Note that additional contact interactions with quark mass insertions are possible [6], such as $D_2 m_\pi^2$ at NLO, $D_{22} m_\pi^2 p^2$ at NNLO, etc. However, to distinguish between these operators and the $C_{2n}^{[k]}$ interactions would require additional information beyond NN scattering, such as πd scattering. Therefore, in the present discussion I subsume them into our definitions of the $C_{2n}^{[k]}$'s.

4 Application of the Method

4.1 An Explicit Example with the Toy Pion

Our first example will utilize the toy pions of Eq. (10). We choose the short-distance physics to also be modeled by a delta-shell so an analytic solution is possible: $V = V_S + V_L$,

$$V_S(r) = -g_\rho \frac{m_\rho}{M}\delta\left(r - \frac{1}{m_\rho}\right) , \qquad V_L(r) = -\frac{m_\pi^2}{M\Lambda_\pi}\delta\left(r - \frac{1}{m_\pi}\right) , \tag{33}$$

with $\Lambda_\pi = 500$ MeV. The exact phase shift is then just given by Eq. (18) with

$$p\cot\widetilde{\delta} = \frac{\xi_\rho - g_\rho \sin\xi_\rho \cos\xi_\rho}{g_\rho \sin^2\xi_\rho} , \tag{34}$$

and $\xi_\rho = p/m_\rho$. Taking $m_\rho = 770$ MeV, we choose $g_\rho = 0.915$ to give the scattering length $a = -23$ fm. The contact interactions $\widehat{C}_{2n}^{[k]}$ were worked out in general in Eqs. (29, 30), so all that is left to do is implement the procedure outlined in §3.3 to fix these constants. The EFT calculation at LO only depends on $\widehat{C}_0^{[-1]}$, which as seen from Eq. (29) is fixed by the experimental scattering length.

At NLO, pions begin to contribute, and the perturbative correction $\widehat{C}_0^{[0]}$ is chosen to ensure the scattering length does not shift. This specifies α_2, which coincides with the analytic result given in Eq. (27). The contact interaction $\widehat{C}_2^{[-2]}$ also contributes at this order and depends on $\widetilde{r}_0$. We use its exact value, which can be calculated from Eq. (34)

$$\widetilde{r}_0 = \frac{2(1+g_\rho)}{3g_\rho m_\rho} = 0.36\ \text{fm}\,, \tag{35}$$

to test our fit. Afterwards, we will discuss how to find this value when the short-distance physics is not known *a priori*.

At NNLO, choosing $\widehat{C}_0^{[1]}$ so the scattering length does not shift produces α_3 as given by Eq. (27). The other new contact interactions $\widehat{C}_2^{[-1]}$ and $\widehat{C}_4^{[-3]}$ are fully determined by already specified quantities Eq. (30), so there are no new constants to fix at this order as already discussed.

The results are shown in Fig. 2. On the left, the phase shift as a function of momentum for the exact result (solid line) is compared to the EFT calculation at NLO (dashed) and NNLO (dot-dashed). Note that the results improve as more orders are added, even when there are no additional free parameters (as occurs at NNLO). The right plot of Fig. 2 shows the improvement in the observable $p\cot\delta$ by plotting its error at each order (solid lines).

If the exact value for $\widetilde{r}_0$ were not known, as is the case in real data, we could treat $\widetilde{r}_0$ as a free parameter of the effective theory, varying its value to achieve the best global fit to data over a representative momentum range such as $[1/a, \Lambda_\pi]$. Doing this at both NLO and NNLO produces the dashed lines in the right plot of Fig. 2, coinciding quite well with the exact result. Another indication of this agreement can be found by comparing the fit values for $\widetilde{r}_0$ with the exact value Eq. (35). The EFT with a global fit gives $\widetilde{r}_0 = 0.41$ fm at NLO and $\widetilde{r}_0 = 0.37$ fm at NNLO, showing convergence to the exact result $\widetilde{r}_0 = 0.36$ fm of Eq. (35), differing only by higher order contributions which the fit cannot resolve.

The point at which this fitting procedure breaks down can also be determined. The EFT should only work for momentum below a scale associated with underlying physics not explicitly accounted for in the Lagrangian, which is m_ρ for this two delta-shell model. Up to this point, we have computed the phase shift by expanding the expression [6]

$$\delta = \frac{1}{2i}\ln\left[1 + i\frac{Mp}{2\pi}\left(\mathcal{A}_{-1} + \mathcal{A}_0 + \ldots\right)\right] \tag{36}$$

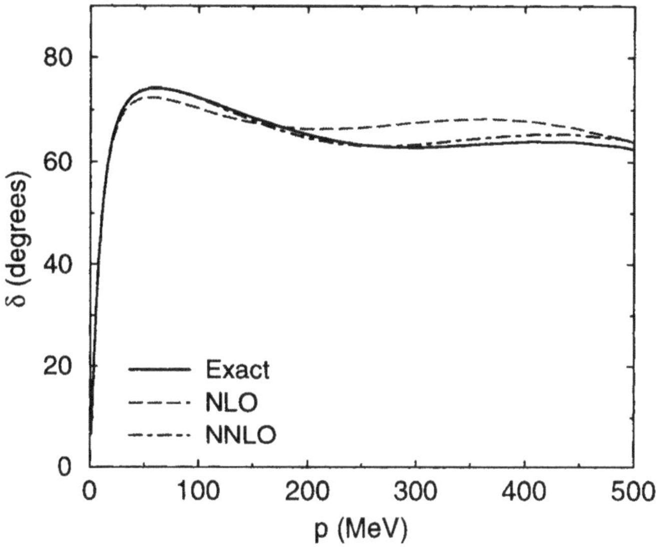

Figure 2: The algorithm of §3.3 for the KSW expansion applied to the two delta-shell model. The plot shows that as more orders are added in the KSW expansion, the agreement with the exact phase shift improves. The result of the Weinberg expansion to NLO would be barely distinguishable from the exact result in this plot.

$$\simeq \frac{1}{2i} \ln\left(1 + i\frac{Mp}{2\pi}\mathcal{A}_{-1}\right) + \frac{Mp}{4\pi}\left(\frac{\mathcal{A}_0}{1 + i\frac{Mp}{2\pi}\mathcal{A}_{-1}}\right) + \ldots, \tag{37}$$

but we could also have kept the full expression Eq. (36). These two expressions are equivalent up to terms that are higher order in the expansion. Those extra terms are an estimate of the corrections to the actual result and depend on the radius of convergence. The breakdown of the KSW expansion can be identified as the point at which the two expressions Eqs. (36, 37) diverge from each other. Doing this exercise for the two delta-shell model using the results of the fit to NLO and NNLO reveals the breakdown to be near m_ρ as expected [10].

The delta-shell model for the pion allows us to obtain analytical expressions for the scattering amplitude and to check the algorithm for determining the $\widehat{C}_{2n}^{[k]}$ coefficients of the EFT as described in §3.3. However, this toy model can be criticized as being too simple to provide evidence that the KSW expansion will work for actual NN scattering. We therefore consider models where pion exchange is represented by the correct OPE potential for the $^1S_0$ channel.

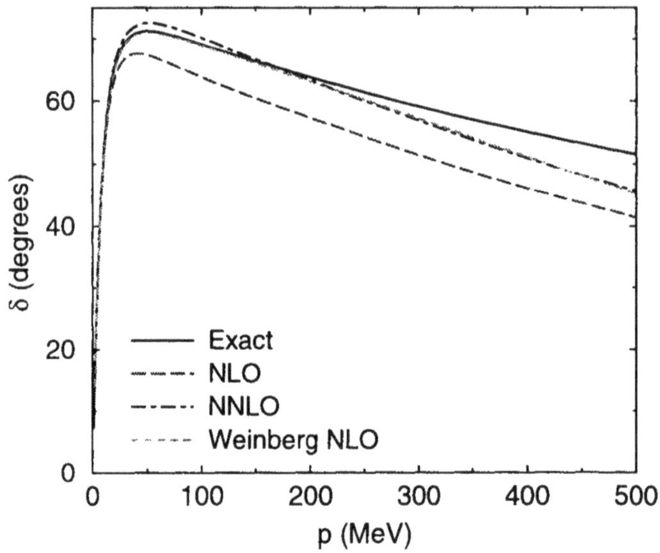

Figure 3: The algorithm of §3.3 for the KSW expansion applied to the two-Yukawa model. NLO and NNLO results from the EFT are compared to the exact result. The breakdown scale is $p \sim 400$ MeV.

4.2 The Two-Yukawa Model

Although the algorithm of §3 was motivated by the model of toy pions, we now apply it to the more sophisticated model with a long-distance potential given by the Yukawa part of the $^1S_0$ OPE potential, and a short-distance Yukawa potential characterized by the rho mass

$$V = V_\pi + V_S\ , \qquad V_\pi = -\alpha_\pi \frac{e^{-m_\pi r}}{r}\ , \qquad V_S = -\alpha_\rho \frac{e^{-m_\rho r}}{r}\ . \tag{38}$$

The pion coupling is taken to be $\alpha_\pi = g_A^2 m_\pi^2/(16\pi f_\pi^2) \simeq 0.075$, as in the real world, and the rho coupling $\alpha_\rho = 1.05$ is tuned to give a large scattering length $a = -23$ fm, as observed in data for the $^1S_0$ partial wave of NN scattering. Analyzing scattering from the short-range V_S potential alone then yields

$$\widetilde{a} = -1.17 \text{ fm}\ , \qquad \widetilde{r}_0 = 0.765 \text{ fm}\ , \tag{39}$$

and the effective range expansion for the phase shift $\widetilde{\delta}$ works well up to $p \simeq m_\rho/2$, as expected from analyticity considerations.

The EFT amplitude with potential pions has been calculated up to NNLO in Refs. [14,15]. We make use of that amplitude and fix the contact interactions according to the algorithm described in §3.3. This gives a LO amplitude that depends upon a, and NLO and NNLO amplitudes that also depend upon $\widetilde{r}_0$, the short-distance effective range parameter. The results obtained for the actual value $\widetilde{r}_0 = 0.765$ fm are shown in Fig. 3 for NLO and NNLO. The corresponding values for the full effective range are $r_0 = 2.1$ fm and $r_0 = 1.3$ fm respectively, compared to the exact value 1.6 fm.

The result from using the Weinberg approach (short dashed line) is also shown. Since it is fit at low-momentum, the agreement there is better than the KSW expansion. Both expansion schemes coincide at large momentum, which is in agreement with the idea that the Weinberg approach includes what the KSW expansion considers to be subleading terms.

Instead of using the actual value for $\widetilde{r}_0$, we could also treat it as a free parameter and use a one-parameter global fit to data. This gives $\widetilde{r}_0 = 0.400$ fm at NLO and $\widetilde{r}_0 = 0.671$ fm at NNLO. The deviations from the true value of $\widetilde{r}_0$ by 47% and 12% respectively are roughly what is expected from the estimates $\mathcal{O}(\alpha_\pi/\alpha_{crit})$ and $\mathcal{O}([\alpha_\pi/\alpha_{crit}]^2)$, where $\alpha_{crit} \simeq 0.25$ is the critical pion coupling that would lead to a bound state in the OPE potential.

The convergence of the KSW expansion to the true result in this model shows that perturbative inclusion of the pions is a good approximation. This is a significant result, as this model includes realistic single pion exchange, unlike the toy model with δ-shell pions in §4.1. However, the two Yukawa model is still unrealistic in that it does not correctly reproduce the large effective range measured in NN scattering. The next example remedies this situation, with major consequences.

4.3 The Three-Yukawa Model

The two-Yukawa model is unable to reproduce the large effective range $r_0 = 2.63$ fm of the $^1S_0$ channel for any choice of couplings in Eq. (38). So as a final example, we add one more Yukawa interaction in order to have the proper effective range. We take for our potential $V = V_\pi + V_S$, , where

$$V_\pi = -\alpha_\pi \frac{e^{-m_\pi r}}{r}, \qquad V_S = -\alpha_\sigma \frac{e^{-m_\sigma r}}{r} + \alpha_\rho \frac{e^{-m_\rho r}}{r}, \tag{40}$$

with the physical values for m_π, m_ρ, and α_π, and with $m_\sigma = 500$ MeV. We then determine the final two parameters in order to obtain $a = -23$ fm and $r_0 = 2.6$ fm, yielding $\alpha_\sigma = 7$ and $\alpha_\rho = 14.65$. This three-Yukawa model is reminiscent of the Bonn potential, which is known to model the data well.

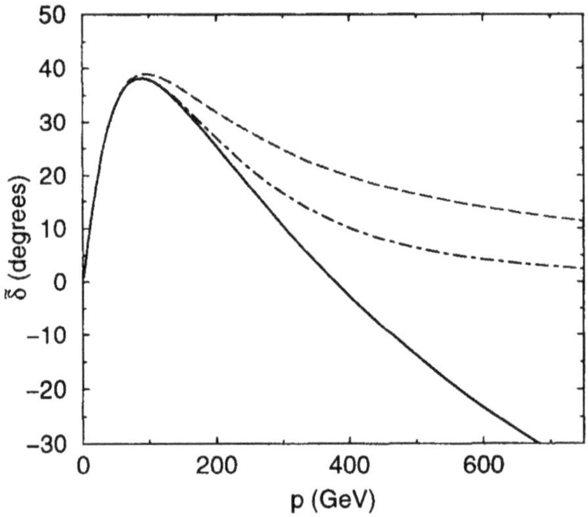

Figure 4: The phase shift $\tilde{\delta}$ due to scattering from the combined ρ and σ contributions to the potential $V(r)$ in Eq. (40). The exact result (solid) is compared with the effective range expansion to order p^2 (dashed) and p^4 (dot-dashed).

Examining scattering from the short-range potential V_S alone, we find

$$\tilde{a} = -3.3 \text{ fm} , \qquad \tilde{r}_0 = 2.59 \text{ fm} , \tag{41}$$

which immediately suggests that we will encounter serious problems due to the large size of $\tilde{r}_0$. Recall that $\tilde{r}_0$ is assumed to be $\mathcal{O}(1/\Lambda_S)$ in all effective field theory approaches proposed to date. For this model, however, the scale Λ_S corresponds to $m_\sigma/2 = 250$ MeV, whereas $\tilde{r}_0 \simeq 1/(80 \text{ MeV})$ corresponds to a much smaller scale.

Performing an effective range expansion for $\tilde{\delta}$, we obtain the result shown in Fig. 4. The short-distance effective range expansion, if carried to high enough order, does indeed have a radius of convergence set by $p \sim m_\sigma/2$. However, the expansion at $\mathcal{O}(p^2)$ deviates from the exact result already at momenta below m_π. This suggests that an EFT expansion to the same order will also fail at momenta below the pion mass, even with the pion included explicitly.

Applying the EFT to the three-Yukawa model reveals this to be the case. The results for NLO and NNLO using the exact $\tilde{r}_0 = 2.59$ fm are shown in Fig. 5. The EFT with a global fit instead gives $\tilde{r}_0 = 1.68$ fm at NLO and $\tilde{r}_0 = 3.71$ fm at NNLO, also showing no signs of convergence. Similarly, the full effective range gives $r_0 = 3.7$ fm at NLO and $r_0 = 1.1$ fm at NNLO, both

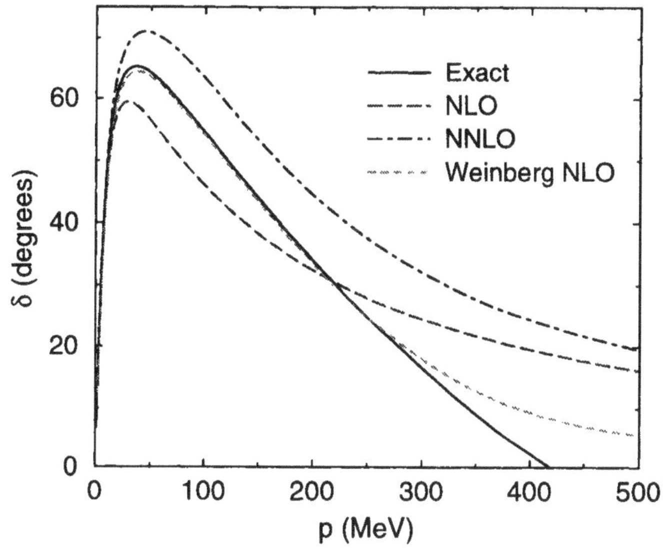

Figure 5: The algorithm of §3.3 for the KSW expansion applied to the three-Yukawa model.

equally far from the exact value 2.6 fm.

The Weinberg approach result does better overall, only deviating from the exact result around 300 MeV. This indicates that the subleading order terms which are accounted for in this case are larger than anticipated from the KSW power counting.

5 Conclusions

The purpose of this exercise was to develope a systematic way to determine the C couplings of the effective theory in order to avoid simply accruing free parameters and using them to curve fit. Curve fitting to the phase shifts hides whether the expansion works well or not, and gives one an exaggerated confidence in the smallness of errors to expect in other processes. Furthermore, the algorithm we develop allows one to better evaluate the quality of the KSW expansion.

It is clear that the KSW expansion to converges to the exact answer in the chiral limit; furthermore, the C coefficients one obtains are directly related to the scattering parameters of the short distance physics. Explicit examples with δ-shell toy pions, and a more realistic two Yukawa model show that the KSW expansion can converge well even for realistic pion mass — there is nothing intrinsically nonperturbative about real pion interactions. Nevertheless, in the three Yukawa example — which, of the models considered here, most resembles reality — the KSW expansion fails rather badly. Why is this?

One way to see the problem is to look at the effective range as computed in the EFT. Even when one sums up C_0 and OPE to all orders, one finds that this pionic contribution by itself only accounts for about one half of the measured value for r_0 [11]. Thus, short range physics must account for the other half. So far, this is consistent with the KSW expansion, which treats both the pionic and nonpionic leading contributions to r_0 as $O(1)$. However, not only is r_0 rather large, but also pions tend to screen the effects of the short distance effective range, $\tilde{r}_0$, which is why in the three Yukawa model, $\tilde{r}_0$ has to be so enormous (2.6 fm).

At NNLO in the KSW expansion for Yukawa pions one finds

$$r_0 = (\text{pion } \& \text{ C}_0 \text{ contribution}) + \left(1 - 2\frac{(m_\pi - 1/a)}{\Lambda_{NN}}\right)\tilde{r}_0 \,. \tag{42}$$

Note that the factor of $[1 - 2(m_\pi - 1/a)/\Lambda_{NN}] \simeq .1$ multiplying $\tilde{r}_0$ exhibits a nearly perfect cancellation between NLO and NNLO contributions. So what exactly is the problem? Is it that one cannot include pions perturbatively, as suggested by Gegelia [16]? Or that there is an unfortunate cancellation between NLO and NNLO contributions, as suggested by Mehen and Stewart [15]? Or is the problem that r_0 is too large for the expansion to work, as suggested by Cohen and Hansen [17,18]? Apparently, these suggestions are all correct: in Eq. (42), if m_π/Λ_{NN} had been small, or if the coefficient of "2" had been a "1", for example, then the near cancellation between NLO and NNLO terms would not have occured, and the short range contribution $\tilde{r}_0$ could have made itself felt. On the other hand, this cancellation was present in the two Yukawa model as well, but it caused no great problems there since in that model pions account for most of r_0 and the screening of $\tilde{r}_0$ by pions at NNLO has little effect on the quality of the expansion. I think it is fair to conclude that the failure of the KSW expansion exhibited in Fig. 5 follows from the confluence of having relatively strongly coupled pions, an unlucky coefficient at NNLO, and an unpleasantly large $\tilde{r}_0$, and that it cannot be ascribed to any one of these factors alone.

The Weinberg expansion result for the three Yukawa model, Fig. 5, works much better than the KSW expansion largely because at NLO, the $\tilde{r}_0$ contributions are summed to all orders. However, one must still note that the power counting arguments that Weinberg invoked to justify his expansion have failed, since the assumption of approximate chiral symmetry requires $m_\pi\tilde{r}_0 \ll 1$, when in fact $m_\pi\tilde{r}_0 \simeq 1.8$. Numerical success without a theoretical justification is what EFT was supposed to supercede.

There are two ways out of this problem, as I see it. The first is the possibility that the three Yukawa model is a poor representation of reality, due to large contributions to r_0 in real NN scattering from radiation pions[19]; however, this seems unlikely. The other way out would be to develop an alternative power counting which gets one to sum up $\tilde{r}_0$ to all orders at NLO. This might be accomplished along the lines of [20], without having to include pions nonperturbatively. However, one must concede that at least certain features of the NN interaction appear to be far from the chiral limit.

References

1. *Nuclear Physics with Effective Field Theory*, ed. R. Seki, U. van Kolck, and M. J. Savage (World Scientific, 1998);
2. U. van Kolck, "Effective Field Theory of Nuclear Forces," nucl-th/9902015, and references therein.
3. S. Weinberg, Phys. Lett. B **251**, 288 (1990); Nucl. Phys. B **363**, 3 (1991); Phys. Lett. B **295**, 114 (1992).
4. D.B. Kaplan and J.V. Steele, nucl-th/9905027.
5. C. Ordonez and U. van Kolck, Phys. Lett. B **291**, 459 (1992); C. Ordonez, L. Ray, and U. van Kolck, Phys. Rev. Lett. **72**, 1982 (1994); Phys. Rev. C **53**, 2086 (1996); U. van Kolck, Phys. Rev. C **49**, 2932 (1994).
6. D. B. Kaplan, M. J. Savage, and M. B. Wise, Phys. Lett. B **424**, 390 (1998); Nucl. Phys. B **534**, 329 (1998); D. B. Kaplan, lecture delivered at *Baryons '98*, Bonn, Sept. 22, 1998, nucl-th/9901003.
7. H. A. Bethe, Phys. Rev. **76**, 38 (1949).
8. M. C. Birse, J. A. McGovern, and K. G. Richardson, hep-ph/9807302.
9. T. S. Park, D. P. Min, and M. Rho, Phys. Rev. Lett. **74**, 4153 (1995); Nucl. Phys. A **596**, 515 (1996).
10. J. V. Steele and R. J. Furnstahl, Nucl. Phys. A **637**, 16 (1998).
11. D. B. Kaplan, M. J. Savage, and M. B. Wise, Nucl. Phys. B **478**, 629 (1996).
12. D.B. Kaplan, nucl-th/9804061, in ref. [1], p. 27.
13. T. Mehen and I. W. Stewart, Phys. Lett. B **445**, 378 (1999).
14. G. Rupak and N. Shoresh, nucl-th/9902077.
15. T. Mehen and I. W. Stewart, nucl-th/9906010; *this volume.*
16. J. Gegelia, nucl-th/9802038; hep-th/9804189; nucl-th/9805008; nucl-th/9806028.
17. T.D. Cohen and J.M. Hansen, Phys. Rev. C **59**, 3047 (1999).
18. T.D. Cohen, nucl-th/9904052; *this volume.*

19. T. Mehen and I.W. Stewart, nucl-th/9901064.
20. D. B. Kaplan, Nucl. Phys. B **494**, 471 (1997).

Discussion during D. Kaplan's presentation

Chair: *M. Luke (University of Toronto)*
Notes taken by *D. R. Phillips (University of Washington)*

David Kaplan opened the session with a presentation of the basics of "Q-counting". Part of the talk involved results of studies he and James Steele had performed of the comparative efficacy of this counting scheme as against that of Weinberg in a toy model. Some discussion centered around what these two different types of counting meant, with Michael Birse pointing out that all orders in Q appear in the amplitude if Weinberg's Λ-counting is employed for the NN potential. Birse was particularly concerned about the validity of this power-counting in the presence of an infra-red (quasi-) fixed-point in the RG flow. Peter Lepage defended Λ-counting, on the grounds that such power-counting in the action is supported by arguments of Wilson, Symanzik, and Zimmerman.

Bira van Kolck asked what happened if the value of the parameter Λ_L, in the Kaplan-Steele toy model's long-distance potential, was lowered, thus increasing the strength of the coupling of the "pion". Kaplan pointed out that the value of Λ_L was chosen to reproduce the behavior of NN phase shifts, but that if one did try to make Λ_L smaller then the agreement between exact results in the toy model and the Q-counting expansion thereof worsened.

Kaplan spent some time in his talk discussing details of fitting the amplitude at NNLO. Tom Cohen suggested that redesigning Q-counting to avoid shifting the position of the deuteron pole was unnecessary if one performed Q-counting in $p \cot \delta$ rather than in the amplitude, T. Kaplan agreed.

Cohen also asked why the subtraction point μ was always chosen to be of order m_π in the actual calculations, and not of order p, the external momentum, as it is formally in Q-counting. Kaplan pointed out that the amplitude was explicitly μ-independent in Q-counting, provided that one worked to a given order. Mark Wise added that the choice $\mu \sim p$ is only necessary to establish power-counting and is not actually needed for calculation. Matthias Lutz then advocated a power-counting scheme using dispersion relations.

PERTURBATIVE PIONS AND THE EFFECTIVE RANGE EXPANSION

T. D. COHEN
Department of Physics, University of Maryland,
College Park, MD 20742
E-mail cohen@physics.umd.edu

In this talk the Q counting scheme to implement effective field theory is discussed. It is pointed out that there are two small mass scales in the problem m_π and $1/a$ with $1/a \ll m_\pi$. It is argued that while the expansion based on $1/a$ being small compared to the underlying short distance scales works well, the chiral expansion may not. The coefficients of the effective range expansion are sensitive to the chiral physics and are very poorly described in Q counting at lowest nontrivial order. A "shape function" is introduced which again is sensitive to pionic physics and insensitive to fitting procedures. It is also poorly described in Q counting.

1 Introduction

The Q counting scheme introduced just over one year ago by Kaplan, Savage and Wise (KSW) [1] represents an important advance in the development of effective field theory techniques for nuclear physics. The approach is systematic, it builds in approximate chiral symmetry and chiral power counting, it solves Weinberg's [2] "unnatural scattering length problem", and, in principle, it provides *a priori* estimates of errors for observables since one works to fixed order in Q/Λ. This last feature is extremely important since in Weinberg's initial formulation[2] it was unclear how to make such error estimates. [a] To a considerable practical extent the major advantage to using EFT technology as opposed to unsystematic models is the ability to specify the accuracy of one's predictions. Thus, in this talk, I will focus entirely on Q counting and not on the many beautifully accurate calculations based on Weinberg's approach implemented with a finite cutoff[3].

The scheme introduced by KSW may be divided up into two parts. The first is Q counting and the second is a set of technical tricks to implement

[a] Of course, the principal difference between Q counting and the Weinberg scheme is that in Weinberg's approach the potential is iterated to all orders whereas in Q counting only the leading term is iterated to all orders; all other operators are treated perturbatively. If Q counting is valid, a scheme like Weinberg's if it can be systematically implemented will, at worst, add uncontrolled higher order contributions which do not spoil the systematic error estimates. Thus even if one is using Weinberg's approach, if one wants to make simple error estimates one may resort to Q counting for the error estimate, provided the system is in a regime where it is valid. This view of error estimation in the Weinberg scheme is featured prominently in the discussions at this workshop.

the Q counting. These technical tricks are rather unusual; they are based on the so-called PDS scheme for doing subtractions in dimensional regularization. Apart from the very peculiar prescriptions required (subtracting the poles as $d = 3$!) the formalism is not completely transparent in terms of the physics. Of course, provided the scheme is consistent and we are in the regime for which Q counting is valid, one should get the same results for any scheme which implements the Q counting. Two other approaches to Q counting have been tried—the OS scheme with dimensional regularization of Mehan and Stewart[4] and a cutoff scheme in configuration space [5]. All the schemes give the same results at fixed order in Q counting. Thus, ultimately the physics turns on whether or not the Q counting is working.

The Q counting scheme is straightforward:

$$\begin{aligned} Q &\sim 1/a \\ Q &\sim m_\pi \\ Q &\sim p \end{aligned} \tag{1}$$

where a is the scattering length–either singlet or triplet, and p represents external momenta. For partisans of the OS scheme, you can just as well use $\gamma = \sqrt{MB}$ where B is the magnitude of the binding energy, in place of $1/a$. All other scales are assumed to be heavy and will collectively be denoted as Λ. The expansion implied by Q counting is in Q/Λ. Now the key point of Q counting is that since p, m_π and $1/a$ are all of the same order, at any order in Q counting we have $1/(pa)$ and p/m_π to all orders. One important feature of Q counting is that while one needs to iterate the lowest order contact term to all orders to get a consistent result, all higher order contributions, including those from the pion can be treated perturbatively.

Q counting has been used to calculate a number of observables [1,4,6] and generally seems to have real predictive power. At first sight this would appear to rule out the possibility that $m_\pi \sim \Lambda$. However, most "vanilla" observables principally test the $1/(a\Lambda)$ part of the theory. Clearly, it is important to identify observables which are principally sensitive to the m_π/Λ parts of the theory and to rigorously test the chiral expansion. Recall that only the chiral part of the Q counting is really understood in terms of QCD. In Q counting, the small value of $1/a$ is treated as an essential fact of life that we cannot ignore. ¿From the QCD level, however, this fact of life is seen as essentially an accident. In contrast, the chiral physics is understood directly in terms of the small quark masses in the QCD Lagrangian along with spontaneous symmetry breaking. The central theme of this talk is that the effective range expansion—namely the expansion of $p\cot(\delta)$ as a power series in energy—is a good place to test whether the chiral part of Q counting is under control.

The effective range expansion (ERE) is a good place to look at pionic effects for a number of reasons. The expansion may be written:

$$p\cot(\delta) = -\frac{1}{a} + \frac{1}{2}r_e\, p^2 + v_2 p^4 + v_3 p^6 + v_4 p^8 + \ldots \tag{2}$$

Simple Q counting shows that the scattering length term is order Q^1 while all other terms in the ERE are at least order Q^2. When one includes pions explicitly the same counting holds; all terms except the first are $\mathcal{O}(Q^2)$. It is important to note the distinction between the Q expansion and the momentum expansion in the ERE. They differ precisely because the Q expansion has k/m_π and $1/(ka)$ to all orders while in the momentum expansion they are multiplied out. One immediately deduces that $v_n \sim Q^{-2n+2}$. Moreover all of the v_i coefficients in the expansion diverge in the chiral limit of $m_\pi \to 0$. Hence they are pion dominated quantities and should provide a test of the chiral part of Q counting.

2 Scales in Nuclear Physics

Before coming to the effective range expansion in this approach, it is useful to look a bit more closely at the various scales underlying Q counting. Formally, there are two light scales intrinsic to the problem, $1/a$ and m_π. In Q counting they are both formally of the same order namely $\mathcal{O}(Q)$. But empirically for both the triplet and singlet channel,

$$m_\pi \gg 1/a \tag{3}$$

with $m_\pi a \approx 4$ for the triplet channel and $m_\pi a \approx 16$ for the singlet channel. This raises the following logical possibility;

$$\begin{aligned} m_\pi &\sim \Lambda \\ 1/a &\ll \Lambda \end{aligned} \tag{4}$$

i.e., that there is no scale separation between pionic scales and the "short distance scales" but there is a good scale separation between them and $1/a$.

Of course, an immediate prejudice is that the first relation in eqs. (4) must be wrong; chiral scales are intrinsically long compared to typical hadronic scales. However, what is relevant here for Λ is not hadronic scales, but *nuclear* scales. The relationship of nuclear scales to QCD is quite obscure, but it is certainly true that typical nuclear mass scales are much lower than typical hadronic scales. If conditions in eqs. (4) turn out be true one would expect that the parts of the theory which depend on $1/(a\Lambda)$ will work quite well, while the parts which depend on m_π/Λ will converge slowly or not at all.

Note, if it turns out that $m_\pi \sim \Lambda$, there is nothing in principle wrong with the Q counting formalism and PDS. It would simply not be useful for real world situations. Of course, one could play God and consider a world in which the pion is much lighter than in nature and then one would have real predictive power. In principle, if lattice technology improves, one could calculate properties in such an artificial world from first principles and could use the Q counting technology to make predictions for this world.

Before looking at explicit calculations, we should ask whether it is reasonable to suppose that $m_\pi \sim \Lambda$. Ultimately, this question comes down to whether $1/a$ and m_π are the only light scales in nuclear physics. The answer appears to be "no". Numerically:

$$\frac{1}{m_\pi} \approx 1.5\text{fm} \;\; a^s \approx -23\text{fm} \;\; a^t \approx 6\text{fm}$$

where the superscript s (t) refers to the singlet (triplet) channel. Compare these with the effective ranges:

$$r_e^s \approx 2.7\text{fm} \;\; r_e^t \approx 1.6\text{fm}$$

It is apparent that $m_\pi r_e \sim 1$; if $r_e \sim 1/\Lambda$ there is a serious potential problem.

Of course, it is possible that the large numerical size of r_e is itself a reflection of chiral physics. For example, if $r_e \sim 1/m_\pi$ there would be no problem. One can use Q counting itself to answer the question of how r_e behaves. At leading nonvanishing order it is given by [4,5]

$$r_e = \mathcal{O}(\Lambda) + \frac{g_A^2 M}{4\pi f_\pi^2 a^2 m_\pi^2} - \frac{g_A^2 M}{3\pi f_\pi^2 a m_\pi} = \mathcal{O}(Q^0) \tag{5}$$

Note that although there *is* a chiral enhancement—the last two terms diverge in the chiral limit—it is compensated for by factors of the scattering length in the denominator. Thus, one expects in the context of Q counting the effective range to be a short distance scale. In practice, however, it is larger than $1/m_\pi$. This suggests, but does not prove, that the chiral scale is not well separated from "short distance" scales.

There is another way to see that "short distance" scales may be comparable to the pion mass scale. Consider the typical scales in so-called realistic N-N potential models, *i.e., those which fit the scattering data.* If you look at the non–one-pion-exchange part of the potential it is, in fact, larger than the OPEP potential for a distance less than $\sim 1.5 - 2$ fm. Since $1/m_\pi$ is comparable to, or shorter than, this distance we again appear to have evidence that non-chiral supposedly short distance scales are comparable to $1/M_\pi$. One might argue that the central potential contains two-pion-exchange physics (suitably reparameterized) in the potential model. However, if Q counting is valid, that contribution is chirally suppressed.

This argument appears to be model dependent as it is based on "typical" potential models. There is a model independent way to constrain the short distance physics.[7] Consider *any* nonrelativistic potential, including possible non-local potentials. Write the potential as the sum of an OPEP potential and some short distance potential with the constraint that the short distance potential vanishes beyond some distance R:

$$V(\vec{r},\vec{r'}) = V_{\rm OPEP}(\vec{r})\,\delta(\vec{r}-\vec{r'}) + V_{\rm short}(\vec{r},\vec{r'})$$

$$\text{with} \quad V_{\rm short}(\vec{r},\vec{r'}) = 0 \quad \text{for} \quad r, r' > R. \tag{6}$$

Now suppose that this potential is inserted to a Schrödinger equation and used to solve for singlet phase shifts. A remarkable theorem can then be proved, namely that if the short distance potential fits the scattering length and effective range there is a minimum value for R. For real world values one can deduce the $R >1.1$ fm [7]. Moreover the derivation of this bound shows that it is unsaturatable so one expects R to be significantly more than than 1.1 fm. From this one deduces that substantial contributions to the scattering come from "short distance" contributions which come from separations of greater than 1.1 fm. Recalling that $1/m_\pi \approx 1.4$ fm, we see immediately that there is no significant scale separation between m_π and the scales fixing the overall range of the nonpionic part of the potential.

While the preceding arguments do not decisively prove that $\Lambda \sim m_\pi$ they certainly show that it is not implausible.

3 Q Counting and Cutoffs

Now the problem comes down to computing $p\cot(\delta)$. This can be done in PDS as in ref. 1. For the present purpose it is instructive to consider the cutoff calculation and we take our discussion from ref. 5. The essential physical idea in this approach is to implement the separation of long distance physics from short distance physics directly in configuration space. A radius, R, is introduced as a matching point between long and short distance effects; renormalization group invariance requires that physical quantities must be independent of R. It is important, however, that R be chosen large enough so that essentially all of the effects of the short distance physics is contained within R. The potential is divided into the sum of two pieces, a short distance part which vanishes for $r > R$ and a long distance part which vanishes for $r < R$. At

R, the information about short distance effects is entirely contained in the energy dependence of the logarithmic derivative (with respect to position) of the wave function at R. Thus, provided we can parameterize this information systematically, we can formulate the problem in a way which is insensitive to the details of the short distance part of the potential. This insensitivity to the details of the short distance physics is at the core of why effective field theory works. For $r > R$, the Schrödinger equation is solved subject to the boundary conditions at R. For s wave scattering, the wave function at R may be parameterized as $A\sin(kr+\delta_0)$; the energy dependence of the logarithmic derivative is independent of A and can be expressed in terms of an expansion similar to an effective range expansion:

$$p\cot(\delta_0) = -1/a_{\text{short}} + 1/2\, r_e^0 p^2 + v_2^0 p^4 + v_3^0 p^6 + v_4^0 p^8 + \ldots \qquad (7)$$

Power counting in Q for s wave scattering can be implemented straightforwardly. All of the coefficients in the preceding expansion are assumed to be order Q^0 except the first term ($-1/a_{\text{short}}$) which will be taken to be order Q^1 to reflect the unnaturally large scale of the scattering length. Power counting for the long range part of the potential simply follows Weinberg's analysis [2], with the proviso that the potentials are only used for $r > R$. At order Q^2 in $p\cot(\delta)$, only the simple one-pion-exchange contribution to the V_{long} contributes. The power counting also justifies an iterative solution of the Schrödinger equation for $r > R$ along the lines of a conventional Born series. It differs from the usual Born series in that the boundary conditions at R are implemented. Finally, Q counting is used in expanding out the final expression for $k\cot(\delta)$.

Carrying out this program gives the following expression for $k\cot(\delta)$ at order Q^2 for the ${}^1S_0$ channel

$$\begin{aligned} p\cot(\delta) &= -\frac{1}{a_0} + m_\pi^2\left[d + \frac{g_A^2 M}{16\pi f_\pi^2}\left(\gamma + \ln(m_\pi R)\right)\right] \\ &+ \frac{1}{2} r_e^0 p^2 - \frac{g_A^2 M}{64\pi a_0^2 f_\pi^2}\left(\frac{m_\pi^2}{p^2}\right)\ln\left(1+\frac{4p^2}{m_\pi^2}\right) \\ &+ \frac{g_A^2 m_\pi M}{16\pi a_0 f_\pi^2}\left(\frac{m_\pi}{p}\right)\tan^{-1}\left(\frac{2p}{m_\pi}\right) + \frac{g_A^2 m_\pi^2 M}{64\pi f_\pi^2}\ln\left(1+\frac{4p^2}{m_\pi^2}\right) \end{aligned} \qquad (8)$$

The convention used here has $f_\pi = 93$ MeV. Apart from well-known parameters from pionic physics, there are three parameters—a_0, d and r_e^0, where $1/a_{\text{short}}$ from eq. (7) is rewritten as $1/a_0 + dm_\pi^2$ with $1/a_0 \sim Q$, and $dm_\pi^2 \sim Q^2$. These parameters fix the energy dependence of the wave function at the matching scale R; renormalization group invariance requires d to depend on R logarithmically. This form is precisely equivalent to the calculation in PDS, provided that the following identifications are made between the coefficients used above and those used in PDS with the notation of ref. 1.

$$\frac{4\pi}{M}\frac{1}{-\mu+1/a_0} = C_0$$

$$\frac{1}{2}r_e^0 = \frac{C_2 M}{4\pi}\left(\mu^2 - \frac{2\mu}{a_0} + \frac{1}{a_0^2}\right)$$

$$m_\pi^2 \quad \left[d + \frac{g_A^2 M}{16\pi f_\pi^2}\left(\gamma + \ln(m_\pi R)\right)\right] =$$

$$\frac{g_A^2 M}{16\pi f_\pi^2}\left(m_\pi^2 \ln\left(\frac{m_\pi}{\mu}\right) - m_\pi^2 + \frac{1}{a_0^2} - 2\frac{\mu}{a_0} + \mu^2\right)$$

$$+ \quad \frac{D_2 M}{4\pi}\left(m_\pi^2\mu^2 - \frac{2m_\pi^2\mu}{a_0} + \frac{m_\pi^2}{a_0^2}\right) \tag{9}$$

Formally, this is encouraging in the sense that it explicitly demonstrates the scheme independence of physical quantities. At the same time, there is an important hint of trouble which may lie ahead. In doing the derivation the matching scale, R was taken to scale as Q^0 and the quantity $m_\pi R$ as order Q^1. To obtain the final expression only the leading term in $m_\pi R$ is kept. If $R \sim 1/m_\pi$ this is clearly problematic, and from the previous discussion about potentials, we see that $m_\pi R \sim 1$.

It should also be stressed that the quantity $p\cot(\delta)$ is an extremely useful observable to work with in Q counting. Unlike the amplitude itself, there are no poles near $p = 0$; thus the issues of reorganizing the expansion as in OS [4] do not come up. Moreover, the expression is valid near $p = 0$ (assuming that Q counting holds) so it should be useful for ultra low energy scattering.

4 Low Energy Theorems

One difficulty with eq. (8) is that it is given in terms of a_0 which is the scattering length for the short distance potential only; as such it is not an observable. However, one can express everything in terms of physical observables and in doing so develop "low energy theorems" [8]. The trick is to rel ate a_0 to the physical scattering length as follows:

$$-\frac{1}{a} = -\frac{1}{a_0} + m_\pi^2 \left[d + \frac{g_A^2 M}{16\pi f_\pi^2} \left(\gamma + \ln(m_\pi R)\right)\right] + \frac{g_A^2 M}{16\pi f_\pi^2} \left(\frac{2m_\pi}{a_0} - \frac{1}{a_0^2} \right)$$

$$= -\frac{1}{a_0} + \mathcal{O}(Q^2/\Lambda) \tag{10}$$

Therefore in all of the $\mathcal{O}(Q^2)$ terms in eq. (8) one can replace a_0 by the physical a; the error in doing this is $\mathcal{O}(Q^3)$ which is one order beyond the order at which I am working. One gets the following expression originally derived in ref. 8.

$$\begin{aligned} p\cot(\delta) = & -\frac{1}{a_0} + m_\pi^2 \left[d + \frac{g_A^2 M}{16\pi f_\pi^2} \left(\gamma + \ln(m_\pi R)\right)\right] \\ & + \frac{1}{2} r_e^0 p^2 - \frac{g_A^2 M}{64\pi a^2 f_\pi^2} \left(\frac{m_\pi^2}{p^2} \right) \ln\left(1 + \frac{4p^2}{m_\pi^2}\right) \\ & + \frac{g_A^2 m_\pi M}{16\pi a f_\pi^2} \left(\frac{m_\pi}{p} \right) \tan^{-1}\left(\frac{2p}{m_\pi} \right) + \frac{g_A^2 m_\pi^2 M}{64\pi f_\pi^2} \ln\left(1 + \frac{4p^2}{m_\pi^2}\right) \end{aligned} \tag{11}$$

One can expand this as a Taylor series in p to obtain ERE coefficients. They are given by:

$$\begin{aligned} v_2 &= \frac{g_A^2 M}{16\pi f_\pi^2} \left(-\frac{16}{3a^2\, m_\pi^4} + \frac{32}{5a\, m_\pi^3} - \frac{2}{m_\pi^2} \right) \\ v_3 &= \frac{g_A^2 M}{16\pi f_\pi^2} \left(\frac{16}{a^2\, m_\pi^6} - \frac{128}{7a\, m_\pi^5} + \frac{16}{3m_\pi^4} \right) \\ v_4 &= \frac{g_A^2 M}{16\pi f_\pi^2} \left(-\frac{256}{5a^2\, m_\pi^8} + \frac{512}{9a\, m_\pi^7} - \frac{16}{m_\pi^6} \right) \\ & \cdots \end{aligned} \tag{12}$$

The preceding low energy theorems are valid to leading nontrivial order in Q counting; corrections are of *relative* order Q/Λ.

Several features of these expressions are notable. The first is that they are true predictions, independent of choices made in fitting, to this order in Q counting. As such they are low energy theorems. One consequence of this is that different schemes (*eg.* fitting out the pole as in OS rather than using $1/a$) will only give corrections at relative order Q/Λ. In this sense these predictions can be considered "low energy theorems " which become exact in the limit

Table 1: A comparison of the predicted effective range expansion coefficients, v_i, for the $^1S_0$ and $^3S_1$ channels with coefficients extracted from the Nijmegen partial wave analysis.

	v_2 (fm^3)	v_3 (fm^5)	v_4 (fm^7)
δ ($^1S_0$ channel)			
low energy theorem	-3.3	17.8	-108.
partial wave analysis	-.48	3.8	-17.
δ ($^3S_1$ channel)			
low energy theorem	-.95	4.6	-25.
partial wave analysis	.04	.67	-4.0

$(m_\pi, 1/a)/\Lambda \rightarrow 0$. The second significant fact is that all terms for all the expressions for the v_i coefficients diverge in the chiral limit of $m_\pi \rightarrow 0$. This implies that these quantities are dominated by pionic effects and hence are a good place to test whether the pionic parts of Q counting are working.

If the pionic parts of Q counting were under control one would expect that these predictions would work well. In practice, however, they work quite poorly. This can be seen in Table (1) where the prediction from the low energy theorems are compared with coefficients extracted from the Nijmegen partial wave analysis (PWA). The prediction from the low energy theorems are typically off by a factor of

5 or so. This suggests that pionic parts of the Q counting are failing rather badly.

One possible difficulty with the comparison of the v_i coefficients from the low energy theorems with the experimental data is that there is no experimental data. The coefficients extracted from the Nijmegen PWA were based on a fit to the smoothed "best fit". In principle one should do this fit including an error analysis based on the uncertainties. This cannot be done from the published data of the Nijmegen group as they did not publish information about correlated errors. Thus, one might wonder whether it is meaningful to extract high derivatives which are presumably rather sensitive to errors. A simple error estimate in ref. 4 concludes that the errors are likely to be too large to get any quantitative information about the v coefficients. From this one might conclude the disagreement between the low energy theorems and the "data" in Table (1) is due to an inability to extract the v from the scattering data in a reliable way. This is almost certainly not the case, however.

Table 2: A comparison of the effective range expansion coefficients, v_i, for the $^3S_1$ and $^3S_1$ channel predicted from the low energy theorem with coefficients extracted from the partial wave analysis and with three potential models—Nijmegen I, Nijmegen II and Reid 93—which were fit directly to the scattering data.

δ ($^3S_1$ channel)	v_2 (fm^3)	v_3 (fm^5)	v_4 (fm^7)
low energy theorem	-.95	4.6	-25.
partial wave analysis	.040	.672	-3.96
Nijm I	.046	.675	-3.97
Nijm II	.045	.673	-3.95
Reid93	0.033	.671	-3.90

The Nijmegen group made several independent fits to the scattering data. One was the PWA. The others were various potential models which were fit directly to the data (*i.e.* not to the PWA). These fits had a χ^2 per degree of freedom of essentially unity. Thus they can be regarded as independent fits to the data [10]. As the potential models have different forms from each other, they clearly have different systematic errors. Moreover in doing the least squares fit different models make different compromises in fitting individual data points so that they tend to explore the statistical errors. Thus, one might expect that the spread in the coefficients extracted in the different fits gives a reasonable feel for the scale of the uncertainty. Table (2) shows the v_i coefficients for these fits for the triplet channel [11], and it is manifestly clear that the spread in the effective range parameters as extracted from the three is vastly smaller than the difference with the predictions from the low energy theorems.

5 Re-summing the Effective Range Expansion

The effective range expansion parameters discussed in the previous section provide a dramatic way to see that the pionic parts of Q counting may have serious problems with convergence. However, there are a number of drawbacks with looking at the v_i coefficients. As noted above, there are ambiguities in the extraction from the data and it is hard to get reliable error bars. Moreover, the effective range expansion itself has a very limited radius of convergence. Because of the pion cut one expects the effective range expansion to converge only for $p < m_\pi/2$.

Of course, all of the low energy theorems for the v_i coefficients are contained in eq. (11). We can study this directly without expanding as a function

Table 3: A comparison of the shape function $\mathcal{S}(p^2) = p\cot(\delta) + 1/a - 1/2r_e p^2$ for the $^3S_1$ channel extracted from the Nijmegen partial wave analysis with the prediction by the low energy theorem.

lab energy (MeV)	$\mathcal{S}$ extracted (Mev)	$\mathcal{S}$ low energy theorem (Mev)
Deuteron Pole	-0.0017 ± 0.0125	-0.743
1	-0.00095 ± 0.00721	-0.0258
5	0.0428 ± 0.0194	-0.535
10	0.245 ± 0.047	-1.78
25	2.18 ± 0.14	-7.54
50	11.03 ± 0.24	-20.10

of k. In effect, this amounts to re-summing the effective range expansion and using $p\cot(\delta)$ as our fundamental quantity. There are two obvious advantages to doing this: First, one can avoid the problem of extracting the v_i coefficients from noisy data and instead we can compare directly with the partial wave analysis (which includes error estimates). Second, one is no longer restricted to $p < m_\pi/2$ since the re-summed expression is valid over the entire domain of Q counting. Unfortunately, $p\,cot(\delta)$ does not isolate the pionic contributions from the rest, and at low p it is dominated by $1/a$ physics and the fitting procedure which gives the effective range. There is a clean way to finesse this problem, however. Rather than study $p\cot(\delta)$ directly, one can study the following "shape function"[12]:

$$\mathcal{S}(p) = p\cot(\delta) - (-1/a + 1/2r_e p^2) \tag{13}$$

which is just the re-summed effective range expansion with the first two terms subtracted off. This has the advantage of removing completely the sensitivity to $1/a$ and the fitting of the effective range. The quantity is completely pion dominated since in a theory with pions integrated out $\mathcal{S}(p)$ is $\mathcal{O}(Q^3)$, while in a theory with explicit pions it is $\mathcal{O}(Q^2)$.

In Table (3) we see the low energy theorem prediction for $\mathcal{S}(p)$ compared with values extracted from the Nijmegen partial wave analysis for the triplet channel. Note the disagreement is quite pronounced. Moreover, note that error estimates are given for the results extracted from the scattering data. It is manifestly clear that the discrepancies are *not* due to uncertainties in the data. Again this suggests that the pionic parts of Q counting are not predictive at leading nontrivial order—at least not for this observable.

6 Conclusions

The development of Q counting is an extremely important step in our understanding of effective field theory for nuclear physics. However, as stressed in this talk Q counting involves two small mass parameters, $1/a$ and m_π. While there is every indication that the part of the theory based on expanding in $1/(a\Lambda)$ is working well, the chiral counting is far more problematic. The expansion has no predictive power for the effective range parameters and the shape function at leading nontrivial order (NLO). As both of these quantities are chirally sensitive this failure suggests that the chiral expansion may not be well under control.

There are a number of possibilities. The most optimistic one is simply that the NLO calculation is not adequate, and if one works at higher order all will be well. The most pessimistic possibility is that that the chiral expansion is not converging and that this failure is general. Clearly, the way to resolve the situation is to work at higher order. It is important when doing so, however, to focus on observables which are highly sensitive to the pion physics.

Acknowledgments

Most of the work discussed in this talk was done in collaboration with James Hansen. The research was funded by the U.S. Department of Energy under grant no. DE-FG02-93ER-40762.

References

1. D.B. Kaplan, M.J. Savage and M.B. Wise, *Phys. Lett.* B **424**, 390 (1998);*Nucl. Phys.* A **534**, 329 (1998).
2. S. Weinberg, *Phys. Lett.* B **251**, 288 (1990); *Nucl. Phys.* B **363**, 3 (1991); *Phys. Lett.* B **295**, 114 (1992).
3. Examples of such calculations include C. Ordonez and U. van Kolck, *Phys. Lett.* B **291**, 459 (1992); C. Ordonez, L. Ray and U. van Kolck, *Phys. Rev. Lett.* **72**, 1982 (1994) ; *Phys. Rev.* C **53**, 2086 (1996) ; U. van Kolck, *Phys. Rev.* C **49**, 2932 (1994); T.-S. Park, D.P. Min and M. Rho, *Phys. Rev. Lett.* **74**, 4153 (1995); *Nucl. Phys.* A **596**, 515 (1996); T.-S. Park, K. Kubodera, D.-P. Min, M. Rho *Nucl. Phys.* A **646**, 83 (1999).
4. T. Mehen and I. W. Stewart, *Phys. Lett.* B **445**, 378 (1999); T. Mehen and I. W. Stewart, `nucl-th/9809095`; `nucl-th/9901064`.
5. T. D. Cohen and J. M. Hansen, *Phys. Lett.* B **440**, 233 (1998).

6. D. B. Kaplan, M. J. Savage and M. B. Wise, *Phys. Rev.* C **59**, 617 (1999); J.-W. Chen, H. W. Griesshammer, M. J. Savage, R. P. Springer, *Nucl. Phys.* A **644**, 221 (1998); *Nucl. Phys.* A **644**, 245 (1998); M. J. Savage, R. P. Springe *Nucl. Phys.* A **644**, 235 (1998); J.-W. Chen, `nucl-th/9809095`; M. Savage, K. A. Scaldeferri, M. B. Wise, `nucl-th/9811029`.
7. K. A. Scaldeferri, D.R. Phillips, C.-W. Kao and T. D. Cohen, *Phys. Rev.* C **56**, 679 (1997).
8. T. D. Cohen and J. M. Hansen, *Phys. Rev.* C **59**, 13 (1999).
9. V. G. J. Stoks, R. A.M. Klomp, M. C. M. Rentmeester, and J. J. de Swart, *Phys. Rev.* C **48**, 792 (1993); on line at http:nn-online.sci.kun.nl
10. V. G. J. Stoks, R. A.M. Klomp, C. P. F. Terheggen, and J. J. de Swart, *Phys. Rev.* C **49**, 2950 (1994).
11. J. J. de Swart, C. P. F. Terheggen, V. G. J. Stoks, nucl-th/9509032, invited talk at "Dubna Deuteron 95".
12. T. D. Cohen and J. M. Hansen, `nucl-th/9901065`, to be published in Phys. Rev C.

Discussion during T. Cohen's presentation

Chair: *M. Luke (University of Toronto)*
Notes taken by *D. R. Phillips (University of Washington)*

Tom Cohen presented results derived with James Hansen on low-energy theorems for the parameters $v_2, v_3, \ldots$ which multiply the terms of order p^4 and beyond in the effective range expansion. These quantities were calculated up to their leading non-vanishing order. The results are then exact at this order in Q-counting. Thus, they are apparently valid provided that $m_\pi R \ll 1$, where R is the range of the short-distance piece of the NN potential which is not pionic in character. The drastic failure of these low-energy theorems at this order led Cohen to seriously question the validity of Q-counting for the "real world", where, he claimed, estimates of R from various physical arguments suggest that $m_\pi R \ll 1$ does not hold.

Cohen's talk provoked much discussion. Ulf Meißner questioned the usefulness of arguments about the convergence of results when only their leading-order piece has been calculated. As pointed out by Kevin Scaldeferri, Meißner, and Tom Mehen, this issue is especially acute for these observables, since a large error in the v_n does not necessarily lead to any noticeable problem in $p \cot \delta$. Empirically, this is because the effective range expansion is dominated by the first two terms. Moreover, formal arguments suggest that in an NNLO calculation a term will be present that will cancel the error noticed by Cohen and Hansen in v_2 almost completely.

Cohen countered by pointing out that quantities such as v_n and $S(p^2)$ should have Q-counting expansions of their own, and so the dramatic failure of Q-counting to predict them at leading order was troubling. After all, the relative size of corrections to the results he displayed should be Q/Λ_{NN}, which, we would hope, would be small. While not disputing this argument, David Kaplan noted that the Q-expansion of all the v_ns begins at negative powers, with $v_n \sim Q^{-2n+2}$, $n \geq 1$. He mentioned that in the toy model he investigated with Jim Steele the quantities v_n tended to oscillate strongly until positive powers in their Q-expansion were reached, at which point they settled down to a convergent series. Numerical results supporting this statement were later presented by Steele.

Christoph Hanhart asked how we knew the behavior of am_π in the chiral limit. Cohen pointed out that in Q-counting $m_\pi a$ is of order 1, and so stays finite in the chiral limit. He also made a plausibility argument for the case of the ${}^1S_0$ channel. There the one-pion exchange potential goes to zero as $m_\pi \rightarrow 0$, and is, in fact, quite weak even for the physical value of m_π. Thus, in that channel, at least, it seems that a is a quantity which is largely determined by short-distance, non-chiral, physics.

Martin Savage asked about the status of these controversial observables v_n in the NNLO calculations of Rupak & Shoresh and Mehen & Stewart.

Finally, Bira van Kolck mentioned that the recent Nijmegen phase-shift analysis of Rentmeester et al. has found that the inclusion of two-pion exchange allows the radius of the truly short-distance piece of the NN potential to be decreased, thus perhaps allowing $m_\pi R \ll 1$ to indeed be valid.

CONVERGENCE ISSUES IN NUCLEON-NUCLEON EFFECTIVE FIELD THEORY

J. V. STEELE
Department of Physics, The Ohio State University
Columbus, OH 43210-1106, USA
E-mail: jsteele@mps.ohio-state.edu

The origin of the breakdown scale in an effective field theory treatment of nuclear forces is investigated. Different organizational schemes used in the nonperturbative calculation of nucleon-nucleon scattering seem to break down for different physical reasons. It is argued that the same physics is actually responsible, namely strong two-pion exchange as also observed in the pion scalar form factor. The prospect of extending the applicability of the effective field theory beyond this breakdown point is discussed.

1 Introduction

Strong interactions at low energies can be described by an effective field theory (EFT) of pions and nucleons with the nucleons treated nonrelativistically.[1] Chiral symmetry prescribes how pions interact. All other interactions are parameterized by the most general local Lagrangian consistent with the symmetries of quantum chromodynamics (QCD).

The coefficients of these general terms can either be predicted directly from QCD, which may be possible in future lattice calculations, or fit from existing data. I will focus on nucleon-nucleon (NN) scattering in the following. The data here is quite good and should lead to an excellent determination of these constants. However, since nuclear forces produce bound states, a nonperturbative treatment of the EFT is required, which adds a new twist to this procedure.[1]

Any EFT will break down at a scale associated with the underlying physics not taken into account in the original Lagrangian. For nuclear EFTs, an optimist may believe this will occur when the center of mass momentum of the nucleons excite either the exchange of the next heaviest meson after the pion, $p \sim m_\rho$, or the nucleon-Δ mass difference $p = \sqrt{M(M_\Delta - M)} \sim 525$ MeV. In practice, the breakdown scale seems to be much lower, $p \sim 300$ MeV. This could be due to anything from quark substructure to subtleties in the fitting.

Furthermore, two different organizational schemes for calculations in the EFT seem to imply different physical reasons for the breakdown. This would be unfortunate, since alternative ways to organize the corrections should have no effect on the physics. By using some models, I will show what actually does

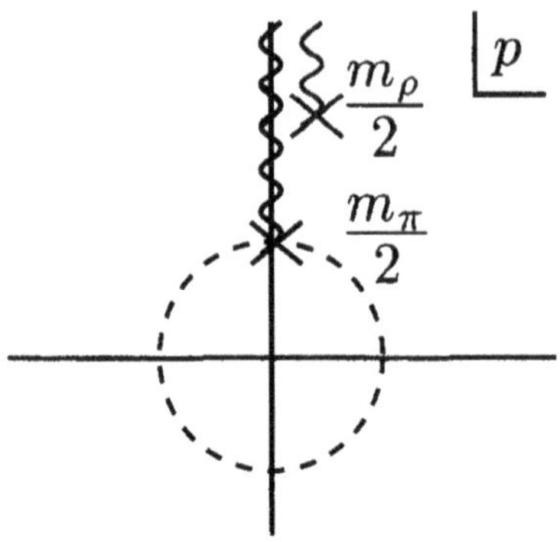

Figure 1: The analytic structure of an observable for the two-Yukawa model of Eq. (1) in the complex momentum plane.

set this scale and then use this to make connections to the real world.

Extending the applicability of the EFT to the scale associated with the Δ seems like a small achievement, but could make calculations of nuclear matter converge much faster. In that case, the momentum of interest is on the order of the Fermi momentum at saturation $p_F \sim 280$ MeV, which currently is too close to the breakdown scale for an EFT analysis to be useful.

2 Removing vs. Weakening the Pion

Assume, for now, that interactions between two nucleons are exactly given by potential exchange of both a long- and short-ranged meson with masses $m_\pi = 140$ MeV and $m_\rho = 770$ MeV respectively,

$$V_{\rm exact}(r) = -\alpha_\pi \frac{e^{-m_\pi r}}{r} - \alpha_\rho \frac{e^{-m_\rho r}}{r} \,. \tag{1}$$

The pion coupling is taken to be $\alpha_\pi = g_A^2 m_\pi^2/(16\pi f_\pi^2) \simeq 0.075$, as in the real world, and the rho coupling $\alpha_\rho = 1.05$ is tuned to give a large scattering length $a = -23.4$ fm, as observed in data for the $^1S_0$ partial wave of NN scattering. The following discussion does not depend on the value of the potential at the origin, so I will take $V_{\rm exact}$ to be zero there for simplicity.

This potential generates observables with analytic structure as shown in Fig. 1, specified by a cut at both $m_\pi/2$ and $m_\rho/2$. The existence of these cuts indicates that a low-momentum expansion of an observable will not converge for momenta above $m_\pi/2$. An EFT can only go beyond this point if it either exactly reproduces or sufficiently approximates the analytic structure of the pion.

This can be illustrated by focusing on the most general Lagrangian without pions (I will restrict myself to the $^1S_0$ channel below)

$$\mathcal{L}_{\rm EFT} = N^\dagger \left[i\partial_t + \frac{\nabla^2}{2M} \right] N - C_0 (N^\dagger N)^2 + \frac{1}{2} C_2 \left[(N^\dagger \nabla N)^2 + h.c. \right] + \dots , \quad (2)$$

which corresponds to a nonrelativistic potential between two nucleons of

$$V_{\rm EFT}(p, p') = C_0 + \frac{1}{2} C_2 \left(p^2 + p'^2 \right) + \dots , \quad (3)$$

where p and p' are the center of mass momentum of the incoming and outgoing nucleons. Since these interactions depend on powers of momentum, a naive calculation of the NN scattering phase shifts from $\mathcal{L}_{\rm eff}$ leads to ultraviolet divergences. This just reflects our ignorance of the high-energy behavior of this theory.

Two possible procedures to regulate these divergences are: i) adding a cutoff to all integrals to eliminate or suppress the high momentum states and ii) using dimensional regularization, which amounts to throwing out any integrals that diverge like powers of momentum. The second procedure may seem a bit counterintuitive at first, but is simpler to implement since it does not break any symmetries of the underlying theory and therefore does not require additional counterterms. Perturbative calculations in chiral perturbation theory have confirmed the equivalence of these regularization procedures.

Nonperturbative calculations, on the other hand, show that a naive implementation of dimensional regularization is incomplete. A generalization to account for integrals that linearly diverge is required. This regularization is called power divergence subtraction [2] (PDS).

After regulating the theory, the EFT can be organized to give predictions order-by-order in powers of an expansion parameter. Since the low-energy constants C_{2n} are not determined by the symmetries of QCD, they must be fit to data. In the process, they will typically pick up the next scale Λ of physics not explicitly accounted for in the EFT, leading to $C_{2n}p^{2n} \sim (4\pi/\Lambda^2)(p/\Lambda)^{2n}$. Therefore interactions with more derivatives are suppressed by additional powers of p/Λ. Power counting in this parameter allows Eq. (2) to be truncated to the first few terms, with more accurate results obtained by keeping more terms in the momentum expansion.

There are two prevalent ways of implementing power counting in nonperturbative effective field theories. The first is to count powers of Λ in the potential $V_{\rm EFT}$ and then solve the Lippmann-Schwinger equation. Such a procedure sums an infinite number of Feynman diagrams to generate the nonperturbative amplitude $\mathcal{A}$, related to the phase shift δ by

$$\mathcal{A} = \frac{4\pi/M}{p\cot\delta - ip}\,. \tag{4}$$

This method of power counting is called Λ-counting[3] and is in the spirit of the Hamiltonian theory of Wilson, first applied to nuclear EFTs by Weinberg.[4] An alternative is to expand $\mathcal{A}$ in powers of $Q \equiv (p, m_\pi, 1/a)$, motivated by chiral perturbation theory, which expands in p and m_π. The nonperturbative nature of the interactions identifies $1/a$ as a small parameter as well, justifying its inclusion in the expansion. This method of power counting is called Q-counting.[2,5]

For a theory without pions, such as Eq. (2), the underlying scale Λ is just m_π. Both power counting procedures work as long as momenta are less than this scale, reproducing the effective range expansion

$$p\cot\delta = -\frac{1}{a} + \frac{1}{2}p^2\sum_{n=0}^{\infty} r_n\frac{p^{2n}}{\Lambda^{2n}}\,. \tag{5}$$

The effective ranges, which are determined from NN scattering data,[6] also exhibit this scale $r_n \sim 1/\Lambda \sim 1/m_\pi$. To extend the EFT to higher momenta, pions need to be explicitly added to the effective Lagrangian. This hopefully will bring Λ up to the QCD scale of m_ρ or possibly even 1 GeV.

The inclusion of pions nonperturbatively is different in the two power counting procedures. Every iteration of one-pion exchange contributes to the pion cut in Fig. 1. In general, n iterations are proportional to[a]

$$n\text{-Pion Exchange} \propto \left[\frac{m_\pi}{\Lambda_{\rm NN}}\frac{m_\pi^2}{4p^2}\ln\left(1+\frac{4p^2}{m_\pi^2}\right)\right]^n, \tag{6}$$

where $\Lambda_{\rm NN} \simeq 300$ MeV is just another way of expressing the strength of the interaction via the relation $\alpha_\pi = m_\pi^2/(M\Lambda_{\rm NN})$. Λ-counting includes this long-range part of one-pion exchange fully at leading order, removing the effect of the pion cut. Q-counting pushes one-pion exchange to subleading order and iterated one-pion exchange to even higher orders. Therefore, each order in the Q-counting expansion weakens the pion cut by an additional power of $m_\pi/\Lambda_{\rm NN}$, as seen in Eq. (6). The topic of the next section is whether this simpler implementation of power counting, also called perturbative pions, can be as successful as Λ-counting.

[a] This can be seen by studying the exact solution of an exponential potential,[8] which has the same cut structure as a Yukawa or by looking at the analytic expressions of higher Yukawa loops, which exhibit the cut structure through Polylogarithms.[9]

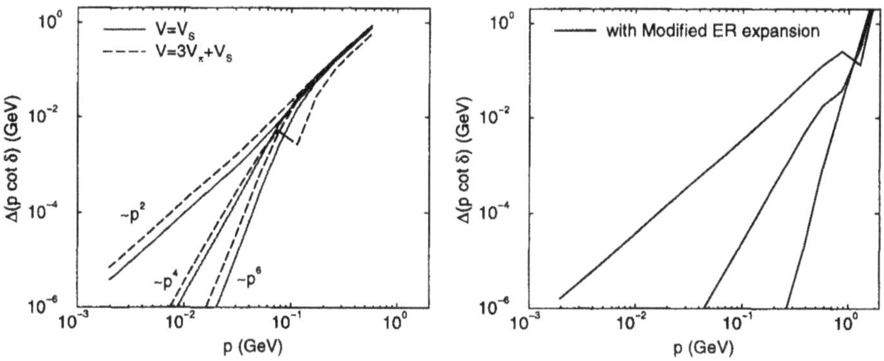

Figure 2: The error in $p\cot\delta$ for a pion coupling three times stronger than the real world. The left plot is from a normal fit to the low-energy constants both without (solid) and with (dashed) explicit pions. The right plot uses the modified effective range expansion to fit.

3 New Fitting Procedures

The NN EFT is expected to work best at low-momentum. So a natural procedure to fix the low-energy constants C_{2n} is to calculate an observable, like $p\cot\delta$, and fit near threshold. However, there are some complications originating from the nonperturbative nature of the problem. These must be cleared up before being able to make a statement about the breakdown scale.

3.1 Λ-counting and the Modified Effective Range Expansion

Most calculations that implement Λ-counting require a numerical treatment (although there are exceptions[5]). A nice way to show the systematic improvement expected of an EFT in this case is to use error plots[3,7,8] as in Fig. 2. As more and more constants are added to the effective potential Eq. (3), the error in $p\cot\delta$ improves systematically by powers of p^2, reflected in the improving slopes. The point at which all errors are of the same order is where the theory is expected to break down.[7] The solid lines of the left plot of Fig. 2 show the case without explicit pions. The breakdown point, or radius of convergence, of the EFT is around $m_\pi/2$, as expected from the discussion in Section 2.

Adding pions to the short-ranged Lagrangian Eq. (2) as dictated by chiral symmetry should improve the radius of convergence.[10] However, one needs to be careful when implementing this with cutoff regularization. If the pion coupling were three times stronger, for example, adding it to the Lagrangian

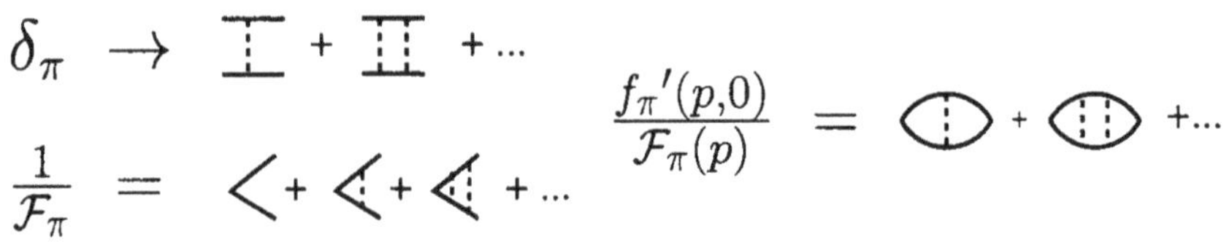

Figure 3: Symbolic definition of quantities in the modified effective range function in Eq. (7), which sums all the pion contributions.

would *not* improve the breakdown scale. [8] This is illustrated by the dashed lines in the left plot of Fig. 2. Chiral symmetry helps in that the coupling is small, but an EFT should work regardless of the strength of the potential being added. What went wrong?

Note that the effective ranges r_n of Eq. (5), which are used to fix the values of the C_{2n}, exhibit the scale m_π. If this scale were completely removed from the fitting procedure, then the constants should take on the proper scale m_ρ. One way to do this is to construct a function similar to $p\cot\delta$ which is analytic all the way to the rho cut in Fig. 1. In proton-proton scattering analyses, such a procedure is used to remove the threshold cuts coming from photon exchange. Called the modified effective range expansion, [11] it can be used here to remove the effect of pion exchange by defining

$$K(p) \equiv \frac{p\cot(\delta-\delta_\pi)}{|\mathcal{F}_\pi(p)|^2} + \mathrm{Re}\left[\frac{f'_\pi(p,0)}{\mathcal{F}_\pi(p)}\right] \equiv -\frac{1}{\widetilde{a}} + \frac{1}{2}p^2\sum_{n=0}^{\infty}\widetilde{r}_n\frac{p^{2n}}{\widetilde{\Lambda}^{2n}} \,. \qquad (7)$$

All possible insertions of one-pion exchange are taken into account in this formula, as seen pictorially in Fig. 3. For the model potential Eq. (1), this function is analytic up to $m_\rho/2$ and has modified effective ranges $\widetilde{r}_n \sim 1/\widetilde{\Lambda} \sim 1/m_\rho$. Fitting the C_{2n}'s using the function $K(p)$ leads to much better results, shown in the right plot of Fig. 2. The radius of convergence is now on the order of m_ρ as expected.

Using a cutoff to regulate the theory mixes long- and short-distance physics, which is what led to the premature breakdown. [8] The modified effective range expansion removes the long-distance physics from the observable, allowing a clean fit of the low-energy constants to the short-distance physics. In principle, this can also be achieved by the addition of more counterterms, but the modified effective range expansion is a simpler alternative. Luckily, one-pion exchange in the real world is weak enough that an improper treatment of this

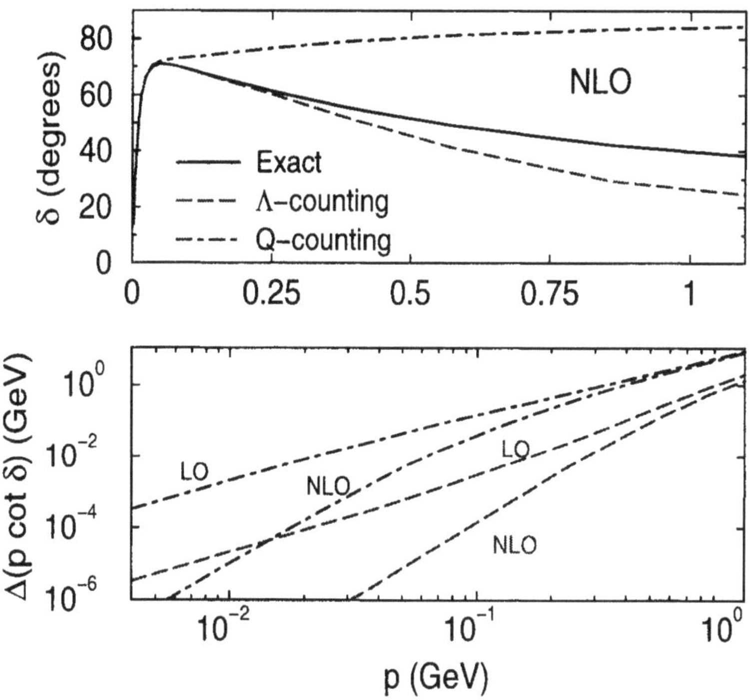

Figure 4: Comparison of the two power counting procedures for the two-Yukawa model Eq. (1) using a fit at low momentum.

malady does not make a difference.[8] However, this might not be the case for two-pion exchange as will be discussed in Section 4.

3.2 Q-counting and a Global Fit

A low-momentum fit to data using Q-counting and perturbative pions also breaks down much earlier than expected.[8] In Fig. 4, the leading order (LO) and next-to-leading order (NLO) calculations using Q-counting are compared to those using Λ-counting. The Λ-counting result has a radius of convergence around m_ρ after the low-energy constants are matched to the modified effective range expansion.

Implementing the modified effective range expansion is inconsistent with the idea of perturbative pions, and so the Q-counting result breaks down around m_π. However, it is important that the low-energy constants only pick up the (unknown) short-distance physics, and matching to the modified effective range expansion is the most convenient way to do this. Using PDS, this would lead to the expressions [12]

$$C_0 = \frac{4\pi/M}{-\mu + 1/\tilde{a}}, \qquad C_2 = \frac{MC_0^2}{4\pi}\frac{1}{2}\tilde{r}_0, \qquad \ldots, \tag{8}$$

which only depend on the short-distance physics.[b] One important modification to this would be to match the pole of the amplitude $\mathcal{A}$ exactly [13]

$$C_0 = \frac{4\pi/M}{-\mu + 1/a}. \tag{9}$$

Otherwise, terms attempting to correct the position of the pole would enter as double poles at higher orders messing up the power counting. This modification requires perturbative corrections to C_0 that compensate the pion contributions at each order. However, this still leaves the other C_{2n}'s dependent on the modified effective ranges $\tilde{r}_n$, which cannot be obtained using perturbative pions.

A method must therefore be devised to carry out the matching of Eq. (8). It is reasonable to assume that when the modified effective ranges are set to the correct values, the error of the EFT will be less than when the ranges are at some random values. This can be checked by using a model. Replacing the Yukawas in Eq. (1) by delta-shells

$$-\alpha_i \frac{e^{-m_i r}}{r} \longrightarrow -g_i\, \delta\left(r - \frac{1}{m_i}\right), \tag{10}$$

an analytic expression for the exact answer can be obtained. [12] Expanding the exact amplitude in powers of Q, it can be compared to the blind EFT fit of the low-energy constants to data using Q-counting. The results coincide best when the modified effective ranges are varied to globally minimize the χ^2 for data in a representative momentum range like $[m_\pi, m_\rho]$.

The results are shown up to next-to-next-to-leading order (NNLO) in Fig. 5. They suggest that for Q-counting, a constrained global fit of the EFT to data to fix the low-energy constants is better than a low-momentum fit. The constraint is that the modified effective ranges, not the low-energy constants themselves, are varied to achieve the best fit.

One thing to notice about Fig. 5 is that the errors do not show the systematic improvement in slope characteristic of the earlier error plots. Although the

[b] Although the renormalization scale μ appears in these low-energy constants, all observables are independent of it.

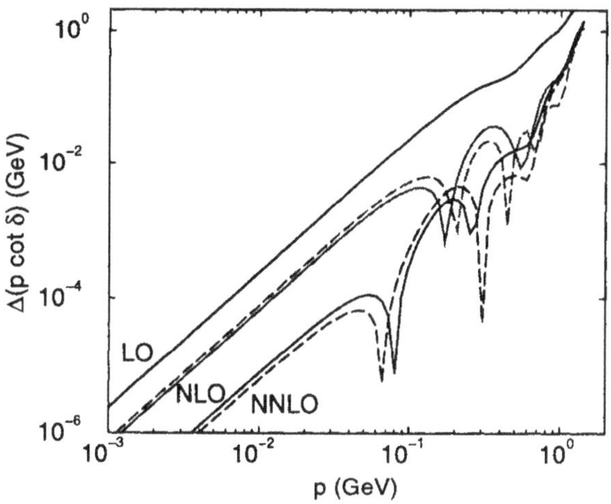

Figure 5: Q-counting expansion of the two delta-shell model. The exact expression (solid) coincides nicely with the EFT (dashed) when a global fit is minimized with respect to the modified effective ranges. The dashed line at LO is on top of the solid line.

scattering length is fit exactly to reproduce the pole in the amplitude Eq. (9), the other terms in the effective range expansion have corrections at each order in the Q-counting. A comparison of Fig. 5 with the right plot of Fig. 2 shows that accuracy is sacrificed near threshold to obtain the global fit needed for this power counting. In addition, the ability to read the EFT radius of convergence from the error plot is gone.

Another way to estimate the breakdown scale is therefore needed. The calculated amplitude $\mathcal{A}$ starts at order Q^{-1} and has a contribution $\mathcal{A}_n$ at each order Q^n in the expansion. Up to this point, I have extracted the Q-counting phase shift by expanding the expression [2]

$$\delta = \frac{1}{2i} \ln\left[1 + i\frac{Mp}{2\pi}(\mathcal{A}_{-1} + \mathcal{A}_0 + \ldots)\right] \tag{11}$$

$$\rightarrow \frac{1}{2i} \ln\left(1 + i\frac{Mp}{2\pi}\mathcal{A}_{-1}\right) + \frac{Mp}{4\pi}\left(\frac{\mathcal{A}_0}{1 + i\frac{Mp}{2\pi}\mathcal{A}_{-1}}\right) + \ldots, \tag{12}$$

but I could also have kept the full expression Eq. (11). These two expressions are equivalent up to terms that are higher order in the Q expansion. Those extra terms, though, are an estimate of the corrections to the actual result and depend on the radius of convergence. The breakdown for Q-counting can

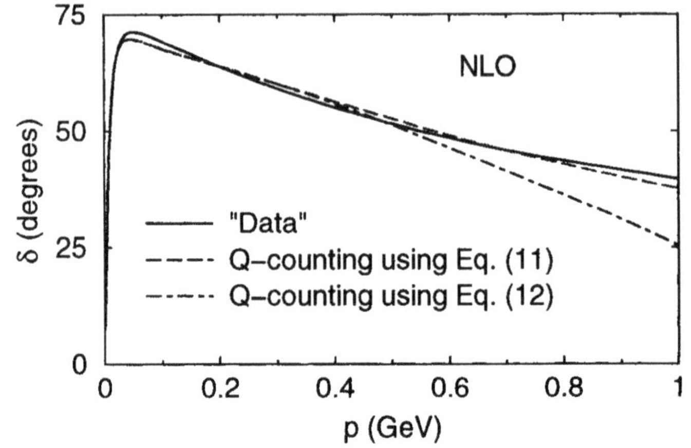

Figure 6: Determination of the Q-counting breakdown scale for the two-Yukawa model.

be identified as the point at which the two expressions Eqs. (11) and (12) diverge from each other. Doing this exercise for the two-Yukawa model results in Fig. 6. There it is easy to see the actual breakdown is somewhere around m_ρ as expected.

4 Data and the Role of Two-Pion Exchange

Now that the two power counting methods are well understood and give the proper radius of convergence for a model, it is appropriate to apply them to actual NN-scattering data. Using the Nijmegen partial wave analysis, [6] the radius of convergence for each of the two different power counting schemes is shown in Fig. 7. The left plot shows that, for Λ-counting with one-pion exchange included as in the previous section, the breakdown is around $p \sim$ 300 MeV. The right plot shows a similar scale for Q-counting with perturbative pions.

This scale is less than expected and seems to be set by physics other than the ρ or Δ. Early calculations in Q-counting predicted a breakdown at $\Lambda_{\rm NN}$, which would coincide with the observed scale. However, recall that in the previous section I showed that Q-counting broke down around m_ρ in the two-Yukawa model. That calculation had identical content with respect to one-pion exchange, indicating that the 300 MeV comes from different physics. A closer study of Λ-counting determines that two-pion exchange should be included at

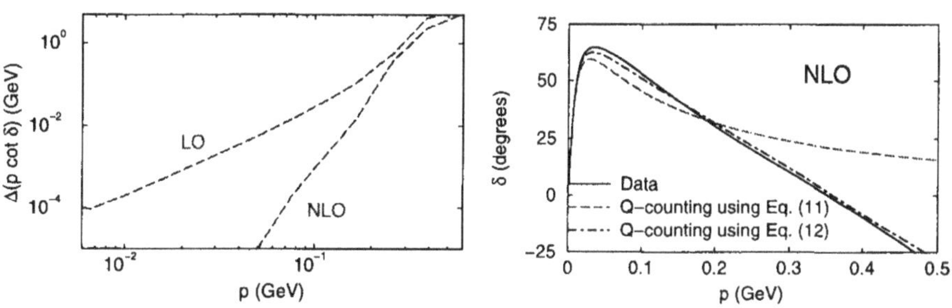

Figure 7: Results for $^1S_0$ NN-scattering data. The breakdown scale is determined by the left plot for Λ-counting and the right plot for Q-counting.

NLO.[10]

The fact that the actual data behaves so much differently from the two-Yukawa model indicates the model was not sophisticated enough. Furthermore, no reasonable choices for the parameters of this model can give as large an effective range $r_0 = 2.63$ fm as seen in the $^1S_0$ channel.[8] Until now I have assumed that m_π and m_ρ are the underlying scales, but nuclear physics wisdom tells us there are actually two short-range Yukawas

$$V(r) = -\alpha_\pi \frac{e^{-m_\pi r}}{r} - \alpha_\sigma \frac{e^{-m_\sigma r}}{r} + \alpha_\rho \frac{e^{-m_\rho r}}{r}, \tag{13}$$

given by an attractive scalar (of mass between 500 and 600 MeV) and repulsive vector particle.[c] Taking the short-distance couplings to be much stronger than the pion $(\alpha_\sigma, \alpha_\rho) = (7, 14.65)$, the modified effective ranges $\tilde{r}_n$ of the data can also be reproduced.[8] This three-Yukawa model is reminiscent of the Bonn potential,[14] which is known to model the data well.

Applying the NN EFT to observables produced from Eq. (13) indeed leads to a breakdown around $m_\sigma/2 \sim 300$ MeV. I can verify the breakdown is not set by the two-pion threshold $2m_\pi$ within this model by varying the pion and scalar masses. The $\sigma(600)$ particle is a common parameterization in nuclear physics for two-pion exchange contributions. The large scalar coupling indicates that including such effects through chiral symmetry might require the modified effective range expansion. Such an analysis could improve the EFT radius of convergence.[d]

[c] The vector particle is normally taken to be the ω meson, but for continuity with the earlier discussion, I will call it the ρ meson.

[d] A preliminary calculation, inspired by discussions at this workshop, seems to indicate this is true.[15]

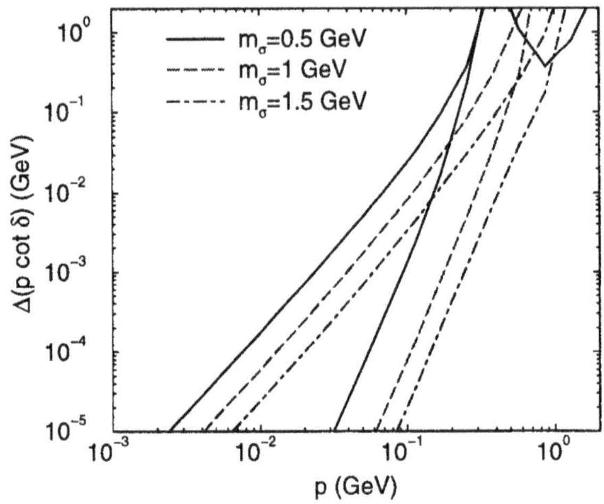

Figure 8: The three-Yukawa model Eq. (13) with different values for the scalar mass, but a, r_0, and $m_\rho - m_\sigma$ fixed. This shows that the radius of convergence tracks with the changing m_σ.

It might seem odd that two-pion effects could be strong, since they should be weak according to chiral symmetry. [3] However, there are other instances where two-pion effects have proven to be strong. The pion scalar form factor, defined by

$$\langle \pi^a(p)\pi^b(p')|\hat{m}\left(\bar{u}u + \bar{d}d\right)|0\rangle = \delta^{ab}\sigma_\pi\Gamma(s), \qquad s = (p+p')^2, \tag{14}$$

has been shown to have poor convergence in chiral perturbation theory, [16] breaking down just above threshold at $\sqrt{s} \simeq 400$ MeV. This comes about from infrared effects that require the summation of $(\ln s)^n$ terms to all orders. Carrying out this summation does lead to improved results and the generation of a "resonance"-like shape in the dispersion analysis just like a scalar resonance around 600 MeV. [16] The same phenomenon could occur in NN scattering and requires further study.

5 Red Herrings

The models developed to understand the breakdown scale better can also be used to shed light on other convergence issues. One feature of the three-Yukawa model is that it can reproduce the somewhat large effective range in NN scattering, which is $r_0 \sim 2/m_\pi$. Any reasonable choices of parameters for the two-Yukawa model can only give r_0 around $1/m_\pi$. Therefore, it is possible that the lower radius of convergence of the three-Yukawa model might have nothing to do with two-pion effects, but come from some type of correlation with the effective range r_0 instead. Choosing different parameters in Eq. (13), I can keep a, r_0, and the mass difference $m_\rho - m_\sigma$ fixed while changing m_σ from 500 MeV to 1500 MeV. The results are shown in Fig. 8, confirming that the radius of convergence is set by $m_\sigma/2$ alone.

Recently, the convergence of Q-counting has been challenged within the context of a low-energy theorem. [5] The point can be made by defining the observable

$$S(p^2) \equiv p \cot\delta - \left(-\frac{1}{a} + \frac{1}{2} r_0 p^2\right) , \tag{15}$$

and taking $p = |1/a|$. Then Q-counting has two types of corrections:

$$\frac{1}{a\Lambda_{\rm NN}} \ll 1 , \qquad\qquad \frac{m_\pi}{\Lambda_{\rm NN}} \sim \frac{1}{2} . \tag{16}$$

It was shown that the Q-counting prediction for $S(p^2)$ at NLO is very far from the data.

This can also be seen in a model. For an analytic understanding, I will use the two delta-shell potential of Section 3. Expanding in Q,

$$S\left(\frac{1}{a^2}\right) = (-22.9 + \underset{\rm NLO}{40.2} - \underset{\rm NNLO}{23.3} + \underset{\dots}{7.9} + \dots) \times 10^{-7} m_\pi , \tag{17}$$

one finds the NNLO term is larger than the NLO term. Only after a few more orders does the series start to converge.

Taking the exact answer, I can also expand in Q to see what is happening analytically. These expressions are messy and can be simplified by observing that even though $m_\pi a = \mathcal{O}(1)$ in Q-counting, it is actually of order 10 numerically. Expanding $S(1/a^2)$ in $(m_\pi a)^{-1}$ then should simplify the expression immensely without blurring its meaning, leading to

$$S\left(\frac{1}{a^2}\right) = \frac{m_\pi}{(m_\pi a)^4}\left[-\frac{m_\pi}{3\Lambda_{\rm NN}} + \frac{m_\pi^2}{\Lambda_{\rm NN}^2}\left(\frac{13}{15} + \frac{8\Lambda_{\rm NN}}{9m_\rho}\right) + \dots\right] \tag{18}$$

$$+ \frac{m_\pi}{(m_\pi a)^5}\left[\frac{4m_\pi}{15\Lambda_{\rm NN}} - \frac{m_\pi^2}{\Lambda_{\rm NN}^2}\left(\frac{68}{45} + \frac{4\Lambda_{\rm NN}}{9m_\rho}\right) + \dots\right] + \dots . \tag{19}$$

Terms in the second bracket and higher are suppressed by $(m_\pi a)^{-1}$ and can be neglected. Focusing on the first bracket of terms, it is clear that the expansion is in $m_\pi/\Lambda_{\rm NN}$ as expected for Q-counting. However, the leading term (which happens to be NLO for this particular observable) has a numerical factor of $\frac{1}{3}$, which suppresses it down to the order of the NNLO term. Therefore, the Q expansion of $S(p^2)$ works, but is slowly converging due to some inconvenient numerical factors as well as the size of the expansion parameter, Eq. (16).

6 Summary

Nonperturbative effective field theories are subtle to implement. Techniques used successfully in chiral perturbation theory, such as naive dimensional regularization and low-momentum fitting, do not produce the expected results in nonperturbative calculations. After a careful study using solvable models, new methods have been developed to deal with these complications.[2,8,12]

Pions enter the calculation differently depending on whether power counting is done in the QCD scale (Λ-counting) or the chiral scale (Q-counting). In both cases a proper determination of the low-energy constants from the short-distance physics requires more than just a low-momentum fit. Utilizing the modified effective range expansion to remove the effect of the long-distance pions from the process of fitting, the Λ-counting radius of convergence moves beyond the pion scale.[8] A constrained global fit minimized with respect to the modified effective ranges is required to achieve the same effect in Q-counting.[12]

Applying the new fitting procedures to $^1S_0$ NN scattering data results in a breakdown of about 300 MeV. Models can be used to confirm this breakdown is associated with the same physics for both power counting procedures. Using a three-Yukawa model, it was argued that including two-pion exchange will extend the EFT radius of convergence. It is not unreasonable to assume this effect is strong since a similar situation occurs in the pion scalar form factor.[16]

Finally, the convergence of Q-counting was investigated using an exactly solvable model. It was shown that some observables could have inconvenient numerical factors, such as occur in the low-energy theorem,[5] which suppress the leading orders and mask the behavior of the expansion. However, analytical results show that this expansion does converge, albeit slowly.

Acknowledgments

The work presented in this talk was done in collaboration with R. J. Furnstahl and D. B. Kaplan. I would like to thank the organizers, P. Bedaque, M. J. Savage, R. Seki, and U. van Kolck, for an enjoyable and productive workshop and the Institute of Nuclear Theory at the University of Washington for its hospitality. This work was supported by the National Science Foundation under Grants No. PHY–9511923 and PHY–9258270.

References

1. *Nuclear Physics with Effective Field Theory*, ed. R. Seki, U. van Kolck, and M. J. Savage (World Scientific, 1998);
   U. van Kolck, "Effective Field Theory of Nuclear Forces," `nucl-th/9902015`, and references therein.
2. D. B. Kaplan, M. J. Savage, and M. B. Wise, *Phys. Lett.* B **424**, 390 (1998); *Nucl. Phys.* B **534**, 329 (1998).
3. G. P. Lepage, "How to Renormalize the Schrodinger Equation," `nucl-th/9706029`;
   G. P. Lepage, *this volume.*
4. S. Weinberg, *Phys. Lett.* B **251**, 288 (1990); *Nucl. Phys.* B **363**, 3 (1991).
5. T. D. Cohen and J. M. Hansen, `nucl-th/9808006`; `nucl-th/9808038`; `nucl-th/9901065`;
   T. D. Cohen, *this volume.*
6. V. G. J. Stoks, R. A. M. Klomp, M. C. M. Rentmeester, and J. J. de Swart, *Phys. Rev.* C **48**, 792 (1993).
7. J. V. Steele and R. J. Furnstahl, *Nucl. Phys.* A **637**, 46 (1998).
8. J. V. Steele and R. J. Furnstahl, *Nucl. Phys.* A **645**, 439 (1999).
9. M. Binger, "Multiloop Diagrams in Nonrelativistic Field Theories and the Deuteron Quadrupole Moment," `nucl-th/9901012`;
   G. Rupak, *this volume*;
   T. Mehen, *this volume.*
10. U. van Kolck, Ph.D. thesis, U. Texas, Aug. 1993;
    C. Ordonez, L. Ray, and U. van Kolck, *Phys. Rev. Lett.* **72**, 1982 (1994); *Phys. Rev.* C **53**, 2086 (1996).
11. H. van Haeringen and L. P. Kok, *Phys. Rev.* A **26**, 1218 (1982);
    A. M. Badalyan, L. P. Kok, M. I. Polikarpov, and Y. A. Simonov, *Phys. Rep.* **82**, 31 (1982).
12. D. B. Kaplan and J. V. Steele, in preparation;
    D. B. Kaplan, *this volume.*
13. T. Mehen and I. W. Stewart, *Phys. Lett.* B **445**, 378 (1999).
14. R. Machleidt, *Adv. Nucl. Phys.* **19**, 189 (1989).
15. J. V. Steele and R. J. Furnstahl, in preparation.
16. J. Gasser and U.-G. Meissner, *Nucl. Phys.* B **357**, 90 (1991).

Discussion during J. Steele's presentation

Chair: *M. Luke (University of Toronto)*
Notes taken by *D. R. Phillips (University of Washington)*

Jim Steele's talk focused on the issue of where the power countings—both Q-counting and Weinberg's Λ-counting—break down. He discussed the importance of using the modified effective range expansion for doing fits, since that procedure avoids contaminating short-distance constants with physics of pion range. He pointed out that if the matching conditions were not chosen carefully this could be an issue even in the case of a perturbatively weak one-pion-exchange interaction. At this point Noam Shoresh raised the issue of pionic contamination of the scattering length in Steele's calculation. Steele explained that if a fitting procedure using the modified effective range expansion was employed then the parameters C_{2n} were fit only to "observables" which were short-distance quantities.

There was further discussion of Steele's fitting procedure when results for the two different power-countings in a toy model were displayed. In particular, Tom Cohen asked whether Steele had used a low-energy fit in Λ-counting and a global fit in Q-counting. He suggested that such a fitting procedure did not really make for a fair comparison of the two power-countings.

Steele presented results which showed large differences in the success of the Q-expansion for both the toy model "data" and actual $^1S_0$ data, depending upon whether the expansion was performed for δ or $p \cot \delta$. Cohen asked why the latter procedure fared worse than the former in when applied to the actual $^1S_0$ phase shifts. Matthias Lutz claimed that this was because the imaginary part of the amplitude was wrong. Some discussion ensued about the exact nature of Steele's fitting procedure. Steele explained that a global fit had been performed over the range $1/a < p < m_\pi$.

Steele then went on to discuss in more detail the case of real NN data. There—unlike in the toy model–the effective range is unnaturally large. At this point, Cohen asked about the scaling of r_e. Steele replied that it was irrelevant for the concerns he was dealing with here. He went on to show that a resummation of pion loop effects into a "σ-" type meson might assist the convergence. Mannque Rho suggested that such a σ-meson might appear in the NN potential if Weinberg's power-counting were employed and the calculation pursued to some sufficiently high order. Later Ulf Meißner suggested that two-pion exchange, and not the σ-meson, would set the scale of breakdown of Steele's calculation. Matthias Lutz then suggested that to get the two-pion exchange cut structure correct at NNLO one would need to calculate all of the two-pion exchange graphs. Steele replied that iterated one-pion exchange does not set the scale of breakdown, but it transpired that Lutz was talking about "genuine" two-pion exchange diagrams due to "radiation" pions. He claimed that these had a large impact on the results in calculations he has performed. Mark Wise said he would be shocked if this were actually the case.

Christoph Hanhart wondered how it was determined that the breakdown scale was not a function of the large scattering length. Steele answered that this had been explored numerically and it had been found that the breakdown scale was independent of a.

The last question before lunch came from Cohen, who asked if the Q-counting results converged if the limit $m_\pi \to 0$ was taken. Steele replied that they did.

SESSION 2

Session Chair: *B. Holstein (University of Massachusetts)*
Notes taken by: *S. Fleming (University of Toronto)*

NUCLEON-NUCLEON SCATTERING IN EFFECTIVE FIELD THEORY

G. RUPAK[a]
Department of Physics, University of Washington,
Seattle, WA 98195, USA
E-mail: grupak@u.washington.edu

I outline the effective field theory (EFT) calculation of nucleon-nucleon scattering which was recently carried out to next-to-next-to-leading order (NNLO) by Noam Shoresh and myself. In this calculation only potential pion contributions are included. These are the leading contributions from pions. A toy model was also considered in which potential pions are the only pion effects. The effective range expansion (ERE), which is valid for low energies, is used to derive matching conditions on some of the EFT couplings. Renormalization group flow analysis is used to fix the rest of the couplings leaving no free parameters at NNLO.

1 Introduction

Effective Field Theory (EFT) is useful for systems with a clear separation of scales. This is the case in nucleon-nucleon scattering, where the light pions have a mass of $m_\pi \sim 140\text{MeV}$, and the next particle to consider, the ρ meson, only comes in at $m_\rho \sim 770\text{MeV}$. In describing the two-nucleon system, however, special care is needed because of the appearance of a new low energy scale, namely the singlet channel scattering length. The Kaplan-Savage-Wise (KSW) power counting scheme [1] was formulated to address this issue.

In the EFT where the pions are included explicitly, the contribution of loops containing pions can be separated into (a) potential pions and (b) radiation and soft pions. Potential pions, corresponding to instantaneous pion exchanges, arise when the nucleons in the loop are put on-shell. Since both nucleons can be put simultaneously on-shell these contributions are enhanced. The radiation and soft pions enter when the *pions* are put on shell and describe retardation pion effects.

In this talk I describe the contribution of the potential pions alone [2]. It is argued below, Subsection 8.2, that this is sufficient for a next-to-next-leading order (NNLO) calculation of nucleon-nucleon $^1S_0$ scattering amplitude at external energies close to the pion production threshold. At energies much below this threshold, there could be contributions from radiation and soft pions. Nevertheless, the calculation with only potential pions serves as a probe for a better understanding of the KSW power counting, the appropriate fitting

[a] Work presented here has been done in collaboration with N. Shoresh.

procedure and the role of renormalization group flow analysis. It is worthwhile to consider a model in which only potential pions exist, and one does not have to worry about the effects of the unaccounted for radiation and soft pions. Such is the two-Yukawa model described in Section 2. The effective Lagrangian with potential pions for center of mass momentum $p \leq m_\rho/2$ is described in Section 3. KSW power counting is briefly reviewed in Section 4. The next-to-leading order (NLO) and NNLO scattering amplitude is calculated in Section 5 and 6 respectively. In Subsection 7.1, some of the EFT couplings are determined from low energy matching conditions for the amplitudes. Renormalization group (RG) flow analysis in Subsection 7.2 fixes the rest of the couplings. Results for the toy model are presented in Subsection 8.1, whereas in Subsection 8.2 I have included the corresponding results for real $^1S_0$ scattering.

There are also relativistic corrections to the amplitude at NNLO due to the finite mass of the nucleons. These effects are very small compared to the potential pion contributions and can be neglected.

2 The Toy Model

We consider a toy model in which non-relativistic nucleons interact via two instantaneous Yukawa potentials:

$$V(r) = -g_\pi \frac{e^{-m_\pi r}}{4\pi r} - g_\rho \frac{e^{-m_\rho r}}{4\pi r} \tag{1}$$

(r is the distance between the nucleons). This model is complex enough in the sense that the interactions involve two well separated scales. The Yukawa potentials describe the instantaneous exchange of the π and a scalar ρ meson (without the complications of retardation effects [3]).

The realistic $^1S_0$ one pion exchange amplitude (here given in momentum space) is

$$\left(-\frac{g_A^2}{2f_\pi^2}\right)\frac{p^2}{p^2+m_\pi^2} = -\frac{g_A^2}{2f_\pi^2} + \frac{g_A^2}{2f_\pi^2}\frac{m_\pi^2}{p^2+m_\pi^2}, \tag{2}$$

where the first term is a contact interaction, and the second is the Yukawa part. The contact interaction is not included in the toy model. g_π is chosen to reproduce the Yukawa part. Comparing Eq. (2) and Eq. (1), we set

$$g_\pi = \frac{g_A^2 m_\pi^2}{2f_\pi^2}. \tag{3}$$

The nucleon mass is taken to be $M = 940$ MeV, as well as $m_\pi = 140$ MeV, $m_\rho = 770$ MeV, $f_\pi = 132$ MeV and $g_A = 1.25$. g_ρ is tuned to give a large scattering length ($g_\rho = 13.5 \rightarrow a = -24.7$ fm) by numerically solving the Schroedinger equation with this potential.

The potential pion contribution to the field theory amplitude in this toy model and the real ${}^1S_0$ nucleon-nucleon scattering are the same. One might wonder about the contact piece in Eq. (2) that was neglected. As will be explained in Section 5, in the field theory amplitude the contact piece of the one-pion-exchange (OPE) can be absorbed in the definition of one of the 4-nucleon couplings.

3 The Effective Lagrangian

The effective Lagrangian for the toy model appropriate for center of mass energies much smaller than m_ρ is presented here. The effects of the short range force, ρ exchanges, are reproduced in the EFT by contact interactions. It is convenient [4] to separate the effective Lagrangian in terms of the number of nucleon fields present:

$$\mathcal{L} = \mathcal{L}_1 + \mathcal{L}_2 + \ldots,$$

where $\mathcal{L}_n$ describes the propagation or scattering of n nucleons. $\mathcal{L}_1$ is the usual kinetic energy term

$$\mathcal{L}_1 = N^\dagger(i\partial_0 + \vec{\nabla}^2/2M)N, \tag{4}$$

where the isodoublet field N represents the nucleons:

$$N = \begin{pmatrix} p \\ n \end{pmatrix}.$$

Next, we write $\mathcal{L}_2^{({}^1S_0)}$, the part of $\mathcal{L}_2$ that has non-vanishing matrix elements between ${}^1S_0$ states. The matrices P_k [4] are used to project onto the ${}^1S_0$ state,

$$P_k \equiv \frac{1}{\sqrt{8}}\sigma_2 \otimes \tau_2\tau_k,$$

where the σ matrices act on the nucleon spin space and the τ matrices act on the nucleon isospin space. The result is [1,2,4]:

$$\begin{aligned}\mathcal{L}_2^{({}^1S_0)} = & -\left(\frac{\mu}{2}\right)^{4-D}(C_0 + D_2 m_\pi^2 + D_4 m_\pi^4 + \cdots)(N^T P_i N)^\dagger (N^T P_i N) \\ & +\left(\frac{\mu}{2}\right)^{4-D}(\frac{1}{8}C_2 + \frac{1}{8}D_4^{(2)} m_\pi^2 + \cdots)[(N^T P_i N)^\dagger (N^T P_i (\overleftrightarrow{\nabla})^2 N) + h.c] \\ & + \cdots\end{aligned} \tag{5}$$

(summation over the repeated isospin index i is implied.)

We use dimensional regularization where the Lagrangian is given in D space-time dimensions and $\mu/2$ is an arbitrary mass scale introduced so that the couplings C_{2n}, $D_{2n}^{(2m)}$ have the same units in any dimension D. C_{2n} are the coefficients of contact interactions containing $2n$ derivatives and $D_{2n}^{(2m)}$ are coefficients of contact interactions involving $2m$ derivatives and $2(n-m)$ powers of the pion mass. We use the convention $D_{2n} \equiv D_{2n}^{(0)}$, $C_{2n} \equiv D_{2n}^{(2n)}$. The ellipses indicate operators with higher powers of derivatives or pion mass insertions.

The interaction with the pions is described by a non-local term in the action:

$$\begin{aligned} S_\pi &= \int d^D x\, d^D y \left(N^T(x) P_i N(x)\right)^\dagger g_\pi \frac{e^{-m_\pi |\mathbf{x}-\mathbf{y}|}}{4\pi|\mathbf{x}-\mathbf{y}|} \delta(x^0 - y^0) \left(N^T(y) P_i N(y)\right) \\ &= \int d^D p\, d^D p' \left(N^T(p) P_i N(p)\right)^\dagger \frac{g_A^2}{2f^2} \frac{m_\pi^2}{|\mathbf{p}-\mathbf{p}'|^2 + m_\pi^2} \left(N^T(p') P_i N(p')\right). \quad (6) \end{aligned}$$

Once the effective Lagrangian has been described, one needs a power counting scheme, i.e. a set of rules that determine the relative sizes of diagrams contributing to the scattering process.

4 The Power Counting

KSW power counting is motivated by the large scattering length a. It can be better understood by looking at the singlet channel amplitude at very low energy where even the pions can be treated as heavy particles and hence integrated out. The effective field theory amplitude is related to the phase shift by

$$\mathcal{A} = \frac{4\pi}{M} \frac{1}{p\cot\delta - ip}, \quad (7)$$

where $p = \sqrt{ME}$ is the center of mass momentum and δ is the singlet channel phase shift. For low energies one can write [5] the effective range expansion (ERE):

$$\begin{aligned} p\cot\delta &= -\frac{1}{a} + \frac{1}{2}\sum_{n=0}^{\infty} r_n p^{2(n+1)} \\ &= -\gamma + \frac{1}{2}\sum_{n=0}^{\infty} s_n \left(p^2 + \gamma^2\right)^{n+1}. \quad (8) \end{aligned}$$

Here the scattering length a can be arbitrarily large, while the other coefficients r_0, r_1, ... are assumed to be of natural sizes, $r_n \sim 1/\Lambda^{2n+1}$, where $\Lambda \sim m_\pi$ is the cut-off for this low energy theory. p is taken to be small compared to the high energy cut-off Λ but not compared to $1/a \ll \Lambda$. The coefficients in the two different ERE expansions are related perturbatively. The first two terms are

$$\begin{aligned}\frac{1}{a} &= \gamma - \frac{1}{2}\gamma^2 s_0 + \mathcal{O}(\gamma^4)\\ r_0 &= s_0 + 2\gamma^2 s_1 + O(\gamma^4).\end{aligned} \tag{9}$$

The difference between γ and $1/a$ enters at NLO, whereas the difference between s_0 and r_0 is only important at orders higher than NNLO.

Substituting Eq. (8) in Eq. (7), it can be seen that the amplitude has a pole at $p = i\gamma$ and this sets the radius of convergence for the Taylor expansion about $p = 0$. Formally identifying γ, $p \sim Q$, where Q/Λ is the expansion parameter, the amplitude Eq. (7) can be expanded about the pole in powers of p/Λ and γ/Λ,

$$\begin{aligned}\mathcal{A} &= -\frac{4\pi}{M}\frac{1}{\gamma+ip}\left[1+\frac{s_0}{2}\frac{p^2+\gamma^2}{\gamma+ip}+\left(\frac{s_0}{2}\right)^2\frac{(p^2+\gamma^2)^2}{(\gamma+ip)^2}\right.\\ &\qquad\left.+\left(\frac{s_0}{2}\right)^3\frac{(p^2+\gamma^2)^3}{(\gamma+ip)^3}+\frac{s_1}{2}\frac{(p^2+\gamma^2)^2}{\gamma+ip}+\cdots\right]\\ &= -\frac{4\pi}{M}\frac{1}{\gamma+ip}\left[1+\frac{s_0}{2}(\gamma-ip)+\left(\frac{s_0}{2}\right)^2(\gamma-ip)^2\right.\\ &\qquad\left.+\left(\frac{s_0}{2}\right)^3(\gamma-ip)^3+\frac{s_1}{2}(\gamma-ip)^2(\gamma+ip)+\cdots\right]\\ &\equiv \sum_{n=-1}^{\infty}\mathcal{A}_n, \qquad \mathcal{A}_n \sim Q^n.\end{aligned} \tag{10}$$

In KSW power counting the field theory amplitude is organized as an expansion in Q, Eq. (10). It has been shown[6] that the leading term in expansion Eq. (10) is obtained by summing up the bubble chains in Fig. 1 with insertions of C_0, giving the leading order (LO) amplitude:

$$\mathcal{A}_{-1} = \frac{-C_0}{1+iC_0 L}, \tag{11}$$

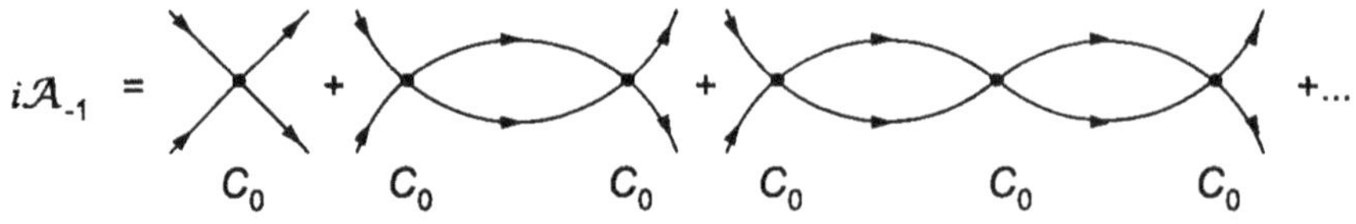

Figure 1: Leading order contribution $\mathcal{A}_{-1}$ from non-perturbative insertions of C_0.

where L is the loop integral

$$L = \left(\frac{\mu}{2}\right)^{4-D} \int \frac{d^D q}{(2\pi)^D} \frac{i}{\frac{E}{2} - q_0 - \frac{q^2}{2M} + i\epsilon} \frac{i}{\frac{E}{2} + q_0 - \frac{q^2}{2M} + i\epsilon}$$
$$= -iM\left(\frac{\mu}{2}\right)^{4-D} \int \frac{d^{D-1}\mathbf{q}}{(2\pi)^{D-1}} \frac{1}{q^2 - EM - i\epsilon} = -i\frac{M}{4\pi}(X_{sub} + ip). \quad (12)$$

A subtraction scheme that reproduces the KSW power counting is the power divergence subtraction (PDS) scheme that was introduced in [1]. In PDS one subtracts the poles of L in both $D = 4$ and $D = 3$ dimension (which gives $X_{sub} = \mu$ in Eq. (12)). Setting $\mu \sim p$, all the diagrams contributing to the bubble chains are rendered equal in size ($\sim 1/p$). This justifies summing up the bubble graphs at LO.

In the theory with pions, pion effects are assumed to be perturbative. From the PDS scheme follows a power counting scheme - KSW - in which:

- The expansion parameter is Q/Λ, where $p, \mu, m_\pi \sim Q$.
- $\mathcal{A} = \sum_{n=-1}^{\infty} \mathcal{A}_n$, $\mathcal{A}_n \sim Q^n$.
- The renormalized coupling $C_0 = C_0(\mu)$ scales as $1/\mu \sim 1/Q$, and more generally the renormalized couplings C_{2n}, $D_{2n}^{(2m)}$ scale as $1/Q^{n+1}$.

Some immediate consequences are:

- The loop integral L scales like p or μ, i.e. $\mathcal{O}(Q)$. Therefore, adding a C_0 and a loop L to a diagram does not change its order since the powers of Q cancel. It is now required to dress each diagram in the theory by C_0 to all orders. (This is diagrammatically represented by the "blob", Fig. 2).
- When the nucleons are put on shell, nucleon propagators scale as $1/Q^2$ and loop integrals $\int dq^4$ scale as Q^5.

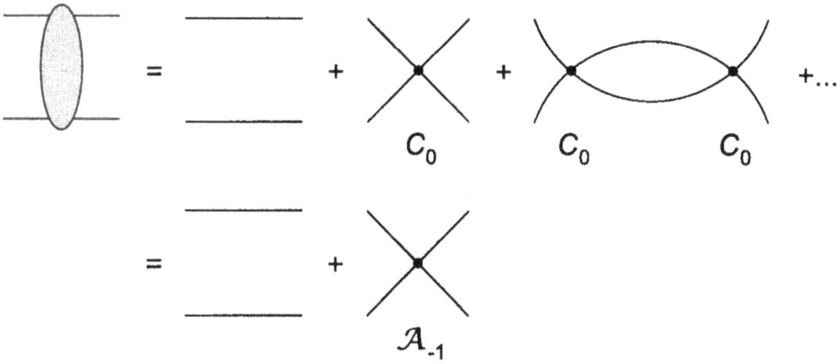

Figure 2: Definition of the "blob".

This theory should correctly describe the physics at momentum below $\Lambda \sim m_\rho/2$, which is the scale associated with the cut in the amplitude due to ρ exchange. A calculation to order Q^n is expected to be accurate up to an error of $\sim (Q/\Lambda)^{n+1}$.

It is a key feature of the KSW power counting that the LO amplitude has the right pole, or the correct scattering length. However, at higher orders there are contributions that move the position of the pole in the amplitude and counterterms must be introduced to cancel these effects. This is done by expanding the EFT couplings in powers of Q [2,7,8]:

$$\begin{aligned} C_0 &= C_{0,-1} + C_{0,0} + C_{0,1} + \cdots \\ C_2 &= C_{2,-2} + C_{2,-1} + \cdots \\ C_4 &= C_{4,-3} + C_{4,-2} + \cdots \\ &\vdots \end{aligned} \tag{13}$$

where the second subscript denotes the scaling with powers of Q.

In the last three sections, all the components of the computational framework of the EFT for nucleons have been identified: the effective Lagrangian (Feynman rules), the regularization and subtraction methods, and the power counting scheme that organizes the diagrams of the theory in a perturbative series. In the following sections we apply this formalism to the calculation of the $^1S_0$ scattering amplitude.

5 The Amplitude at NLO

In the theory with truly dynamical pions, the contact piece of the pion exchange can be re-absorbed in the definition of $C_{0,0}$. Hence, there is no difference in the potential pion amplitude for this toy model and the real N-N scattering amplitude. The potential pion contribution at NLO has already been calculated before [1]. I include the calculation here for completeness and clarity.

At NLO, we need to consider local operators that make up the following four-nucleon vertices, classified here according to their Q counting:

$$\begin{aligned} \mathcal{O}(Q^{-1}) &\quad : \quad -C_{0,-1} \\ \mathcal{O}(Q^{0}) &\quad : \quad -\tfrac{(\mathbf{q}_1^2+\mathbf{q}_2^2)}{2} C_{2,-2}\,; -C_{0,0}\,; -m_\pi^2 D_{2,-2} \end{aligned}$$

with $\mathbf{q}_1$, $\mathbf{q}_2$ the single nucleon incoming and outgoing momenta, in the center of mass coordinates. Only the linear combination $C_{0,0} + m_\pi^2 D_{2,-2}$ appears in the amplitude, and it is denoted by $B_{0,0}$. From this point on we adopt the convention that the second subscript is omitted from leading couplings, so that, for instance, C_0 stands for $C_{0,-1}$, C_2 is really $C_{2,-2}$, etc.

The diagrams contributing to the NLO amplitude are shown in Fig. 3. The amplitude is

$$\mathcal{A}_0 = \mathcal{A}_0^{(0\pi)} + \mathcal{A}_0^{(\pi)}$$

where $\mathcal{A}_0^{(0\pi)}$ includes all the diagrams with no pion exchange, and $\mathcal{A}_0^{(\pi)}$ contains all the diagrams with a single pion exchange. It is convenient to express Fig. 3(b) and Fig. 3(c) in terms of the loop integrals P_1 and P_2, so that

$$\begin{aligned} \mathcal{A}_0^{(0\pi)} &= (1 + i\mathcal{A}_{-1}L)^2(-B_{0,0} - C_2 p^2) \\ \mathcal{A}_0^{(\pi)} &= \frac{g_A^2}{2f^2}\frac{m_\pi^2}{4p^2}\ln(1+\frac{4p^2}{m_\pi^2}) + 2\,(\mathcal{A}_{-1})P_1 + i(\mathcal{A}_{-1})^2 P_2, \end{aligned} \tag{14}$$

with the definitions

$$\begin{aligned} P_1 &\equiv i\frac{g_A^2}{2f^2}m_\pi^2\left(\frac{\mu}{2}\right)^{4-D}\int\frac{\mathrm{d}^D q}{(2\pi)^D}\frac{i}{\frac{E}{2}+q_0-\frac{\mathbf{q}^2}{2M}+i\epsilon} \\ &\qquad \times\frac{i}{\frac{E}{2}-q_0-\frac{\mathbf{q}^2}{2M}+i\epsilon}\frac{1}{(\mathbf{q}-\mathbf{p})^2+m_\pi^2} \\ &\overset{PDS}{\rightarrow} \frac{1}{8\pi}\frac{g_A^2}{2f^2}\frac{m_\pi^2 M}{p}\left(\arctan(2\frac{p}{m_\pi}) + \frac{i}{2}\ln(1+4\frac{p^2}{m_\pi^2})\right) \end{aligned} \tag{15}$$

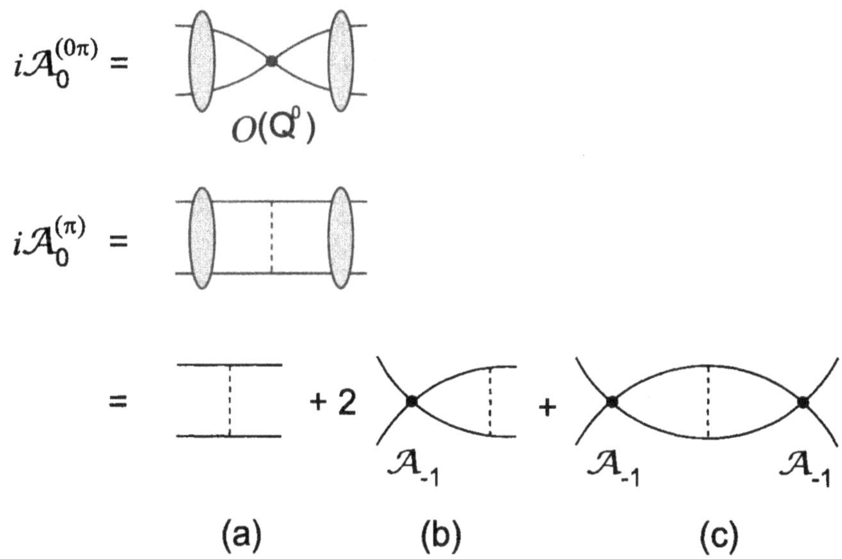

Figure 3: Next-to-leading order contribution $\mathcal{A}_0$.

and

$$\begin{aligned} P_2 &\equiv i\frac{g_A^2}{2f^2}m_\pi^2\left(\frac{\mu}{2}\right)^{8-2D}\int\frac{\mathrm{d}^Dq}{(2\pi)^D}\frac{\mathrm{d}^Dl}{(2\pi)^D}\frac{i}{\frac{E}{2}+q_0-\frac{q^2}{2M}+i\epsilon} \\ &\times \frac{i}{\frac{E}{2}-q_0-\frac{q^2}{2M}+i\epsilon}\frac{i}{\frac{E}{2}+l_0-\frac{l^2}{2M}+i\epsilon}\frac{i}{\frac{E}{2}-l_0-\frac{l^2}{2M}+i\epsilon}\frac{1}{(\mathbf{l}-\mathbf{q})^2+m_\pi^2} \\ &\overset{PDS}{\rightarrow} -i\frac{g_A^2}{2f^2}\frac{m_\pi^2M^2}{32\pi^2}\left(-\gamma_E+\ln(\pi)+1\right. \\ &\qquad\qquad \left.-2\ln\left(\frac{m_\pi}{\mu}\right)-2\ln\left(1-i2\frac{p}{m_\pi}\right)\right). \qquad (16) \end{aligned}$$

Eq. (16), where only the poles have been subtracted, differs from an earlier calculation of P_2 [1] that involved finite subtractions.

6 The Amplitude at NNLO

At this order, there are insertions of three kinds of local operators, listed here according to their Q counting:

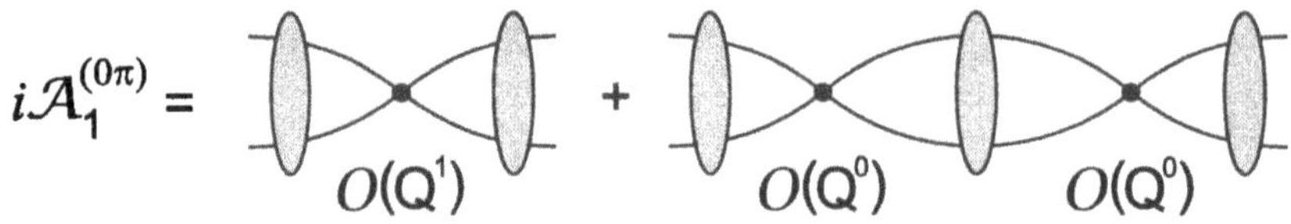

Figure 4: Next-to-next-to-leading order contribution $\mathcal{A}_1^{(0\pi)}$ from diagrams with no pion exchange.

$$
\begin{aligned}
\mathcal{O}(Q^{-1}) &\quad : \quad -C_0 \\
\mathcal{O}(Q^{0}) &\quad : \quad -B_{0,0}\ ; -\tfrac{(\mathbf{q}_1^2+\mathbf{q}_2^2)}{2}C_2 \\
\mathcal{O}(Q^{1}) &\quad : \quad -B_{0,1}\ ; -\tfrac{(\mathbf{q}_1^2+\mathbf{q}_2^2)}{2}B_{2,-1}\ ; -p^4 C_4
\end{aligned}
$$

There are 3 new couplings that enter at this order: $B_{0,1} \equiv C_{0,1} + m_\pi^2 D_{2,-1} + m_\pi^4 D_4$, $B_{2,-1} \equiv C_{2,-1} + m_\pi^2 D_4^{(2)}$ and C_4. The NNLO amplitude is

$$\mathcal{A}_1 = \mathcal{A}_1^{(0\pi)} + \mathcal{A}_1^{(\pi)} + \mathcal{A}_1^{(\pi\pi)}.$$

6.1 $\mathcal{A}_1^{(0\pi)}$

The diagrams in Fig. 4 are generated by two insertions of $\mathcal{O}(Q^0)$ operators or a single insertion of $\mathcal{O}(Q^1)$ operator, and give:

$$
\begin{aligned}
\mathcal{A}_1^{(0\pi)} &= (1+i\mathcal{A}_{-1}L)^2(-C_4p^4 - B_{2,-1}p^2 - B_{0,1}) \\
&\quad + i(1+i\mathcal{A}_{-1}L)^3 L(-C_2p^2 - B_{0,0})^2 .
\end{aligned} \tag{17}
$$

6.2 $\mathcal{A}_1^{(\pi)}$

All the diagrams of Fig. 5 are generated by insertions of local operators of order $\mathcal{O}(Q^0)$ and a single pion exchange. The one pion exchange contribution at this order is

$$
\begin{aligned}
\mathcal{A}_1^{(\pi)} &= 2(1+i\mathcal{A}_{-1}L)^2(-C_2p^2 - B_{0,0})(P_1 + P_2(i\mathcal{A}_{-1})) \\
&\quad + 2(1+i\mathcal{A}_{-1}L)\left(-\frac{C_2}{2}\right)(Q_1 + Q_2(i\mathcal{A}_{-1})),
\end{aligned} \tag{18}
$$

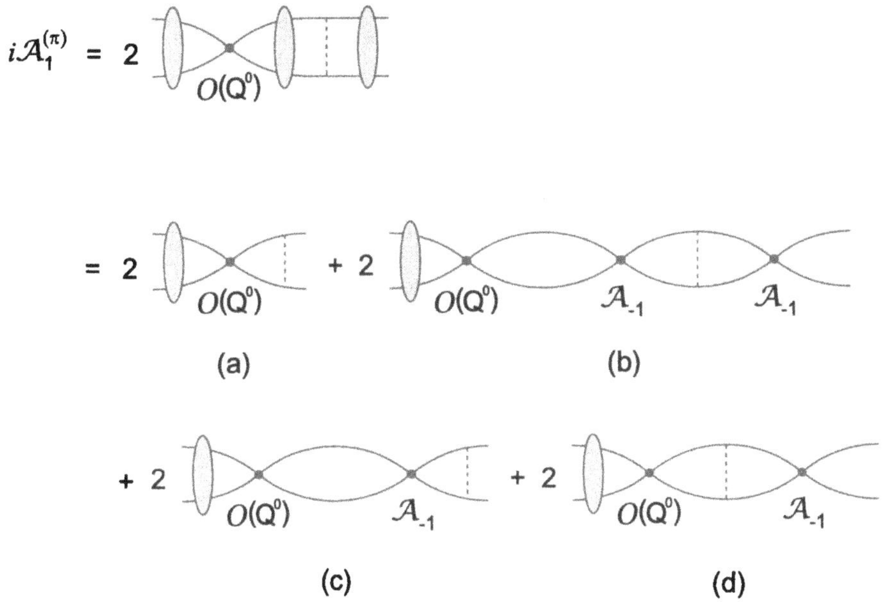

Figure 5: Next-to-next-to-leading order contribution $\mathcal{A}_1^{(\pi)}$ from one pion exchange diagrams.

where

$$\begin{aligned} Q_1 &\equiv \frac{g_A^2}{2f^2} m_\pi^2 M \left(\frac{\mu}{2}\right)^{4-D} \int \frac{\mathrm{d}^{D-1}\mathbf{q}}{(2\pi)^{D-1}} \frac{1}{(\mathbf{q}-\mathbf{p})^2 + m_\pi^2} \\ &\overset{PDS}{\rightarrow} \frac{g_A^2}{2f^2} m_\pi^2 M \left(\frac{\mu - m_\pi}{4\pi}\right), \end{aligned} \tag{19}$$

and

$$\begin{aligned} Q_2 &\equiv -i\frac{g_A^2}{2f^2} m_\pi^2 M^2 \left(\frac{\mu}{2}\right)^{8-2D} \int \frac{\mathrm{d}^{D-1}\mathbf{q}}{(2\pi)^{D-1}} \frac{\mathrm{d}^{D-1}\mathbf{l}}{(2\pi)^{D-1}} \frac{1}{l^2 - p^2 - i\epsilon} \\ &\qquad \times \frac{1}{(\mathbf{q}-\mathbf{l})^2 + m_\pi^2} \\ &= Q_1 L \\ &\overset{PDS}{\rightarrow} -i\frac{g_A^2}{2f^2} m_\pi^2 M^2 (\frac{\mu - m_\pi}{4\pi})(\frac{\mu + ip}{4\pi}), \end{aligned} \tag{20}$$

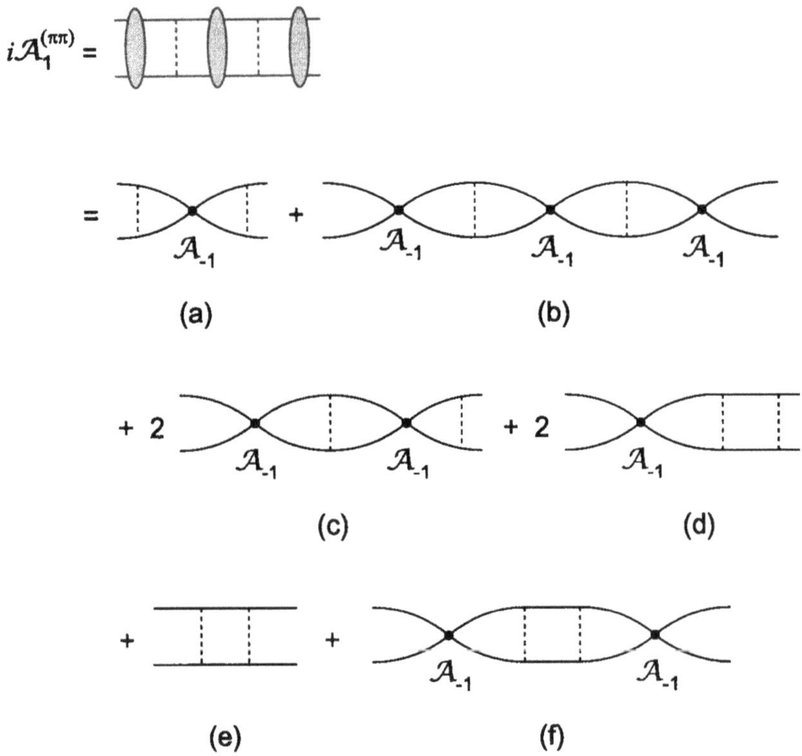

Figure 6: Next-to-next-to-leading order contribution $\mathcal{A}_1^{(\pi\pi)}$ from two pion exchange diagrams.

and $P_{1,2}$ are defined in Section 5. In carrying out the subtractions in Q_1 and Q_2, some care is needed. The poles and the corresponding counterterms should be calculated treating C_0 perturbatively.

6.3 $\mathcal{A}_1^{(\pi\pi)}$

Fig. 6 contains the diagrams of this set. The graphs Fig. 6(e), (d) and (f) can be written in terms of analytic functions $I_1(p)$, $I_2(p)$ and $I_3(p)$ respectively which are presented in the Appendix. The rest of the diagrams can be written in terms of expressions that have already been defined. We find

$$\mathcal{A}_1^{(\pi\pi)} = -iI_1 + 2(\mathcal{A}_{-1})I_2 + i(\mathcal{A}_{-1})^2 I_3$$

$$+P_1(\mathcal{A}_{-1})P_1 + 2i(\mathcal{A}_{-1})^2 P_1 P_2 - (\mathcal{A}_{-1})^3 (P_2)^2. \qquad (21)$$

In the last two sections, we have calculated an expression for the LO+NLO +NNLO amplitude in terms of six independent free parameters C_0, $B_{0,0}$, C_2, $B_{0,1}$, $B_{2,-1}$ and C_4. The table below shows the new free parameters introduced at each order in the expansion.

LO:	C_0		
NLO:	$B_{0,0}$	C_2	
NNLO:	$B_{0,1}$	$B_{2,-1}$	C_4

When the number of parameters is doubled (from NLO to NNLO) it is hard to consider an improvement of the fit as evidence for the convergence of the EFT expansion. Therefore, it is desirable to find theoretically viable ways to reduce the number of independent parameters. This is the subject of the next section.

7 Determining the Effective Field Theory Couplings

7.1 Matching conditions

In this subsection I discuss a method for deriving conditions that reduce the number of free parameters. The guiding principle is simple: the leading couplings are defined in such a way that certain features of the low energy phase shift (like the effective range expansion parameters) are reproduced exactly. Conditions are then imposed on the sub-leading couplings so that these features are not modified at higher orders. Similar matching conditions were introduced in [7].

From Eq. (7), the Q expansion of $p\cot\delta$ in terms of the amplitude $\mathcal{A} = \mathcal{A}_{-1} + \mathcal{A}_0 + \mathcal{A}_1 + \cdots$, is

$$p\cot\delta = ip + \frac{4\pi}{M\mathcal{A}_{-1}} - \frac{4\pi}{M\mathcal{A}_{-1}}\frac{\mathcal{A}_0}{\mathcal{A}_{-1}} + \frac{4\pi}{M\mathcal{A}_{-1}}\left(\frac{\mathcal{A}_0^2}{\mathcal{A}_{-1}^2} - \frac{\mathcal{A}_1}{\mathcal{A}_{-1}}\right) + \cdots . \qquad (22)$$

At low energies, the same quantity is given by expansion Eq. (8):

$$p\cot\delta = -\gamma + \frac{1}{2}s_0\left(p^2+\gamma^2\right) + \frac{1}{2}s_1\left(p^2+\gamma^2\right)^2 + \cdots . \qquad (23)$$

where $p\cot\delta$ has been expanded about $p^2 = -\gamma^2$ (or $p = \pm i\gamma$).

The LO amplitude contributes only a constant to $p\cot\delta$. C_0 is determined by requiring

$$ip + \frac{4\pi}{M\mathcal{A}_{-1}} = -\gamma. \qquad (24)$$

This puts the pole in the amplitude at $p = i\gamma$. At NLO and NNLO there are more constant contributions to $p \cot \delta$. The requirements that the position of the pole does not shift at NLO, nor at NNLO, reduce to:

$$\begin{aligned} \text{NLO}: \ & \mathcal{A}_0|_{p=-i\gamma} = 0 \\ \text{NNLO}: \ & \mathcal{A}_1|_{p=-i\gamma} = 0. \end{aligned} \tag{25}$$

These conditions determine the couplings $B_{0,0}$ and $B_{0,1}$. Similarly, one requires that s_0 in Eq. (23) is determined at NLO by C_2:

$$\frac{\partial}{\partial p^2}\left(-\frac{4\pi}{M\mathcal{A}_{-1}}\frac{\mathcal{A}_0}{\mathcal{A}_{-1}}\right)|_{p^2=-\gamma^2} = \frac{1}{2}s_0, \tag{26}$$

and that higher order (NNLO) effects do not change s_0:

$$\text{NNLO}: \ \frac{\partial}{\partial p}\mathcal{A}_1|_{p=-i\gamma} = 0. \tag{27}$$

This condition determines $B_{2,-1}$. (Eqns. (25) and (27) can be more directly obtained by considering the expansion of the amplitude Eq. (10).) Note that at NNLO, fixing γ is not exactly the same as fitting the scattering length a whereas reproducing s_0 *is* essentially identical to fixing the effective range r_0.

The values of γ and s_0 are used to determine C_0 and C_2. One can determine C_4 from s_1, the next free parameter in the expansion Eq. (23). However, as in the theory without pions [1,9], it can be shown using RG analysis that at NNLO, C_4 is completely determined. This is done below.

7.2 Renormalization group flow

The RG equations are derived by requiring that the amplitude be exactly μ independent at each order in Q for any value of p. The solutions to the RG equations for C_0 and C_2 are:

$$\begin{aligned} C_0(\mu) &= \frac{4\pi}{M}\left(\frac{1}{\xi_{C_0}} - \mu\right)^{-1} \\ C_2(\mu) &= \xi_{C_2}\left(\frac{MC_0(\mu)}{4\pi}\right)^2. \end{aligned} \tag{28}$$

The integration constants ξ_{C_0} and ξ_{C_2} are determined by boundary conditions. The other coupling constants, denoted here by X, have the general solution

$$X(\mu) = \xi_X\left(\frac{MC_0(\mu)}{4\pi}\right)^2 + f_X(C_0(\mu), B_{0,0}(\mu), C_2(\mu); \mu) \tag{29}$$

where f_X is known, and the only freedom in X is in the constant ξ_X, determined by boundary conditions. Specifically,

$$\begin{aligned}
B_{0,0} &= \left(\frac{MC_0}{4\pi}\right)^2 \left(\xi_{B_{0,0}} + \frac{g_A^2}{2f^2} m_\pi^2 \ln\left(\frac{\mu}{M}\right)\right) \\
B_{0,1} &= \frac{B_{0,0}^2}{C_0} - \frac{g_A^2}{2f^2} m_\pi^2 \frac{M\mu}{4\pi} C_2 + \xi_{B_{0,1}} \left(\frac{MC_0}{4\pi}\right)^2 \\
B_{2,-1} &= \frac{2B_{0,0}C_2}{C_0} + \xi_{B_{2,-1}} \left(\frac{MC_0}{4\pi}\right)^2 \\
C_4 &= \frac{C_2^2}{C_0} + \xi_{C_4} \left(\frac{MC_0}{4\pi}\right)^2 .
\end{aligned} \tag{30}$$

Examination of the solutions in Eq. (30) shows that the μ scaling of some couplings do not agree with their expected Q counting. To determine the Q counting one also needs to know the Q counting of the RG constants ξ_X. That γ can make up Q counting for some of the ξ_Xs was already shown in the theory without pions, where, for example, $C_{2,-1} \sim \gamma^2/(\mu-\gamma)^3$ [2,8]. It is possible, however, to determine the Q scaling of the leading couplings. We demonstrate this point by discussing directly the coupling C_4, which has power counting of $1/Q^3$. The second term in the solution to the RG equation for C_4 (Eq. (30)) is $\sim \xi_{C_4}/\mu^2$. In order for it to contribute at NNLO, ξ_{C_4} should be $\mathcal{O}(1/Q)$. Since C_4 has a sensible large scattering length ($\gamma \to 0$) limit, the possibility $\xi_{C_4} \sim 1/\gamma$ is ruled out. It is also assumed that C_4 is associated with short distance physics and thus it is fundamentally independent of pion physics. These considerations allow setting $\xi_{C_4} = 0$ at this order, and have C_4 be completely determined by C_0 and C_2. There are in fact no new parameters at NNLO!

In the next section we check how well the EFT reproduces the phase shift of the toy model and the real data when the matching conditions derived in the last two subsections are imposed.

8 Results

The ${}^1S_0$ scattering cross section σ is related to the phase shift and amplitude by

$$\begin{aligned}
\sigma &= \frac{4\pi}{p^2}\sin^2\delta \\
&= \frac{M^2}{4\pi}|\mathcal{A}|^2
\end{aligned} \tag{31}$$

which can be expanded in powers of Q (writing $\sigma = \sigma_{-2} + \sigma_{-1} + \sigma_0 + \cdots$):

$$\begin{aligned}
\sigma_{-2} &= \frac{M^2}{4\pi}|\mathcal{A}_{-1}|^2 \\
\sigma_{-1} &= \frac{M^2}{4\pi}\left(\mathcal{A}_0\mathcal{A}^*_{-1} + \mathcal{A}^*_0\mathcal{A}_{-1}\right) \\
\sigma_0 &= \frac{M^2}{4\pi}\left(\mathcal{A}_{-1}\mathcal{A}^*_1 + \mathcal{A}^*_{-1}\mathcal{A}_1 + |\mathcal{A}_0|^2\right).
\end{aligned} \tag{32}$$

There are several advantages to using σ as opposed to δ in comparing EFT result to data. The experimentally measured cross section σ can be related to the sizes of the absolute value of the amplitude Eq. (32). δ is very sensitive to the value of r_1 and other higher moment coefficients and it is reflected in the value of C_4 that one chooses. This problem is avoided if one plots σ or $\sin^2\delta$ instead, where the effects of r_1 or ξ_{C_4} are truly sub-leading.

In the following subsections we compare the EFT result with the exact phase shift from the toy model, and also do the corresponding comparisons for the real $^1S_0$ phase shift. We consider how well the data is reproduced by the EFT, whether the expansion is converging, and what is the breakdown scale.

8.1 Results for the toy model

For the toy model, $\gamma = -7.743$ MeV and $s_0 = 1.584$ fm and these values are used to determine the couplings. The results are presented in Fig 7. Fig 7 (a) shows a clear improvement of fit as we go to higher order in the expansion. The NNLO result is a prediction as we do not introduce any new free parameter. Fig. 7 (b). shows the errors [10] at each order, $\Delta \equiv |\sin^2(\delta)_{EFT} - \sin^2(\delta)_{exact}|$, on a log-log scale. The breakdown scale, where the errors become comparable, can be read off of the error plot to be $\sim$ 520 MeV which is slightly higher than the expected $m_\rho/2$.

Table 1 shows the couplings at $\mu = m_\pi$. The entries are given in units of fm^2. To make a direct comparison of the couplings easier, we included the momentum factors from the operators C_2p^2, $B_{2,-1}p^2$ and C_4p^4, evaluated at $p = m_\pi$. From the table we see a hierarchy in the sizes of the couplings corresponding to the orders in the expansion. This is a verification of the power counting scheme. A rough estimate of the expansion parameter is $Q/\Lambda \sim m_\pi/(m_\rho/2) \sim 1/3$, which is also suggested by the table.

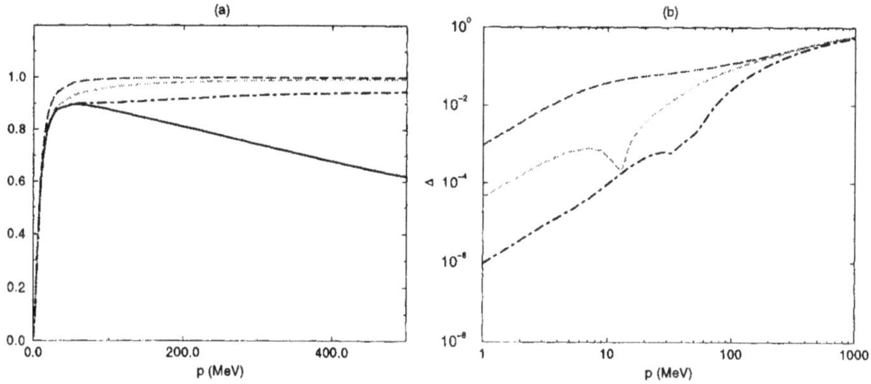

Figure 7: Toy model: (a) $\sin^2(\delta)$ vs. p(MeV); (b) log-log plot of $\Delta \equiv |\sin^2(\delta)_{EFT} - \sin^2(\delta)_{exact}|$ vs. p(MeV). The exact phase shift is described by the solid curve. The long-dashed curves denote the LO phase shift and error, the dashed curves are the NLO results, and the dot-dashed curves describe the NNLO results.

Table 1: Renormalized couplings for the toy model at $\mu = m_\pi$ in units of fm^2.

LO	NLO		NNLO		
C_0	$C_2 m_\pi^2$	$B_{0,0}$	$B_{0,1}$	$B_{2,-1}\ m_\pi^2$	$C_4\ m_\pi^4$
−3.526	0.06912	1.412	0.5620	0.2886	−0.001355

8.2 Results for the real $^1S_0$ scattering

Let us recall that in the KSW power counting the LO amplitude, which is inversely proportional to the momentum p, is calculated in terms of a non-perturbative dimension 6 four-nucleon operator. The other higher derivative operators and pion exchanges are treated perturbatively. The expansion parameter is Q/Λ with p, $m_\pi \sim Q$ and Λ some high energy cut-off, and at $p \sim Q$ the LO amplitude scales as $1/Q$. Pion production, however, at external momentum $p = \sqrt{m_\pi M} \sim 360$ MeV, introduces a new scale, and it is convenient to use a different expansion parameter, $Q_r \sim \sqrt{m_\pi M}$, instead of $Q \sim Q_r^2/M \ll Q_r$ [7]. At $p \sim Q_r$, the LO amplitude scales as $1/Q_r$. The relative importance of the potential, radiation and soft pion contributions depends on the physics associated with different scales determined by the momentum p.

At $p \sim Q_r$, potential and soft pions contribute at Q_r^0 and Q_r^2 respectively, while radiation pions contribute at Q_r^3 [7]. Thus, for an NNLO ($\mathcal{O}(Q_r)$) cal-

Table 2: Renormalized couplings for the real $^1S_0$ channel at $\mu = m_\pi$ in units of fm^2.

LO	NLO		NNLO		
C_0	$C_2 m_\pi^2$	$B_{0,0}$	$B_{0,1}$	$B_{2,-1}\, m_\pi^2$	$C_4\, m_\pi^4$
−3.5226	1.39685	1.41738	0.56156	0.541789	−0.553904

culation, only potential pions contribute at this scale. At $p \sim Q$, potential pions contribute at order Q^0. The power counting for the soft and radiation pion is less clear. Examination of the radiation pion diagrams that have been calculated in [7], shows that some diagrams, which scale as Q_r^3 at high momentum, become order $Q^{1/2}$ at low external momentum. Such contributions seem to violate the presumed expansion of the amplitude in integer powers of Q, but in fact there is a cancellation between the different diagrams, and the leading contribution from radiation pion graphs starts at $\mathcal{O}(Q)$. Furthermore, the remaining terms are highly suppressed because of Wigner's $SU(4)$ spin-isospin symmetry [3] and can be neglected in the NNLO calculation (see Tom Mehen's discussion in these proceedings). We do not see (at the moment) a general reason for such cancellations (of order $Q^{1/2}$), and it is in principal possible that radiation or soft pion graphs that are of high order in Q_r at high momentum, become important at low momentum, and should in fact be included in an NNLO calculation of the amplitude. Here, however, this option will not be investigated, and only the potential pion contributions will be taken into account, admittedly with the possibility of having left out some graphs. The EFT amplitude for the toy model with potential pions is identical to the potential pion contribution to the real $^1S_0$ amplitude, after a redefinition of $C_{0,0} \to C_{0,0} - \frac{g_A^2}{2f^2}$. With these considerations, the EFT amplitude with only potential pions is applied to the real $^1S_0$ scattering. We use $\gamma = -7.889$ MeV ($a = -23.68$ fm) and $s_0 = 2.73$ fm to fix the couplings.

Here again, we have a prediction at NNLO, Fig. 8. By going to higher order we get better result, Fig. 8(a). From Fig. 8(b), the breakdown scale is estimated to be $\sim$ 400MeV. Just like in the toy model, a rough estimate of the breakdown scale is $m_\rho/2$ due to the exchange of vector meson ρ.

Table 2 shows the couplings at $\mu = m_\pi$. The entries are given in units of fm^2. We included the momentum factors from the operators C_2p^2, $B_{2,-1}p^2$ and C_4p^4, evaluated at $p = m_\pi$. The sizes of the couplings at various order in the expansion suggest convergence of the expansion with $Q/\Lambda \sim 1/3$.

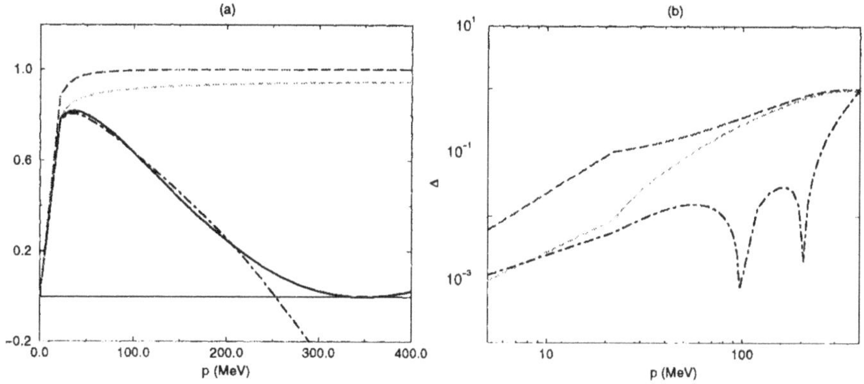

Figure 8: Nijmegen: (a) $\sin^2(\delta)$ vs. p(MeV); (b) log-log plot of $\Delta \equiv |\sin^2(\delta)_{EFT} - \sin^2(\delta)_{Nijmegen}|$ vs. p(MeV). The Nijmegen phase shift is described by the solid curve. The long-dashed curves denote the LO phase shift and error, the dashed curves are the NLO results, and the dot-dashed curves describe the NNLO results.

9 Summary and Conclusion

A NNLO calculation for nucleon-nucleon scattering in the $^1S_0$ channel using potential pions was presented. Radiation and soft pions do not contribute at this order for center of mass momentum $p \sim Q_r$. However, there could be soft and radiation pion contributions at momentum $p \sim Q$ which have not been calculated yet. If for low momentum $p \sim Q$, soft and radiation pion contributions are found to be small at NNLO, then this calculation gives the NNLO nucleon-nucleon scattering amplitude in the $^1S_0$ channel for momentum $p \leq m_\rho/2$.

The calculation was applied to a toy model and the real data. The toy model provided a setup where only potential pions contribute by construction. This allowed us to address some general questions regarding the convergence of KSW power counting, number of free parameters and the role of renormalization group flow, without worrying about unaccounted for radiation and soft pion contributions.

At NNLO there are six independent parameters. In the effective field theory, these parameters are associated with contact operators which encodes information about the high energy physics. However, in the singlet channel the high energy physics conspire to produce a low energy scale $\gamma \sim 8$ MeV. This small energy scale has to be taken into account in using dimensional analysis to determine the sizes of the parameter. RG analysis, which only determines the μ scaling of parameters, is not always enough to determine their Q scaling. Some of the parameters such as ξ_{C_4} can be determined from RG analysis. We introduce a method to determine the rest of the parameters. This is done by matching the EFT amplitude at low momentum to the effective range expansion and all the parameters are given in terms of two experimental inputs, γ and s_0. We find that there are no free parameters at NNLO and we make a prediction for the high momentum behavior of the scattering amplitude.

For the toy model we find that including higher order terms in the amplitude improves the fit to "data". The hierarchy in the sizes of couplings show converges with an expansion parameter of $Q/\Lambda \sim 1/3$. The breakdown scale is estimated to be ~ 520 MeV by comparing the sizes of errors at different order. This is slightly higher than the expected scale $m_\rho/2$.

The fit to real data shows significant improvement at NNLO. We see an hierarchy in the sizes of the couplings which suggests an expansion parameter similar to the toy model $Q/\Lambda \sim 1/3$. The breakdown scale is estimated to be ~ 400 MeV.

Acknowledgments

I thank Noam Shoresh for a rewarding collaboration. I would also like to thank Paulo Bedaque, Harald Grießhammer, David Kaplan, Tom Mehen, Dan Phillips, Martin Savage, James Steele and Iain Stewart for many useful discussions. It is also a pleasure to thank the organizers of the conference and the participants. This work is supported in part by U.S. Department of Energy Grant DE-FG03-97ER41014.

Appendix: The integrals I_1, I_2 and I_3

Fig. 6 (e) is given by:

$$
\begin{aligned}
I_1 &= i\left(\frac{g^2}{2f^2}\right)^2 m_\pi^4 M \bar{I}_1 \\
\bar{I}_1 &= \left\langle \int \frac{d^3\mathbf{q}}{(2\pi)^3} \frac{1}{q^2 - p^2 - i\epsilon} \frac{1}{(\mathbf{q}-\mathbf{p})^2 + m_\pi^2} \frac{1}{(\mathbf{q}-\mathbf{p}')^2 + m_\pi^2} \right\rangle \\
&= \frac{1}{32\pi m_\pi^3 \rho^3} \left(iL(\rho) + M(\rho)\right)
\end{aligned} \tag{33}
$$

$$L(\rho) = \int_0^{1/2} dt \frac{1}{t(1-t)} \ln\left(\frac{(1+4\rho^2 t)(1+4\rho^2(1-t))}{1+4\rho^2}\right)$$
$$= \frac{1}{2}\ln^2(1+4\rho^2)$$
$$M(\rho) = 2\int_0^{1/2} dt \frac{1}{t(1-t)} \left(\tan^{-1}(2\rho\sqrt{1+4\rho^2 t(1-t)}) - \tan^{-1}(2\rho)\right)$$
$$= \frac{i}{2}\pi^2 - i\ln^2(2) + \frac{i}{2}\ln(4)\ln(1-2i\rho) + 2\pi\ln(2-2i\rho)$$
$$-i\ln\left(2(1+i\rho)^2(1-2i\rho)\right)\ln(1+2i\rho)$$
$$-\tan^{-1}(2\rho)\ln\left(\frac{4(1+\rho^2)^2}{1+4\rho^2}\right) - 2\tan^{-1}(\rho)\ln(1+4\rho^2)$$
$$-2i\,Li_2\left(\frac{1}{2}+i\rho\right) - 2i\,Li_2(-1-2i\rho) - 2i\,Li_2(2-2i\rho),$$

where the average is over the outgoing momentum $\mathbf{p}'$. $\rho = p/m_\pi$ and the dilogarithm, $Li_2(z)$, is defined as

$$Li_2(z) = -\int_0^z dt \frac{\ln(1-t)}{t}.$$

Fig. 6 (d) is given by:

$$I_2 = \left(\frac{g^2}{2f^2}\right)^2 m_\pi^4 M^2 \bar{I}_2 \tag{34}$$
$$\bar{I}_2 = \int \frac{d^3\mathbf{q}}{(2\pi)^3}\frac{d^3\mathbf{l}}{(2\pi)^3}\frac{1}{q^2-p^2-i\epsilon}\frac{1}{(\mathbf{q}-\mathbf{p})^2+m_\pi^2}\frac{1}{l^2-p^2-i\epsilon}\frac{1}{(\mathbf{q}-\mathbf{l})^2+m_\pi^2}$$
$$= \frac{i}{128\pi^3 m_\pi^2 \rho}\int_{-\infty}^{\infty} dx \ln\left(\frac{i+x+\rho}{i-x+\rho}\right)\ln\left(\frac{(x+\rho)^2+1}{(x-\rho)^2+1}\right)\frac{1}{x^2-\rho^2-i\epsilon}$$
$$= \frac{1}{128\pi^2\rho^2 {m_\pi}^2}\left(-\pi^2 + 4i\pi\ln(2-2i\rho) + \ln(2)\ln(1+2i\rho)\right.$$
$$+\ln(1-2i\rho)\left[4\ln(1-i\rho) - \ln((1-2i\rho)(1+4\rho^2)/8)\right]$$
$$\left. -Li_2\left(\frac{1}{2}-i\rho\right) + Li_2\left(\frac{1}{2}+i\rho\right) + 4\,Li_2(2-2i\rho)\right).$$

From Fig. 6(f), we define

$$I_3 = -i\left(\frac{g^2}{2f^2}\right)^2 m_\pi^4 M^3 \bar{I}_3$$
$$\bar{I}_3 = \int \frac{\mathrm{d}^3\mathbf{k}}{(2\pi)^3}\frac{1}{k^2-p^2-i\epsilon}I(\mathbf{k},\mathbf{p})^2$$
$$I(\mathbf{k},\mathbf{p}) = \int \frac{\mathrm{d}^3\mathbf{q}}{(2\pi)^3}\frac{1}{q^2-p^2-i\epsilon}\frac{1}{(\mathbf{q}-\mathbf{k})^2+m_\pi^2}. \quad (35)$$

and it gives

$$\bar{I}_3 = \frac{i}{64\pi^3 p}\int_0^{1/2} dt\frac{1}{t(1-t)}\int_0^\infty dse^{-(m-2ip)s}\frac{e^{-i2pst}-1}{s}$$
$$= \frac{i}{64\pi^3\rho m_\pi}\left(\ln\left(\frac{1-i\rho}{1-i2\rho}\right)\ln(i\rho)+Li_2\left(\frac{-i\rho}{1-i2\rho}\right)\right.$$
$$\left.+Li_2(1-i\rho)-Li_2(1-i2\rho)\right). \quad (36)$$

The explicit expressions for I_1, I_2 and I_3 presented above are valid only for $\rho \geq 0$ and small positive imaginary ρ ($|\rho| \ll 1$). For other values of ρ, the integrals have to be done taking into account the poles and cuts in the integrand appropriately.

References

1. D.B. Kaplan, M.J. Savage and M.B. Wise, *Phys. Lett.* B **424**, 390 (1998); *Nucl. Phys.* B **534**, 329 (1998).
2. G. Rupak and N. Shoresh, nucl-th/9902077.
3. T. Mehen and I.W. Stewart, nucl-th/9901064; T. Mehen, I.W. Stewart and M.B. Wise, hep-ph/9902370.
4. D.B. Kaplan, M.J. Savage and M.B. Wise, nucl-th/9804032.
5. H.A. Bethe, *Phys. Rev.* **76**, 38 (1949); H.A. Bethe and C. Longmire, *Phys. Rev.* **77**, 647 (1950).
6. S. Weinberg, *Phys. Lett.* B **251**, 288 (1990); *Nucl. Phys.* B **363**, 3 (1991); *Phys. Lett.* B **295**, 114 (1992).
7. T. Mehen and I.W. Stewart, nucl-th/9809071; nucl-th/9809095.
8. J.W. Chen, G. Rupak and M.J. Savage, nucl-th/9902056.
9. U. van Kolck, *Nucl. Phys.* A **645**, 273 (1999).
10. G.P. Lepage, nucl-th/9706029; J.V. Steele and R.J. Furnstahl, *Nucl. Phys.* A **637**, 46 (1998).

Discussion during G. Rupak's presentation

Chair: *B. Holstein (University of Massachusetts)*
Notes taken by: *S. Fleming (University of Toronto)*

M. Lutz asked what was meant by the comment that relativistic corrections are small. G. Rupak replied that relativistic corrections are numerically small relative to other contributions of the same order in the power counting. I. Stewart pointed out that relativistic corrections are suppressed by a factor of the nucleon mass rather than the range of the theory Λ.

The Lepage analysis G. Rupak presented was only for the observable $\cot\delta$. J. Steele pointed out that one should look at both δ and $\cot\delta$ to determine where the theory breaks down.

N. Shoresh pointed out that if you use a fixed point argument to set $C_4 = C_2^2/C_0$ you miss additional perturbative pieces.

R. Perry asked if the expansion is really converging. G. Rupak replied that the extraction of $1/a$ is approaching the value in the model.

NUCLEON-NUCLEON EFFECTIVE FIELD THEORY AT NNLO: RADIATION PIONS AND $^1S_0$ PHASE SHIFT

T. MEHEN[a]
California Institute of Technology, Pasadena, CA, 91125
E-mail: mehen@theory.caltech.edu

Low energy phenomena involving two nucleons can be successfully described using effective field theory. Because of the relatively large expansion parameter, it is only at next-to-next-to-leading order (NNLO) where one can expect to see agreement with experiment at the few percent level. The first part of this talk will focus on radiation pion effects, which first appear at NNLO. The power counting for radiation pions is simple for center of mass momentum $p \sim \sqrt{Mm_\pi} \equiv Q_r$, the threshold for pion production. We explain how graphs calculated with the Q_r power counting scale for $p \sim m_\pi$. The Q_r^3 radiation pion contributions to nucleon-nucleon scattering are suppressed by inverse powers of the S-wave scattering lengths. However, we point out that order Q_r^4 radiation contributions might give a NNLO contribution for $p \sim m_\pi$. In the second part of the talk, results for the potential pion and contact interaction part of the NNLO $^1S_0$ phase shift are presented. We emphasize the importance of eliminating spurious poles in the expression for the amplitude at each order in the perturbative expansion. Doing this leaves a total of three free parameters at NNLO. We obtain a good fit to the $^1S_0$ phase shift.

1 Introduction

This talk focuses on higher order calculations in the low energy effective field theory for two-nucleon systems. In particular, we will be discussing nucleon-nucleon scattering at next-to-next-to-leading order (NNLO) in the expansion recently proposed by Kaplan, Savage and Wise (KSW).[1,2] Observables are expanded in powers of Q/Λ, where Q is either p, the three-momentum of a nucleon in the center of mass frame, or m_π. Λ is the range of the effective theory. The nuclear S-wave scattering lengths (denoted by a) are very large so that powers of pa have to be summed to all orders at each order of the Q expansion. This requires a novel power counting in which the leading 4-nucleon operator with no derivatives is treated nonperturbatively. To make this power counting manifest in dimensional regularization it is necessary to use subtraction schemes such as the power divergence subtraction (PDS)[1,2] or the off-shell momentum subtraction(OS)[3,4], but predictions of the theory are manifestly scheme and scale independent order by order in Q. Higher derivative operators and pion exchange are treated perturbatively.

Various estimates of the range of the theory exist. One estimate comes from examining pion exchange ladder graphs, where each additional loop gives

[a]Work presented here has been done in collaboration with I. W. Stewart.

a contribution of order $p \times Mg_A^2/(8\pi f^2) \equiv p/\Lambda_{NN}$, where $\Lambda_{NN} = 300\,\text{MeV}$. Another possibility is that m_ρ or the threshold for Δ production sets the scale for the breakdown of the effective theory, implying a range for S-wave scattering $\sim 700\,\text{MeV}$. It has been suggested [5,6] that two pion exchange contributions to the nucleon-nucleon potential may be become important for momenta of order 400 MeV. These considerations point to a range somewhere between 300 MeV and 700 MeV. Therefore, for $p \sim m_\pi$, the expansion parameter, Q/Λ, is between 1/2 and 1/5. Because the expansion parameter of the theory is rather large, low order calculations in the effective theory cannot be expected to reproduce phase shift data as accurately as potential models with many parameters.

Many observables have been computed to next-to-leading order (NLO) in the KSW expansion, including nucleon-nucleon phase shifts,[1,7] Coulomb corrections to proton-proton scattering,[8] electromagnetic form factors of the deuteron,[9] deuteron polarizabilities,[10] proton-proton fusion,[11] $np \to d\gamma$,[12] Compton deuteron scattering,[13] and $\nu d \to \nu d$.[14] Some of these calculations are reviewed in the talk by Martin Savage in this volume.[15] One typically finds errors of order 30 – 40% at leading order and 10% at NLO. This is consistent with $Q/\Lambda \approx 1/3$, or $\Lambda \approx 400\,\text{MeV}$. This suggests that at NNLO, effective field theory calculations of low energy processes in the two body sector should agree with data at the few percent level, approaching an accuracy comparable to that of potential models. It is for this reason that extending calculations to this order is an important part of the effective field theory program.

In the first half of this talk, we will discuss radiation pion effects, which first appear at NNLO in calculations of nucleon-nucleon scattering. The power counting of KSW has to be modified in the presence of pion radiation because a new scale, $Q_r = \sqrt{Mm_\pi}$, appears.[16] For power counting radiation pions, it is simpler to take $p \sim Q_r$ and count powers of Q_r rather than Q. We give a procedure for determining how a Q_r^n correction scales with Q for $p \sim m_\pi$. The order Q_r^3 radiation pion contribution to nucleon-nucleon scattering turns out to be suppressed by powers of $1/a$. This is actually a consequence of the invariance of the leading order theory under Wigner's $SU(4)$ spin-isospin symmetry.[17] Wigner symmetry is discussed in detail in the talk by Mark Wise in this volume.[18] The order Q_r^4 radiation pion corrections can give an order Q contribution. In the second half of the talk, we present results of a NNLO calculation of nucleon-nucleon scattering in the $^1S_0$ channel. It is emphasized that coupling constants of the theory must be treated in a Q expansion. The parameter space is constrained by the requirement that perturbative corrections do not shift the location of the pole in the amplitude and by the solutions of renormalization group equations. Once these constraints are imposed, a three parameter fit to the $^1S_0$ phase shift is demonstrated which has $< 2\%$ accuracy at $p \sim m_\pi$ and also reproduces the data well for higher momenta. A similar calculation is discussed in the talk by Gautam Rupak[19,20].

2 Radiation and Soft Pions

The Lagrangian for the theory of nucleons and pions is

$$\begin{aligned}
\mathcal{L}_\pi &= \frac{f^2}{8}\mathrm{Tr}\,(\partial^\mu \Sigma\, \partial_\mu \Sigma^\dagger) + \frac{f^2\omega}{4}\,\mathrm{Tr}(m_q\Sigma + m_q\Sigma^\dagger) + \frac{ig_A}{2}\, N^\dagger \sigma_i (\xi \partial_i \xi^\dagger - \xi^\dagger \partial_i \xi) N \\
&+ N^\dagger \left(iD_0 + \frac{\vec{D}^2}{2M} \right) N - C_0^s (N^T P_i^s N)^\dagger (N^T P_i^s N) \qquad (1) \\
&+ \frac{C_2^s}{8}\left[(N^T P_i^s N)^\dagger (N^T P_i^s \overleftrightarrow{\nabla}^2 N) + h.c. \right] \\
&- D_2^s\, \omega\, \mathrm{Tr}(m^\xi)(N^T P_i^s N)^\dagger (N^T P_i^s N) + \dots ,
\end{aligned}$$

where operators relevant at NLO are included (and isospin violation is neglected). Here $g_A = 1.25$ is the nucleon axial-vector coupling, $\Sigma = \xi^2$ is the exponential of pion fields, $f = 131\,\mathrm{MeV}$ is the pion decay constant, $m^\xi = (\xi m_q \xi + \xi^\dagger m_q \xi^\dagger)/2$, where $m_q = \mathrm{diag}(m_u, m_d)$ is the quark mass matrix, and $m_\pi^2 = w(m_u + m_d)$. The matrices P_i^s project onto states of definite spin and isospin, and the superscript s denotes the partial wave amplitude mediated by the operator. This talk will be concerned only with S-wave scattering, so $s = S$ (for $^1S_0$) or T (for $^3S_1$). This notation will be omitted when it is not necessary to distinguish between the two channels.

In the KSW power counting, the C_0 operator is treated nonperturbatively and graphs with a single pion exchange (dressed with C_0 bubbles) first appear at NLO. Loop graphs with pions contain three different kinds of contributions, which are called potential, radiation and soft. The three kinds of pion are characterized by different energy (q_0) and momentum ($\vec{q}$):

potential	$q_0 \sim \vec{q}^{\,2}/M$
radiation	$q_0 \sim \lvert\vec{q}\rvert \sim m_\pi$
soft	$q_0 \sim \lvert\vec{q}\rvert \sim Q_r = \sqrt{M m_\pi}$.

As stated earlier, when calculating radiation and soft contributions we take $p \sim Q_r$ rather than $p \sim Q$. The three contributions will differ in size and it is necessary to devise a power counting which correctly takes this into account. Before giving the power counting we will illustrate how these contributions arise with a few illustrative examples.

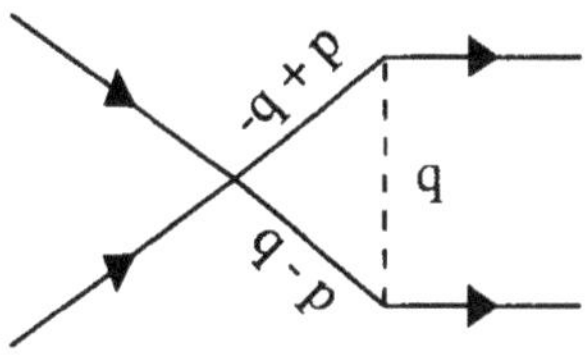

Figure 1: One loop graph with a C_0 and pion that has both potential and radiation contributions

Consider the one loop graph shown in Fig. 1, whose contribution to the amplitude is proportional to

$$\frac{g_A^2 C_0}{2f^2}\int \frac{d^4q}{(2\pi)^4}\frac{1}{\frac{E}{2}+q_0-\frac{(\vec{q}-\vec{p})^2}{2M}+i\epsilon}\,\frac{1}{\frac{E}{2}-q_0-\frac{(\vec{q}-\vec{p})^2}{2M}+i\epsilon}\,\frac{\vec{q}^{\,2}}{q_0^2-\vec{q}^{\,2}-m_\pi^2+i\epsilon}.$$

When the q_0 integral is performed via contour integration, the integral receives contributions from both pion and nucleon poles. If a nucleon pole is taken then $|q_0| = E/2-(\vec{q}-\vec{p})^2/2M = (2\vec{q}\cdot\vec{p}-\vec{q}^{\,2})/2M$, where in the last step we have used $E = \vec{p}^{\,2}/M$. Since $|\vec{p}|, |\vec{q}| \ll M$, $q_0 \ll |\vec{q}|$, we can expand the pion propagator:

$$\frac{1}{q_0^2-\vec{q}^{\,2}-m_\pi^2} = -\frac{1}{\vec{q}^{\,2}+m_\pi^2}-\frac{q_0^2}{(\vec{q}^{\,2}+m_\pi^2)^2}+\ldots.$$

A pion is referred to as a potential pion whenever the energy dependent piece of its propagator is treated perturbatively. In the KSW power counting, $|\vec{p}| \sim m_\pi \sim \mu_R \sim |\vec{q}| \sim Q$. Each nucleon propagator gives a factor of M/Q^2 since $E \sim q_0 \sim Q^2/M$. The measure $d^4q \sim Q^5/M$, and in a scheme with manifest power counting, $C_0 \sim 1/(M\mu_R) \sim 1/(MQ)$. Using this counting it is straightforward to see that this graph is order Q^0, i.e., it is a NLO contribution. The first correction from the expansion in $q_0^2/(\vec{q}^{\,2}+m_\pi^2)$ is suppressed relative to the leading potential contribution by Q^2/M^2, and so is N^{3}LO.

There is also a contribution from the pion pole. In this case $q_0^2 = \vec{q}^{\,2} + m_\pi^2$. In the nucleon propagators, the factors of $(2\vec{q}\cdot\vec{p}-\vec{q}^{\,2})/2M \ll q_0$ and must be treated perturbatively. With the KSW power counting, the nucleon propagators in the graph in Fig. 1 are $\pm 1/q_0 \sim 1/Q$, and the loop measure scales as Q^4, so the graph is $\sim Q$. Therefore, this radiation pion contribution first appears at NNLO.

While KSW power counting works for the graph in Fig. 1, it fails for other graphs with radiation pions. As an example, consider the graph in Fig. 2 which

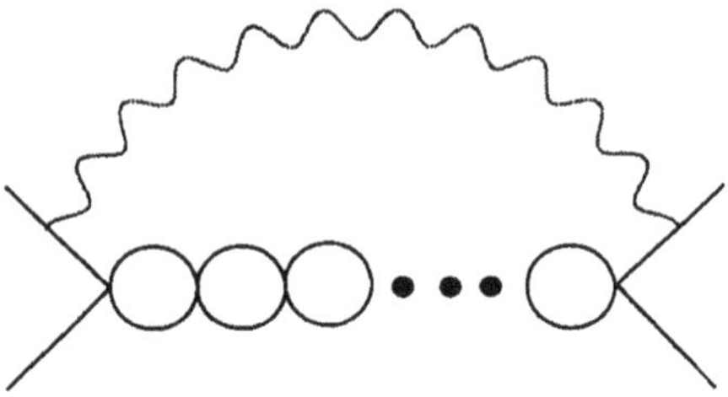

Figure 2: A radiation pion graph with n internal C_0 bubbles.

contains n nucleon bubbles inside a radiation pion loop. The loop integral in this graph vanishes if the pion pole is not taken so there is no potential pion contribution. Emission of the radiation pion in these graphs changes the spin/isospin of the nucleon pair. Therefore, if the external nucleons are in a spin-triplet (singlet) state, then the coefficients appearing in the internal bubble sum are C_0^S (C_0^T). For definiteness, consider nucleon-nucleon scattering in the ${}^1S_0$ channel. The contribution from the graph in Fig. 2 is:

$$\frac{g_A^2}{2f^2}\int\frac{d^4q}{(2\pi)^4}\frac{i}{q_0+i\epsilon}\frac{i}{q_0+i\epsilon}\frac{-i\,\vec{q}^{\,2}}{q_0^2-\vec{q}^{\,2}-m_\pi^2+i\epsilon}[-iC_0^T(\mu_R)]^{n+1}$$
$$\times\left[\int\frac{d^4k}{(2\pi)^4}\frac{i}{q_0-k_0+\frac{E}{2}-\frac{(\vec{k}-\vec{q})^2}{2M}+i\epsilon}\frac{i}{k_0+\frac{E}{2}-\frac{\vec{k}^2}{2M}+i\epsilon}\right]^n .$$

The q_0 integral is closed around the one pion pole above the real axis so $q_0 \sim |\vec{q}| \sim Q$. In the k_0 integrals a nucleon pole must be taken. In the KSW power counting, $k_0 \sim Q^2/M, |\vec{k}\,| \sim Q$. The graph then scales as $(Q/M)^{n+1}$. This suggests that graphs with nucleon bubbles inside the radiation loop are suppressed relative to the one loop radiation pion graph. However, explicitly performing the q_0 and k_0 integrals gives

$$-iC_0^T(\mu_R)\,\frac{g_A^2}{2f^2}\int\frac{d^3q}{(2\pi)^3}\frac{\vec{q}^{\,2}}{(\vec{q}^{\,2}+m_\pi^2)^{3/2}}\left[\int\frac{d^3k}{(2\pi)^3}\frac{-MC_0^T(\mu_R)}{\vec{k}^{\,2}+M(\vec{q}^{\,2}+m_\pi^2)^{1/2}-ME}\right]^n .$$

The size of the loop momenta k in the nucleon bubbles is $\sim\sqrt{Mm_\pi}$ even for $p<\sqrt{Mm_\pi}$. The q integral will be dominated by $\vec{q}\sim m_\pi$ so the graph will scale as

$$\frac{1}{\Lambda_\chi^2}\,\frac{m_\pi^2}{M\mu_R}\left(\frac{\sqrt{Mm_\pi}}{\mu_R}\right)^n .$$

(Recall $\Lambda_\chi = 4\pi f$). For $\mu_R \sim Q \sim m_\pi$, we see that graphs with additional bubbles are enhanced, contrary to the KSW power counting. The sum over graphs with an arbitrary number of bubbles is μ_R independent and the correct estimate for the size is obtained when $\mu_R \sim \sqrt{M m_\pi}$. At this scale, these graphs and their sum are of order $m_\pi^{3/2}/(M^{3/2}\Lambda_\chi^2) = Q_r^3/(M^3\Lambda_\chi^2)$.

In the KSW power counting, one assumes the loop momentum in the nucleon bubbles is dominated by $k \sim Q, k_0 \sim Q^2/M$. However, when the nucleon bubbles are inside a radiation pion loop, the energy flowing into the nucleon bubbles is actually order m_π, and therefore $k^0 \sim m_\pi$ and $k \sim \sqrt{M m_\pi} \equiv Q_r$. This scale corresponds to the threshold for on-shell pion production.

In general, radiation pion graphs will depend on p, m_π and M in a complicated way, making them difficult to power count if one takes $p \sim m_\pi$. However, the natural scale for loop momenta with radiation pions is Q_r, and the power counting simplifies considerably if we consider nucleons scattering at $p \sim Q_r$. Later we will discuss what happens as p is lowered back down to m_π.

Power counting at the scale Q_r is straightforward. We take $p \sim \mu_R \sim Q_r$. The C_{2n} scale with μ_R exactly the same way as in the KSW power counting, $C_{2n}p^{2n} \sim p^{2n}/\mu_R^{n+1} \sim Q_r^{n-1}$. A radiation pion propagator gives M^2/Q_r^4, the pion nucleon coupling gives Q_r^2/M. Nucleon propagators scale like M/Q_r^2. In a radiation loop $q_0 \sim |\vec{q}| \sim m_\pi$, so the loop measure $d^4q \sim Q_r^8/M^4$. The measure of a potential loop scales as Q_r^5/M. Using this power counting it is straightforward to show that all graphs with one radiation pion and an arbitrary number of C_0's scale as $Q_r^3/(M^3\Lambda_\chi^2)$. These graphs are shown in Fig. 3.

It is interesting to examine the result of evaluating some of the diagrams in Fig. 3 explicitly[16] ($\bar{\mu}^2 = \mu^2 \pi e^{-\gamma_E}$):

$$
\begin{aligned}
a) &= -3i\mathcal{A}_{-1}\frac{g_A^2 m_\pi^2}{(4\pi f)^2}\left[\frac{1}{\epsilon} + \frac{1}{3} - \ln\left(\frac{m_\pi^2}{\bar{\mu}^2}\right)\right], \qquad (2)\\
b) &= [\mathcal{A}_{-1}]^2\frac{g_A^2 M m_\pi^2}{(4\pi f)^2}\left\{\frac{3p}{4\pi}\left[\frac{1}{\epsilon} + \frac{7}{3} - 2\ln 2 - \ln\left(\frac{m_\pi^2}{\bar{\mu}^2}\right) - \ln\left(\frac{-p^2}{\bar{\mu}^2}\right)\right]\right.\\
&\qquad \left. + \frac{i\sqrt{M m_\pi}}{4\sqrt{\pi}} I_1\left(\frac{E}{m_\pi}\right)\right\},\\
c) &= \frac{i g_A^2}{\sqrt{\pi} f^2}\left(\frac{m_\pi}{M}\right)^{3/2} I_2\left(\frac{E}{m_\pi}\right).
\end{aligned}
$$

I_1 and I_2 are hypergeometric functions. The $1/\epsilon$ poles are canceled by inser-

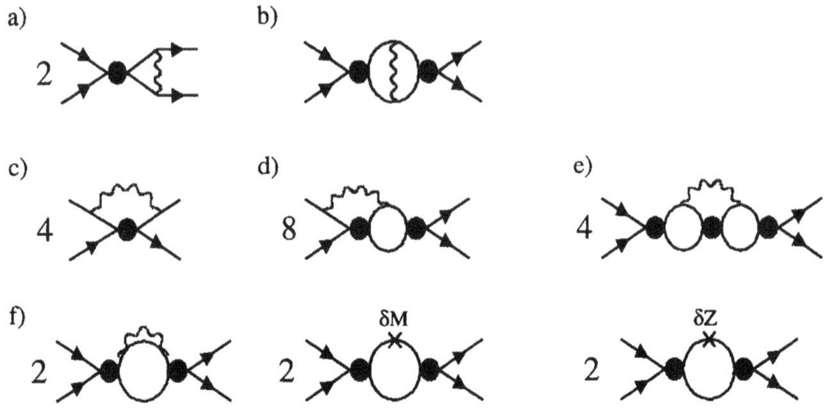

Figure 3: Leading order radiation pion graphs for NN scattering. The solid lines are nucleons, the wavy lines are radiation pions and δM, δZ are the mass and field renormalization counterterms. The filled dot denotes the $C_0(\mu_R)$ bubble chain. There is a further field renormalization contribution that is not shown, but is included in the calculation.[16]

tions of a $D_2 m_\pi^2$ counterterm. The leading order amplitude

$$\mathcal{A}_{-1} = -\frac{4\pi}{M}\frac{1}{1/a^S + ip},$$

scales as $\sim 1/(Mp)$, so we see that Eq. (2) has terms proportional to

$$\left(\frac{m_\pi}{M}\right)^{3/2}, \quad \frac{m_\pi^2}{Mp} \quad \text{and} \quad \frac{m_\pi^{5/2}}{M^{1/2}p^2}.$$

For $p \sim Q_r$ these terms scale as Q_r^3/M^3, as anticipated by the power counting. At $p \sim m_\pi \sim Q$, these terms scale like $(Q/M)^{3/2}, Q/M$, and $(Q/M)^{1/2}$ respectively. Bubble sums which do not appear inside radiation loops will be referred to as external bubble sums. These bubble sums are responsible for the factors of p in the denominators, and thus the enhancement of some terms at $p \sim m_\pi$. Graphs with two external bubble sums have terms that are enhanced by $Q_r^2/Q^2 \sim 1/m_\pi$ (and $Q_r/Q \sim 1/\sqrt{m_\pi}$), while graphs with one external bubble sum have terms enhanced by $1/\sqrt{m_\pi}$. Individual graphs like b) have parts that scale differently with Q. Terms which scale like $Q^{1/2}$ at $p \sim m_\pi$ are actually larger than NNLO in the Q counting. The $Q^{1/2}$ contributions come from graphs $b), e)$ and $f)$, and cancel when the graphs are added together.

Presently it is not known whether this cancellation occurs for some reason or is merely an accident.

The sum of all Q_r^3 graphs is:

$$i\mathcal{A}^{rad} = 6i\mathcal{A}_{-1}^2 \frac{g_A^2 m_\pi^2}{(4\pi f)^2} \frac{M}{4\pi}\left(\frac{1}{a^S}-\frac{1}{a^T}\right)\left[\kappa + \ln\left(\frac{\mu^2}{m_\pi^2}\right)\right]$$
$$+i\mathcal{A}_{-1}^2\left(\frac{M}{4\pi}\right)^2\left(\frac{1}{a^S}-\frac{1}{a^T}\right)^2 \frac{g_A^2}{\sqrt{\pi}f^2}\left(\frac{m_\pi}{M}\right)^{3/2} I_2\left(\frac{E}{m_\pi}\right), \qquad (3)$$

where the $\ln(\mu)$ dependence is canceled by $D_2(\mu)$. (For the $^3S_1$ channel, the result is the same as above with $a_S \leftrightarrow a_T$.) The final answer turns out to be much smaller than anticipated by the power counting. For $p \sim Q_r$, the first term is suppressed by a factor of $1/Q_r(1/a^S - 1/a^T)$, the second by $1/Q_r^2(1/a^S - 1/a^T)^2$. This suppression occurs because the radiation pions couple to a charge of Wigner's $SU(4)$, which is a symmetry of the leading order Lagrangian in the limit $a_S, a_T \to \infty$ (or $a_S = a_T$).[17,18] The order Q_r^3 radiation pion graphs are a tiny correction to the S-wave scattering amplitude.

The next important radiation pion contribution comes from graphs with one insertion of a $C_2 p^2$, $D_2 m_\pi^2$, or G_2 operator [4] or a potential pion, and one radiation pion with an arbitrary number of C_0's. Power counting these graphs gives $Q_r^4/(M^3\Lambda_\chi^2\Lambda)$, i.e. these are suppressed by Q_r/Λ relative to the leading radiation pion graphs in Fig. 3. Note that $Q_r = 360\,\text{MeV}$, so for the most pessimistic estimates of Λ, the Q_r/Λ expansion does not converge. If this is the case then the radiation pion contribution is incalculable. This is true of radiation contributions even when we scale down to $p \sim m_\pi$. However, at the low momenta where the theory would be applicable, the radiation pions could be integrated out. For example, the hypergeometric function $I_2(E/m_\pi) = I_2(p^2/(Mm_\pi))$ could be expanded in a series in p^2 and the effect of each term absorbed[b] into the definition of a C_{2n}. Assuming the radiation pion contribution is computable, the Q_r^4 correction is almost certainly larger than the Q_r^3 correction. Since the $C_2 p^2$, $D_2 m_\pi^2$ operators and potential pion exchange do not respect Wigner's $SU(4)$, there will be no suppression by factors of $1/(a\,Q_r)$.

An important issue which needs to be addressed is how the radiation pion graphs scale as p is lowered from Q_r to m_π. We saw that the Q_r^3 graphs have pieces that scale as $Q^{1/2}, Q, Q^{3/2}, \ldots$, for $p \sim m_\pi$, and that this enhancement can be understood by counting the number of external bubble sums. In order to know which radiation pion graphs to include at a given order in the KSW

[b]Unfortunately, the resulting theory below the scale Q_r would no longer respect chiral symmetry. Operators involving a different number of pion fields could have different coefficients.

power counting, we must know how a Q_r^n correction scales with Q for $p \sim m_\pi$. It turns out that an order Q_r^n calculation is sufficient to determine the order $Q^{n/2-1}$ result.

To see this first consider the Q expansion of $p\cot\delta$:

$$\begin{aligned} p\cot\delta &= ip + \frac{4\pi}{M}\frac{1}{\mathcal{A}} \\ &= ip + \frac{4\pi}{M}\frac{1}{\mathcal{A}_{-1}} - \frac{4\pi}{M}\frac{\mathcal{A}_0}{\mathcal{A}_{-1}^2} - \frac{4\pi}{M}\left(\frac{\mathcal{A}_1}{\mathcal{A}_{-1}^2} - \frac{\mathcal{A}_0^2}{\mathcal{A}_{-1}^3}\right) \\ &\quad - \frac{4\pi}{M}\left(\frac{\mathcal{A}_2}{\mathcal{A}_{-1}^2} - \frac{2\mathcal{A}_0\mathcal{A}_1}{\mathcal{A}_{-1}^3} + \frac{\mathcal{A}_0^3}{\mathcal{A}_{-1}^4}\right) + \dots . \end{aligned}$$

$p\cot\delta$ is real and an analytic function of p^2 near $p = 0$. This will be true order by order in Q so:

$$\begin{aligned} \frac{\mathcal{A}_0}{\mathcal{A}_{-1}^2} &= f_0 \Rightarrow \mathcal{A}_0 = f_0\mathcal{A}_{-1}^2\,, \\ \frac{\mathcal{A}_1}{\mathcal{A}_{-1}^2} - \frac{\mathcal{A}_0^2}{\mathcal{A}_{-1}^3} &= f_1 \Rightarrow \mathcal{A}_1 = f_1\mathcal{A}_{-1}^2 + f_0^2\mathcal{A}_{-1}^3\,, \\ \frac{\mathcal{A}_2}{\mathcal{A}_{-1}^2} - \frac{2\mathcal{A}_0\mathcal{A}_1}{\mathcal{A}_{-1}^3} + \frac{\mathcal{A}_0^3}{\mathcal{A}_{-1}^4} &= f_2 \Rightarrow \mathcal{A}_2 = f_2\mathcal{A}_{-1}^2 + 2f_0f_1\mathcal{A}_{-1}^3 + f_0^3\mathcal{A}_{-1}^4\,, \end{aligned}$$

where the f_n are real functions of p which are analytic about $p^2 = 0$. We see that the general form of a higher order amplitude is powers of $\mathcal{A}_{-1}$ multiplied by functions of p. The crucial point is that the function multiplying the $\mathcal{A}_{-1}^2$ is the only new contribution. The coefficient of $\mathcal{A}_{-1}^n, n > 2$, is determined by lower order amplitudes. In the Q expansion of $p\cot\delta$ the latter contributions will cancel.

This generalizes to the Q_r expansion of radiation pion graphs, the only difference being that the radiation pion contribution starts out at Q_r^3, while the potential pion starts out at Q_r^0. A Q_r^n radiation pion correction to the amplitude will be of the form:

$$\mathcal{A}_n = \mathcal{A}_{-1}^2 f_{n,2} + \mathcal{A}_{-1}^3 f_{n,3} + \dots + \mathcal{A}_{-1}^{n-1} f_{n,n-1}\,.$$

Again, the $f_{n,m}$ are analytic about $p^2 = 0$ and all the $f_{n,m}$ except for $f_{n,2}$ will be determined from lower order amplitudes. Since $\mathcal{A}_n \sim Q_r^n$ and $\mathcal{A}_{-1} \sim 1/(Mp)$, $f_{n,m} \sim Q_r^{n+m}$ for $p \sim Q_r$. To understand how $f_{n,m}$ scales with Q as p is lowered to m_π, note that without loss of generality, $f_{n,m}$ can be written as

$$f_{n,m} = \frac{(\sqrt{Mm_\pi})^{n+m}}{\Lambda_\chi^2\,\overline{\Lambda}^{n-m}}\hat{f}_{n,m}\left(\frac{p}{\sqrt{Mm_\pi}},\dots\right),$$

Figure 4: Example of order Q_r^4 graphs that have three external bubble sums.

where the ellipses denote momentum dependence that involves scales other than Q_r, and $\overline{\Lambda} = \Lambda_\chi$, Λ, or M. For $p \sim m_\pi$ the ellipse denote dependence on the dimensionless variables p/m_π, pa, and $p/\overline{\Lambda}$. For $p \sim m_\pi$, $p/\sqrt{Mm_\pi} \sim (Q/M)^{1/2}$ and the function $\hat{f}_{n,m}$ can be expanded in its first argument:

$$\mathcal{A}_{-1}^m f_{n,m} = \mathcal{A}_{-1}^m \frac{(\sqrt{Mm_\pi})^{n+m}}{\Lambda_\chi^2\, \overline{\Lambda}^{n-m}} \hat{f}_{n,m}(0,\ldots)\left[1 + O\left(\frac{Q}{M}\right)^{1/2}\right].$$

The leading term scales as $\sim Q^{n/2-m/2}$ for $p \sim m_\pi \sim Q$. Since only the $m = 2$ term actually ends up contributing to $p\cot\delta$, the new contribution at Q_r^n scales like $Q^{n/2-1}$ (plus subleading terms) for $p \sim m_\pi$. This is consistent with the result of the Q_r^3 calculation, where the largest contributions from individual graphs scaled as $Q^{1/2}$. A cancellation between graphs resulted in this contribution vanishing. The remaining terms scale as Q, $Q^{3/2}$, ... (counting $1/a \sim Q$). The Q_r^n radiation pion contribution could have a $Q^{1/2}$ contribution from the $\mathcal{A}_{-1}^{n-1} f_{n,n-1}$ term. But this contribution is determined by the Q_r^3 amplitude which vanished; so there will be no $Q^{1/2}$ contribution from any radiation pion graph. For example, at Q_r^4 the terms proportional to $\mathcal{A}_{-1}^3$ comes from graphs such as those shown in Fig. 4. These graphs factorize into two pieces which are lower order and it is easy to see that the same cancellation between the $Q^{1/2}$ pieces of the graphs b), e) and f) will also occur at Q_r^4.

Since $Q^{n/2-1} = Q$ for $n = 4$, the Q_r^4 radiation pion graphs may have a contribution that is NNLO for $p \sim m_\pi$. We have checked that graphs with one insertion of the $D_2 m_\pi^2$ operator do not give rise to such a contribution, but a calculation of the remaining Q_r^4 graphs has not been performed. These graphs need to be computed in order to obtain the complete NNLO amplitude. Unfortunately, graphs with one potential and one radiation pion are numerous. Higher Q_r^n amplitudes may have a contribution which scales as Q for $p \sim m_\pi$ from the $\mathcal{A}_{-1}^{n-2} f_{n,n-2}$ term. But these will cancel in the Q expansion of $p\cot\delta$. Note that a calculation of the order Q_r^5 graphs would be necessary to determine the order $Q^{3/2}$ terms.

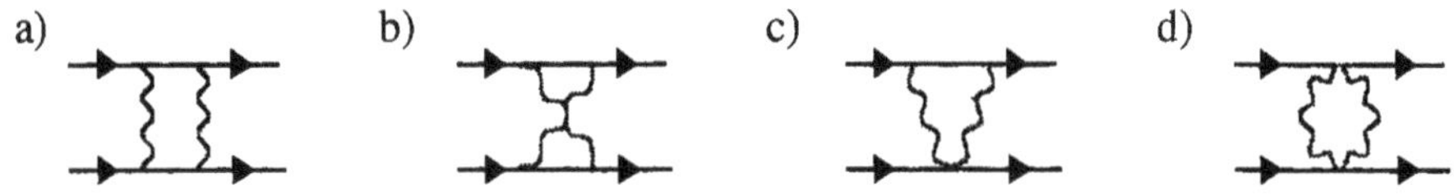

Figure 5: Examples of one-loop graphs which have soft pion contributions. Graphs a)-d) also have a radiation pion contribution, while in addition graph a) has a potential pion contribution.

Finally, we briefly discuss soft contributions. These first arise in box-type diagrams with two pions, such as those shown in Fig. 5. Dressing these graphs on the outside with C_0 bubbles gives further diagrams of the same order. Unlike radiation graphs, which are dominated by loop momenta $|\vec{q}| \sim m_\pi$, the box-like diagrams receive nonvanishing contributions from $|\vec{q}| \sim Q_r$. It is this latter type of contribution which is called soft. In soft loops the measure $d^4q \sim Q_r^4$ instead of Q_r^8/M^4. For nucleon propagators in soft loops, the loop energy is always greater than the nucleon's kinetic energy, so the propagator is static (like heavy quark effective theory propagators, see [21] or [22]). These propagators scale as $1/Q_r$. Pion propagators scale as $1/Q_r^2$ and pion-nucleon vertices give a factor of Q_r. Power counting the graphs in Fig. 5 gives $\sim Q_r^2$ for the soft contribution and $\sim Q_r^4/M^2$ for the radiation contribution. The contribution from each regime can be separated using the method of asymptotic expansions [23,24,25] and explicit evaluation [16] of the graphs verifies the power counting appropriate for each type of contribution. At $p \sim Q_r$, the soft contribution actually dominates all radiation pion graphs we have considered so far. However, for $p \sim m_\pi$ these soft graphs are order Q^2, and therefore are not enhanced by tne scaling down to $p \sim m_\pi$. The leading order soft graphs are next-to-next-to-next-to-leading order (N^3LO) in the KSW power counting.

3 NNLO Calculation of the $^1S_0$ Phase Shift

In this section[c], we present a partial NNLO calculation of the $^1S_0$ phase shift. A more detailed analysis will be given in a future publication.[26] The first piece of the Q_r^3 radiation pion contribution in Eq. (3) is order Q. The second term is order $Q^{3/2}$ and is not included in the NNLO calculation. The relativistic corrections are computed in Ref.[27] and shown to be negligible. This is because they are suppressed relative to the leading order amplitude by $(Q/M)^2$ rather

[c]Work presented in this section was done in collaboration with Sean Fleming.

than $(Q/\Lambda)^2$ and so are smaller than other NNLO corrections. The calculation here includes order Q contact interactions and potential pion graphs. The NNLO calculation is incomplete because of the omission of the Q_r^4 radiation pion graphs, as discussed in the previous section. The pieces of the NNLO amplitude which are included are expected to scale as $Q/(M\Lambda^2)$ while the Q_r^4 radiation pion graphs are expected to scale as $Q/(\Lambda_\chi^2\Lambda)$ for $p \sim m_\pi$. Since $\Lambda < \Lambda_\chi \approx M$, the contribution from Q_r^4 radiation pion graphs may be smaller than what has been included.

Since we are only interested in the ${}^1S_0$ channel, only $s = S$ operators are relevant and this superscript will be omitted in the following discussion. At NNLO, the following terms are added to the Lagrangian in Eq. (1):

$$\begin{aligned}
\mathcal{L} = &-\frac{C_4}{64}\left[(N^T P_i N)^\dagger (N^T P_i \overleftrightarrow{\nabla}^4 N) + h.c. + 2(N^T P_i \overleftrightarrow{\nabla}^2 N)^\dagger (N^T P_i \overleftrightarrow{\nabla}^2 N)\right] \\
&+\frac{E_4}{8}\omega \mathrm{Tr}(m^\xi)\left[(N^T P_i N)^\dagger (N^T P_i \overleftrightarrow{\nabla}^2 N) + h.c.\right] \\
&-\frac{D_4}{2}\omega^2\left\{\mathrm{Tr}^2(m^\xi) + 2\mathrm{Tr}[(m^\xi)^2]\right\}(N^T P_i N)^\dagger (N^T P_i N)\,, \qquad (4)
\end{aligned}$$

where only terms relevant for the phase shift are included and isospin violation is neglected. All calculations presented in this section will be done in the PDS renormalization scheme[1,2] with spin and isospin traces done in four dimensions.

There are six coefficients that appear at NNLO: C_0, which is present in the leading order calculation, C_2 and D_2, which first appear at NLO, and C_4, E_4, and D_4. It is important that the coupling constants are expanded in Q: [3]

$$\begin{aligned}
C_0 &\rightarrow C_0 + C_{0,0} + C_{0,1} \\
C_2 &\rightarrow C_2 + C_{2,-1} \\
D_2 &\rightarrow D_2 + D_{2,-1}\,. \qquad (5)
\end{aligned}$$

The first piece of C_0 is treated nonperturbatively (i.e. $C_0 \sim Q^{-1}$), while $C_{0,0} \sim Q^0, C_{0,1} \sim Q$. Solving the renormalization group equations (RGE) for the couplings perturbatively ensures that the amplitude is μ independent order by order in the expansion. Therefore, theoretical expressions for physical quantities, such as the scattering length or the location of the pole, are always μ independent. Physically, the Q expansion of the couplings is a consequence of the fact that higher order loop graphs with pions can renormalize the short distance operators at different orders in Q, and therefore different values of the couplings will be obtained at different orders in the expansion.

When a coupling is expanded in Q and its RGE solved perturbatively, a new constant of integration is obtained for each term in the expansion. For example, the RGE for C_0 is:

$$\mu\frac{\partial}{\partial\mu}C_0 = \frac{M\mu}{4\pi}\left(C_0 + \frac{g_A^2}{2f^2}\right)^2 . \tag{6}$$

After the perturbative expansion of C_0 this becomes

$$\mu\frac{\partial}{\partial\mu}C_0 = \frac{M\mu}{4\pi}C_0^2 , \qquad \mu\frac{\partial}{\partial\mu}C_{0,0} = 2\frac{M\mu}{4\pi}C_0\left(C_{0,0} + \frac{g_A^2}{2f^2}\right),$$
$$\mu\frac{\partial}{\partial\mu}C_{0,1} = \frac{M\mu}{4\pi}\left[2\,C_0\,C_{0,1} + \left(C_{0,0} + \frac{g_A^2}{2f^2}\right)^2\right],$$

with solutions

$$C_0 = \frac{4\pi}{M}\frac{1}{-\mu+\gamma} , \qquad C_{0,0} = \frac{M\kappa}{4\pi}C_0^2 - \frac{g_A^2}{2f^2} ,$$
$$C_{0,1} = \frac{1}{C_0}\left(C_{0,0} + \frac{g_A^2}{2f^2}\right)^2 + \frac{M\kappa'}{4\pi}C_0^2 . \tag{7}$$

Three constants of integration appear: γ, κ and κ'; one for each term in the Q expansion of C_0. For consistency we must assign these constants a Q counting. For instance, $C_{0,0}$ has a term in its solution $\kappa\, C_0^2 \sim \kappa/\mu^2$. Since $C_{0,0} \sim Q^0$, κ must be of order Q^2. This reflects the fact that κ is intrinsically small. In the theory without pions, $\kappa = \gamma - 1/a$. As we will see below, the values of γ, κ, κ' may be fixed by demanding that the amplitude has the correct pole structure. In this case, $\kappa \approx r_0/(2a^2) \sim Q^2$. The results in Eq. (7) can also be obtained by solving Eq. (6) exactly and expanding the result (including the constant of integration) in Q. Because of the perturbative expansion of the couplings in Eq. (5) there are ten constants of integration at NNLO. However, the NNLO amplitude will depend only on six independent linear combinations of these constants. There are two further constraints on the number of free parameters: 1) at this order, C_4, E_4 and D_4 are determined entirely in terms of lower order couplings; 2) spurious double and triple poles in the NLO and NNLO amplitudes must be canceled in order to obtain a good fit.

The fact that C_4, E_4 and D_4 are determined in terms of lower order couplings is a consequence of solving the RGE's and applying the KSW power counting. For instance[2], the RGE for C_4 is:

$$\mu\frac{\partial}{\partial\mu}C_4 = \frac{M\mu}{4\pi}\left(2\,C_0\,C_4 + C_2^2\right),$$

which has the solution

$$C_4 = \frac{C_2^2}{C_0} + \rho \frac{M}{4\pi} C_0^2 \,,$$

where ρ is a constant of integration. In the theory without pions, ρ is proportional to the shape parameter, which is $\sim Q^0$ in the KSW power counting. It is also reasonable to consider $\rho \sim Q^0$ in the theory with pions, since ρ is a constant of integration in the RGE for the lowest order term in the Q expansion of C_4. Therefore, $C_2^2/C_0 \sim Q^{-3}$, while $\rho C_0^2 M/(4\pi) \sim Q^{-2}$. The second term is subleading in the Q expansion, and should be omitted at NNLO, so $C_4 = C_2^2/C_0$. Similar relations for E_4, D_4 hold at NNLO:

$$E_4 = \frac{2C_2 D_2}{C_0} \,, \qquad D_4 = \frac{D_2^2}{C_0} \,.$$

It is interesting to note that these relations arise even though there are $\ln(\mu^2)$ terms in the amplitude which contribute to the beta functions for E_4 and D_4.

Because of the nonperturbative treatment of C_0, spurious poles arise at higher orders in the expansion. The leading order amplitude $\mathcal{A}_{-1}$ has a simple pole at $p = i\gamma$. The NLO calculation is proportional to $\mathcal{A}_{-1}^2$, and therefore has a double pole, while the NNLO amplitude has terms proportional to $\mathcal{A}_{-1}^2$ and $\mathcal{A}_{-1}^3$. To obtain a good fit at low momentum, parameters need to be fixed so that the amplitude has only a simple pole at each order in the expansion. This requires that $\mathcal{A}_{-1}$ have its pole in the correct location and that the residues of the spurious double and triple poles vanish. This requirement leads to the following good fit conditions: [3]

$$\left.\frac{1}{\mathcal{A}_{-1}}\right|_{p=p^*} = 0 \,, \qquad \left.\frac{\mathcal{A}_0}{\mathcal{A}_{-1}^2}\right|_{p=p^*} = 0 \,, \qquad \left.\frac{\mathcal{A}_1}{\mathcal{A}_{-1}^2}\right|_{p=p^*} = 0 \,, \tag{8}$$

where p^* is the location of the physical pole. The second condition first appears at NLO, the third at NNLO. The residue of the triple pole in $\mathcal{A}_1$ is canceled by the second equation in Eq. (8). The first equation results in $\gamma = -ip^*$, while the other equations give constraints which eliminate two of the remaining parameters. Eq. (8) will also apply in the ${}^3S_1$ channel.

The graphs evaluated in the NNLO calculation are shown in Fig. 6. The final result is surprisingly simple. The amplitude up to NNLO is:

$$\mathcal{A}_{-1} = -\frac{4\pi}{M}\frac{1}{\gamma + ip} \,,$$

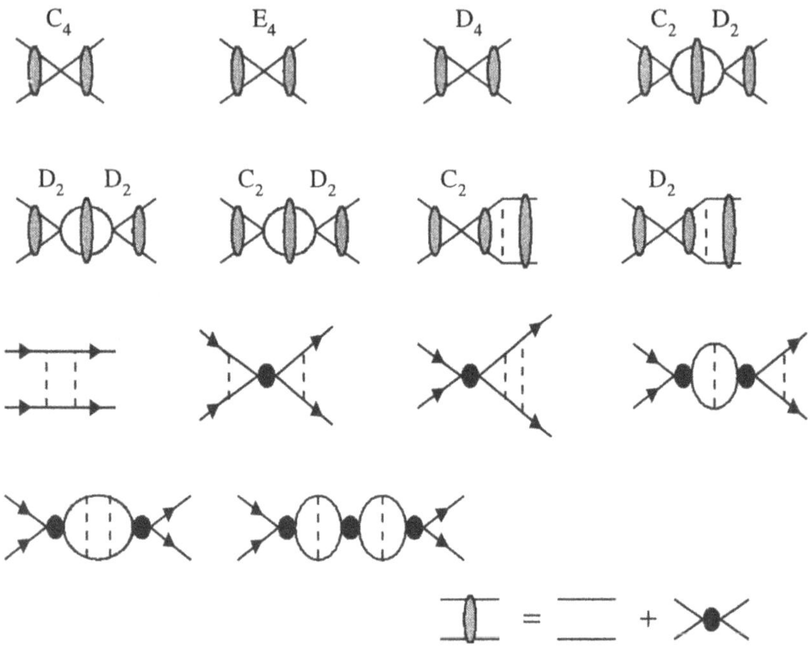

Figure 6: Order Q contact interaction and potential pion graphs for the $^1S_0$ channel. The shaded circle denotes a C_0 bubble sum. At this order the first three graphs do not introduce new parameters as explained in the text.

$$
\begin{aligned}
\mathcal{A}_0 &= -\mathcal{A}_{-1}^2(\zeta_1\, p^2 + \zeta_2\, m_\pi^2) \\
&\quad + \frac{g_A^2}{2f^2}\mathcal{A}_{-1}^2\left(\frac{M m_\pi}{4\pi}\right)^2\left[\frac{(\gamma^2-p^2)}{4p^2}\ln\left(1+\frac{4p^2}{m_\pi^2}\right) - \frac{\gamma}{p}\tan^{-1}\left(\frac{2p}{m_\pi}\right)\right], \\
\mathcal{A}_1 &= \frac{\mathcal{A}_0^2}{\mathcal{A}_{-1}} - \mathcal{A}_{-1}^2\left(\zeta_3\, m_\pi^2 + \zeta_4\, p^2 + \zeta_5\frac{p^4}{m_\pi^2}\right) + \mathcal{A}_0\frac{M g_A^2}{8\pi f^2}\,\frac{m_\pi^2}{p}\left[\frac{\gamma}{2p}\ln\left(1+\frac{4p^2}{m_\pi^2}\right)\right. \\
&\quad \left. -\tan^{-1}\left(\frac{2p}{m_\pi}\right)\right] + \frac{M\mathcal{A}_{-1}^2}{4\pi}\left(\frac{M g_A^2}{8\pi f^2}\right)^2\frac{m_\pi^4}{4p^3}\left\{2(\gamma^2-p^2)\,\mathrm{Im}\,\mathrm{Li}_2\left(\frac{-m_\pi}{m_\pi-2ip}\right)\right. \\
&\quad -4\gamma\, p\,\mathrm{Re}\,\mathrm{Li}_2\left(\frac{-m_\pi}{m_\pi-2ip}\right) - \frac{\gamma\, p\,\pi^2}{3} - (\gamma^2+p^2)\left[\mathrm{Im}\,\mathrm{Li}_2\left(\frac{m_\pi+2ip}{-m_\pi+2ip}\right)\right. \\
&\quad \left.\left. +\frac{\gamma}{4p}\ln^2\left(1+\frac{4p^2}{m_\pi^2}\right) - \tan^{-1}\left(\frac{2p}{m_\pi}\right)\ln\left(1+\frac{4p^2}{m_\pi^2}\right)\right]\right\}.
\end{aligned}
\tag{9}
$$

Using this amplitude it is easy to verify that the S-matrix is unitary to the order we are working. The six linearly independent constants appearing in the amplitude are $\gamma, \zeta_1, \zeta_2, \zeta_3, \zeta_4, \zeta_5$. By definition $\zeta_1 - \zeta_5$ are dimensionless. They are given in terms of coupling constants in the Appendix.

For the ${}^1S_0$ channel, the location of the pole is determined by solving

$$-\frac{1}{a} + \frac{r_0}{2}(p^*)^2 - ip^* = 0\,. \tag{10}$$

Adding the shape parameter correction to Eq. (10) changes the location of the pole by less than 0.01%, so this approximation for the physical pole location is sufficiently accurate. Eq. (10) has a solution for $p^* = -i\,7.877\,\mathrm{MeV}$, which fixes the single LO parameter, $\gamma = -7.877\,\mathrm{MeV}$. The NLO good fit condition relates the constants ζ_1 and ζ_2,

$$\begin{aligned} -\zeta_1\gamma^2 + \zeta_2 m_\pi^2 + \frac{M m_\pi^2}{4\pi}\frac{g_A^2 M}{8\pi f^2}\log\left(1+\frac{2\gamma}{m_\pi}\right) &= 0 \\ \Rightarrow \zeta_2 = \frac{\gamma^2}{m_\pi^2}\zeta_1 + \frac{M}{4\pi}\frac{g_A^2 M}{8\pi f^2}\left[-\frac{2\gamma}{m_\pi} + \frac{2\gamma^2}{m_\pi^2} + \mathcal{O}\left(\frac{\gamma^3}{m_\pi^3}\right)\right]&. \end{aligned} \tag{11}$$

We can use this equation to eliminate ζ_2 in favor of ζ_1, leaving one new parameter in the fit at NLO. This good fit condition gives non-trivial m_π dependence to the perturbative contributions to C_0 (such as κ) as emphasized in Refs. [28,19,6]. ζ_1 is fixed by doing a weighted least squares fit to low momentum data. Note that ζ_1 appears in the good fit condition multiplied by γ^2. Therefore, the value of ζ_2 is insensitive to the value of ζ_1 obtained from the fitting procedure. To a good degree of accuracy we can ignore ζ_1 in Eq. (11), and then we find $\zeta_2 \approx 0.03$. ζ_2 is small because it is proportional to γ/m_π.

At NNLO, $\zeta_5 = 0$ once we impose $C_4 = C_2^2/C_0$. The condition in Eq. (11) must still be satisfied. The NNLO good fit condition involves ζ_3 and ζ_4,

$$\begin{aligned} \zeta_3 &= \frac{\gamma^2}{m_\pi^2}\zeta_4 + \left(\frac{M g_A^2}{8\pi f^2}\right)^2 \frac{M}{4\pi}\frac{m_\pi^2}{\gamma}\left[\mathrm{Re\,Li}\left(\frac{-m_\pi}{m_\pi+2\gamma}\right) + \frac{\pi^2}{12}\right] \\ &= \frac{\gamma^2}{m_\pi^2}\zeta_4 + \left(\frac{M g_A^2}{8\pi f^2}\right)^2 \frac{M m_\pi}{4\pi}\left[2\ln 2 - (1+2\ln 2)\frac{\gamma}{m_\pi} + \mathcal{O}\left(\frac{\gamma^2}{m_\pi^2}\right)\right]. \end{aligned} \tag{12}$$

Since ζ_4 is multiplied by γ^2/m_π^2, this condition basically fixes the value of ζ_3. At this order ζ_1 may change from its value at NLO. We have chosen to fix ζ_1 and ζ_4 by performing a least square fit to the lower momentum data.

The ${}^1S_0$ phase shift is shown in Fig. 7. The solid line is the result of the Nijmegen phase shift analysis[29]. The ${}^1S_0$ phase shift has an expansion in

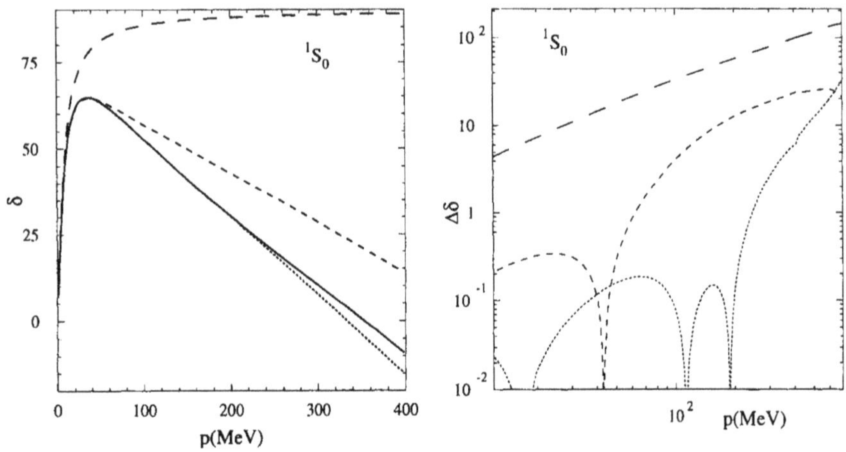

Figure 7: Fit to the $^1S_0$ phase shift δ emphasizing the low momentum region. The solid line is the Nijmegen fit[29] to the data (for $p > 400\,\text{MeV}$ values from the VPI[30] phase shift analysis were used.). The long dashed, short dashed, and dotted lines are the LO, NLO, and NNLO results respectively. $\Delta\delta$ is the difference between these results and the solid line.

powers of Q, and we have plotted the LO, NLO and NNLO results. The LO phase shift at $p \sim m_\pi$ is off by 49%. At NLO, the error is 14%. At NNLO, the error in the $^1S_0$ channel is less than 2% at $p \sim m_\pi$, and the NNLO result gives improved agreement with the data even at $p \sim 400\,\text{MeV}$.

Using $M = 939\,\text{MeV}$ and $m_\pi = 137\,\text{MeV}$, the parameters for our fit in the $^1S_0$ channel are:

$$\begin{aligned} &\text{NLO}: \quad && \zeta_1 = 0.2163; \quad && \zeta_2 = 0.0318; \\ &\text{NNLO}: \quad && \zeta_1 = 0.0777; \quad && \zeta_2 = 0.0313; \quad && \zeta_3 = 0.1831; \quad && \zeta_4 = 0.2447; \end{aligned}$$

Note that $\zeta_3 \sim Q$ is larger than $\zeta_2 \sim Q^0$ because from Eqs. (11) and (12), $\zeta_3/\zeta_2 \sim m_\pi^2/(\gamma\Lambda_{NN})$. The parameter ζ_2 is stable because it is fixed by the NLO good fit condition. On the other hand, ζ_1 changes by a factor of 2.7 going from NLO to NNLO. One expects the value of coupling constants to change at each order in the expansion, but a factor of three difference is somewhat surprising. It is also disturbing that ζ_4 is greater than ζ_1, since, on the basis of the RGE, it is expected that $\zeta_4 < \zeta_1$. At NNLO the RGE for C_2 is: [4]

$$\mu \frac{\partial}{\partial \mu} C_2 = 2 \frac{M\mu}{4\pi} \left(C_0 + \frac{g_A^2}{2f^2} \right) C_2 \, .$$

Expanding C_2 perturbatively results in two equations:

$$\mu\frac{\partial}{\partial\mu}C_2 = 2\frac{M\mu}{4\pi}\,C_0\,C_2\,, \tag{13}$$

$$\mu\frac{\partial}{\partial\mu}C_{2,-1} = 2\frac{M\mu}{4\pi}\left[\left(C_{0,0}+\frac{g_A^2}{2f^2}\right)C_2 + C_0\,C_{2,-1}\right],$$

with solutions

$$C_2 = \zeta_1\,C_0^2\,, \qquad C_{2,-1} = 2\frac{C_2}{C_0}\left(C_{0,0}+\frac{g_A^2}{2f^2}\right) + \zeta_4\,C_0^2\,. \tag{14}$$

The second term in $C_{2,-1}$ has exactly the same form as the leading C_2. $C_{2,-1}$ is supposed to be a perturbative correction to C_2, so $\zeta_4 \sim Q < \zeta_1 \sim Q^0$, and one does not expect this part of $C_{2,-1}$ to be significantly larger than the leading C_2.

Some insight into this puzzle can be obtained by comparing the NLO and NNLO expressions for r_0. Using Eq. (9) we have

$$\begin{aligned} r_0^{NLO} &= \frac{8\pi}{M}\zeta_1 + \frac{2}{\Lambda_{NN}}\left(1-\frac{8\gamma}{3m_\pi}+\frac{2\gamma^2}{m_\pi^2}\right) \\ &= \frac{8\pi}{M}(\zeta_1 + 0.2952)\,, \end{aligned} \tag{15}$$

$$\begin{aligned} r_0^{NNLO} &= \frac{8\pi}{M}\left[\zeta_4 + \zeta_1\left(1-\frac{2m_\pi - 2\gamma}{\Lambda_{NN}}\right) + \frac{\zeta_2}{\Lambda_{NN}}\left(\frac{8m_\pi}{3} - 4\gamma\right) + \ldots\right] \\ &= \frac{8\pi}{M}(\zeta_4 + 0.0138\,\zeta_1 + 1.3508\,\zeta_2 + 0.2104)\,, \end{aligned} \tag{16}$$

where $1/\Lambda_{NN} = (Mg_A^2)/(8\pi f^2)$. Any reasonable fit for the phase shifts will at least approximately reproduce the observed effective range. In Eq. (16), the piece of the NLO correction proportional to ζ_1 is almost exactly canceled by the NNLO correction. This cancellation occurs because $1-2(m_\pi-\gamma)/\Lambda_{NN} \simeq 0.01$ in the ${}^1S_0$ channel. This is simply an unfortunate accident; in the ${}^3S_1$ channel, where $\gamma = 45.7$ MeV instead of -7.88 MeV, the coefficient of ζ_1 in Eq. (16) is ≈ 0.4. Since ζ_2 is small due to the NLO good fit condition, this accidental cancellation forces ζ_4 to make up the observed effective range. Therefore, ζ_4 is much larger than anticipated.

It is important to note that the coupling C_2 is not changing nearly as drastically at each order. If one were to solve the theory exactly, one would find that C_2 had a term $\hat{r}\,C_0^2$, where $\hat{r}$ represents a short distance contribution to the effective range. ζ_1 and ζ_4 can be thought of as the first few terms in an expansion of $\hat{r}$. The theory should eventually converge to the exact $\hat{r}$ but it need not reproduce $\hat{r}$ exactly at low orders in perturbation theory. It is reassuring that $\zeta_1^{NLO} = 0.22$ and $\zeta_1^{NNLO} + \zeta_4^{NNLO} = 0.32$, indicating that the coupling constant C_2 is not changing more than one would expect in a theory with an expansion parameter $\sim 1/3$. Note that ζ_1^{NNLO} must be small in order for $\zeta_1^{NNLO} + \zeta_4^{NNLO}$ to not be much larger than ζ_1^{NLO}. At NLO potential pions make up $\sim 60\%$ of r_0, with short distance physics making up the remaining $\sim 40\%$. At NNLO the situation does not change by very much; potential pions give $\sim 40\%$, short distance physics $\sim 50\%$ and cross-terms make up the rest.

Kaplan and Steele[28,31] have proposed a fitting procedure in which $\hat{r}$ is not expanded in powers of Q. This amounts to imposing the additional condition $\zeta_4 = 0$, so there is no new parameter at NNLO. Only the linear combination $\zeta_1 + \zeta_4$ appears in C_2. However the amplitude depends on ζ_1 and ζ_4 very differently because they appear at different orders in the Q expansion. This is why we treat ζ_1 and ζ_4 as separate parameters. Where it not for the cancellation in r_0^{NNLO} noted above, then the difference between the two methods would be small, i.e., the size of a N^{3}LO correction. In fact, it is impossible to reproduce the observed effective range if one demands $\zeta_4 = 0$. In the $^3S_1$ channel there is a logarithmic divergence[4] at order Q, introducing a $\ln(\mu/K)$ dependence into the coupling $C_{2,-1}$. Since the constant K is undetermined, ζ_4 cannot be set to zero, so there is a new parameter at NNLO. For C_2^S this type of $\ln(\mu)$ dependence occurs at order Q^2 from soft pion graphs.[16] Kaplan and Steele have suggested that the failure of their fitting procedure when applied to models with effective ranges close to that seen in nature may indicate that r_0 is unnaturally large, and that the power counting of the effective theory might need to be modified to take this scale into account. It seems more likely that the failure observed in Ref. [31] may just be the consequence of a numerical accident in the amplitude at NNLO as shown in Eq. (16). This cancellation does not occur in the $^3S_1$ channel at NNLO, nor is it likely to persist at higher orders. For this reason, it seems premature to conclude on the basis of the NNLO amplitude that the expansion is failing due to a large r_0.

In Ref. [20], Rupak and Shoresh use a similar fitting procedure to the one discussed here. Instead of finding ζ_1 and ζ_4 by fitting to the phase shift, they fix these constants by matching onto

$$p\,\cot(\delta) = -\gamma + \frac{s_0}{2}\,(p^2 + \gamma^2) + \dots\,, \tag{17}$$

which is similar to the effective range expansion except $p\cot(\delta)$ is expanded about $p = i\gamma$. At NLO ζ_1 is fixed to give s_0. At NNLO the same value of ζ_1 is used and ζ_4 is again fixed to reproduce s_0. This procedure was applied to data from a two-Yukawa toy model and the convergence of the effective field theory looks similar to that in Fig. 7. In our approach we have not demanded that the exact value of r_0 is reproduced since we know that there will be corrections to r_0 from higher orders in the m_π/Λ expansion.

Finally, we would like to comment on the prediction of higher order terms in the effective range expansion

$$p\cot(\delta) = -\frac{1}{a} + \frac{r_0}{2}p^2 + v_2 p^4 + v_3 p^4 + v_4 p^4 + \dots .$$

Using the NLO expression for $p\cot(\delta)$, Cohen and Hansen[32] obtained predictions for v_2, v_3 and v_4. At NLO, the effective field theory predictions for v_2, v_3, and v_4 disagree with the v_i obtained from a fit to the Nijmegen phase shift analysis. The NNLO predictions for the shape parameters are shown in the table below. The prediction for r_0 is not better at NNLO than at NLO. The NNLO v_i predictions depend on ζ_1 and ζ_2. We see that the NNLO correction substantially reduces the discrepancy between the effective field theory prediction and the fit to the Nijmegen phase shift analysis, but the discrepancy is still quite large. This gives some evidence that the EFT expansion is converging on the true values of the v_i, albeit slowly. Effective field theory predictions for the shape parameters have been studied in toy models where one is able to go to very high orders in the Q expansion.[33] In the toy models, the effective field theory did eventually reproduce the shape parameters, but the observed convergence is rather slow.

$^1S_0$	r_0	v_2	v_3	v_4
Fit[32]	2.73 fm	$-0.48\,\mathrm{fm}^3$	$3.8\,\mathrm{fm}^5$	$-17\,\mathrm{fm}^7$
NLO	2.65 fm	$-3.3\,\mathrm{fm}^3$	$19\,\mathrm{fm}^5$	$-117\,\mathrm{fm}^7$
NNLO	2.63 fm	$-1.2\,\mathrm{fm}^3$	$2.9\,\mathrm{fm}^5$	$-0.7\,\mathrm{fm}^7$

What can we learn from Fig. 7 about the convergence of the KSW expansion? It is pleasing to see a NNLO calculation reproducing the $^1S_0$ phase shift at $p \sim m_\pi$ with an accuracy of a few percent, and giving an improved fit to the data even for larger momenta. It is important to keep in mind that the NNLO calculation of the phase shift is incomplete since there is a possible contribution from order Q_r^4 radiation pion graphs. Many other process involving two nucleons can be examined at this order. Once enough processes are calculated to NNLO, all parameters of the theory appearing at this order can be extracted and it will be possible to make predictions with no free parameters. The accuracy of these predictions will constitute a serious test of the KSW power counting.

Acknowledgments

We would like to thank D. Kaplan, G. Rupak, and N. Shoresh, for useful discussion. This work was supported in part by the Department of Energy under grant number DE-FG03-92-ER 40701. T.M. was also supported by a John A. McCone Fellowship.

Appendix

Here we give the definitions of the constants appearing in the NNLO expression for the amplitude in Eq. (9). After solving the renormalization group equations at this order one finds that all quantities in parentheses and curly brackets are separately μ independent. The quantities in curly brackets vanish at NNLO in the Q expansion.

$$\gamma = \frac{4\pi}{MC_0} + \mu \; ; \qquad \zeta_1 = \left(\frac{C_2}{C_0^2}\right) ; \tag{18}$$

$$\zeta_2 = \left(\frac{D_2}{C_0^2} - \frac{g_A^2}{4f^2}\left(\frac{M}{4\pi}\right)^2\left[1 + \ln\left(\frac{\mu^2}{m_\pi^2}\right)\right]\right) + \left(\frac{g_A^2/(2f^2) + C_{0,0}}{C_0^2\, m_\pi^2}\right) ;$$

$$\zeta_3 = -\frac{g_A^2}{2f^2}\frac{Mm_\pi}{4\pi}\left(\frac{C_2}{C_0^2}\right) + \frac{1}{m_\pi^2}\left(\frac{C_{0,1}}{C_0^2} - \frac{[g_A^2/(2f^2) + C_{0,0}]^2}{C_0^3}\right)$$
$$+ m_\pi^2\left\{\frac{D_4}{C_0^2} - \frac{D_2^2}{C_0^3}\right\} + \left(\frac{D_{2,-1}}{C_0^2} - \frac{2D_2[g_A^2/(2f^2) + C_{0,0}]}{C_0^3} + \frac{g_A^2}{2f^2}\frac{\mu M}{4\pi}\frac{C_2}{C_0^2}\right)$$
$$+ \left(\frac{\Delta D_2(\mu)\, m_\pi^2}{C_0(\mu)^2} - 6\,\frac{g_A^2 m_\pi^2}{(4\pi f)^2}\frac{M}{4\pi}\left(\frac{1}{a^S} - \frac{1}{a^T}\right)\left[\frac{1}{3} + \ln\left(\frac{\mu^2}{m_\pi^2}\right)\right]\right) ;$$

$$\zeta_4 = \left(\frac{C_{2,-1}}{C_0^2} - \frac{2\, C_2\, [g_A^2/(2f^2) + C_{0,0}]}{C_0^3}\right) + m_\pi^2\left\{\frac{E_4}{C_0^3} - \frac{2\, C_2\, D_2}{C_0^2}\right\} ;$$

$$\zeta_5 = m_\pi^2\left\{\frac{C_4}{C_0^2} - \frac{C_2^2}{C_0^3}\right\} .$$

ζ_1 and ζ_4 are short distance constants of integration of the RGE's in Eq. (13). On the other hand, ζ_2 and ζ_3 are sums of constants and variables that appear in the amplitude. Note that the order Q radiation pion contribution from order Q_r^3 graphs appears in ζ_3. $\Delta D_2(\mu)$ is a correction to D_2 which cancels the $\ln(\mu)$ dependence from the Q_r^3 radiation pion graphs.

References

1. D.B. Kaplan, M.J. Savage, and M.B. Wise, *Phys. Lett.* B **424**, 390 (1998).
2. D.B. Kaplan, M.J. Savage, and M.B. Wise, *Nucl. Phys.* B **534**, 329 (1998).
3. J. Gegelia, nucl-th/9802038; T. Mehen, and I. Stewart, *Phys. Lett.* B **445**, 378 (1999).
4. T. Mehen and I. Stewart, *Phys. Rev.* C **59**, 2365 (1999).
5. G.P. Lepage, nucl-th/9706029.
6. J. Steele, *this volume*, nucl-th/9904023.
7. E. Epelbaum and U-G. Meissner, *this volume*; nucl-th/9903046; nucl-th/9902042.
8. X. Kong and F. Ravndal, hep-ph/9903523; *Phys. Lett.* B **450**, 320 (1999).
9. D.B. Kaplan, M.J. Savage, and M.B. Wise, *Phys. Rev.* C **59**, 617 (1999).
10. J.W. Chen, H.W. Griesshammer, M.J. Savage and R.P. Springer, *Nucl. Phys.* A **644**, 221 (1999).
11. X. Kong and F. Ravndal, nucl-th/9904066; nucl-th/9902064.
12. M.J. Savage, K.A. Scaldeferri and M.B. Wise, nucl-th/9811029; D.B. Kaplan, M.J. Savage, R.P. Springer and M.B. Wise, *Phys. Lett.* B **449**, 1 (1999): J.-W. Chen, G. Rupak and M.J. Savage, nucl-th/9905002.
13. J.W. Chen, H.W. Griesshammer, M.J. Savage and R.P. Springer, *Nucl. Phys.* A **644**, 245 (1999); J.-W. Chen, nucl-th/9901012.
14. M. Butler, and J.W. Chen, nucl-th/9905060.
15. M.J. Savage, *this volume*; nucl-th/9905009.
16. T. Mehen and I.W. Stewart, nucl-th/9901064.
17. T. Mehen, I.W. Stewart, and M.B. Wise, hep-ph/9902370.
18. M.B. Wise, *this volume*, nucl-th/9904509.
19. G. Rupak, *this volume*.
20. G. Rupak and N. Shoresh, nucl-th/9902077.
21. M. Neubert, *Phys. Rep.* **245**, 259 (1994).
22. H.W. Griesshammer, *Phys. Rev.* D **58**, 094027 (1998); hep-ph/9810235.
23. M. Beneke and V.A. Smirnov, *Nucl. Phys.* B **522**, 321 (1998).
24. S.G. Gorishny, *Nucl. Phys.* B **319**, 633 (1989); G.B. Pivovarov and F.V. Tkachov, *Int. Journ. Mod. Phys.* A **8**, 2241 (1993); V.A. Smirnov, *Phys. Lett.* B **394**, 205 (1997); A. Czarnecki and V.A. Smirnov, *Phys. Lett.* B **394**, 221 (1997); V.A. Smirnov and E.R. Rakhmetov, hep-ph/9812529.
25. K.G. Chetyrkin, *Teor. Mat. Fiz.* **75**, 26 (1998); *ibid* **76**, 207 (1998); K.G. Chetyrkin and V.A. Smirnov, *Phys. Lett.* B **144**, 419 (1984);

V.A. Smirnov, *Commun. Math. Phys.* **134**, 109 (1990); V.A. Smirnov, *Renormalization and asymptotic expansions*, (Birkhäuser, Basel, 1991).
26. S. Fleming, T. Mehen and I.W. Stewart, work in progress.
27. J.-W. Chen, G. Rupak and M.J. Savage, nucl-th/9902056.
28. D.B. Kaplan, *this volume.*
29. V.G.J. Stoks, R.A.M. Klomp, M.C.M. Rentmeester, and J.J. de Swart, *Phys. Rev.* C **48**, 792 (1993); V.G.J. Stoks, R.A.M. Klomp, C.P.F. Terheggen, and J.J. de Swart, *Phys. Rev.* C **49**, 2950 (1994), nucl-th/9406039. (cf. http://nn-online.sci.kun.nl/NN/)
30. R.A. Arndt and R.L. Workman, *Few Body Syst. Suppl.* **7**, 64 (1994). Data from: http://said.phys.vt.edu/
31. J. Steele and D.B. Kaplan, nucl-th/9905027.
32. T.D. Cohen and J.M. Hansen, *Phys. Rev.* C **59**, 13 (1999); nucl-th/9901065; T.D. Cohen, *this volume.*
33. D.B. Kaplan, *private communication*

Discussion during T. Mehen's presentation

Chair: *B. Holstein (University of Massachusetts)*
Notes taken by: *S. Fleming (University of Toronto)*

During the part of the talk where T. Mehen discussed the role of different pion modes M. Lutz emphasized that the leading pion contribution does not come from static pions. H. Griesshammer, citing the paper on the threshold expansion by Beneke and Smirnov, disagreed, saying that the potential pion contribution is dominant.

During the latter part of talk T. Mehen discussed the NNLO results in the $^1S_0$ channel, and the subtleties of fitting the data. T. Cohen pointed out that in the theory without pions there are no free parameters at NNLO. D. Kaplan commented that in the theory with pions there should be one free parameter at NNLO, and asked if it is fixed by r_0. T. Mehen explained that there it is possible to expand C_2 perturbatively, and have two free parameters at NNLO. The subleading contribution would be fixed by r_0 while the leading parameter could be fixed by v_2.

R. Seki pointed out that the convergence to k cotδ is going to be slow because this quantity diverges as $\delta \to 0$ for a finite k.

POWER COUNTING FOR AND SYMMETRIES OF THE EFFECTIVE FIELD THEORY FOR NN INTERACTIONS

M. B. WISE
California Institute of Technology, Pasadena
CA 91125, USA
E-mail: wise@theory.caltech.edu

The appropriate power counting for the effective field theory of NN interactions is reviewed. It is more subtle than in most effective field theories since in the limit that the S-wave NN scattering lengths go to infinity it is governed by a nontrivial fixed point. The leading two body terms in the effective field theory for nucleon self interactions are scale invariant and invariant under Wigner SU(4) spin-isospin symmetry in this limit. Higher body terms with no derivatives (i.e., three and four body terms) are automatically invariant under Wigner symmetry.

1 Basics

Consider $N(\mathbf{p})N(-\mathbf{p}) \to N(\mathbf{p}')N(-\mathbf{p}')$ scattering in the ${}^1S_0$ channel. Since the spins of the two nucleons are combined anti-symmetrically Fermi statistics implies that this channel is $I = 1$ (similarly the ${}^3S_1$ and ${}^3D_1$ channels are $I = 0$). The energy $E = p^2/M = p'^2/M$ where $p^{(\prime)} = |\mathbf{p}^{(\prime)}|$ and the scattering matrix, S, is related to the scattering amplitude $\mathcal{A}$ by $S = 1 + iMp\mathcal{A}/2\pi$. Since $S = e^{2i\delta}$, where δ is the phase shift,

$$\mathcal{A}^{({}^1S_0)} = \frac{4\pi}{M}\frac{1}{p\cot\delta^{({}^1S_0)} - ip}, \tag{1}$$

where M is the nucleon mass. For $p < m_\pi/2$ the quantity $p\cot\delta$ can be expanded in a power series in p^2

$$p\cot\delta^{({}^1S_0)} = -\frac{1}{a^{({}^1S_0)}} + \frac{1}{2}r_0^{({}^1S_0)}p^2 + \ldots, \tag{2}$$

when a is called the scattering length and r_0 is called the effective range. The scattering length in the ${}^1S_0$ channel is very large,[1] $a^{({}^1S_0)} \simeq -23.7$ fm or $1/a^{({}^1S_0)} \simeq -8.3$ MeV. On the other hand the nuclear potential is characterized by a momentum scale $\Lambda \sim 200$ MeV. The smallness of $|1/a^{({}^1S_0)}|$ compared with this scale is the result of an accidental cancellation which causes a state in the spectrum to be very near zero binding energy. ($a \to -\infty$ as a scattering state approaches zero energy and $a \to \infty$ as a bound state approaches zero binding energy.) Neglecting the small ${}^3S_1 - {}^3D_1$ mixing, formulas analogous to eqs. (1) and (2) hold in the ${}^3S_1$ channel. The scattering length is also large in that case,[1] $a^{(^3S_1)} \simeq 5.4$ fm or $1/a^{(^3S_1)} \simeq 36$ MeV. The bound state in this channel that is near zero binding energy is the deuteron.

2 Expansions of $\mathcal{A}$

The simplest expansion of $\mathcal{A}$ is a momentum expansion. This is analogous to what is done in standard applications of effective field theory, e.g., chiral perturbation theory for $\pi\pi$ scattering. For NN scattering in the $s = {}^1S_0$ or ${}^3S_1$ channels,

$$\mathcal{A}^{(s)} = \frac{4\pi}{M} \frac{1}{\left[-1/a^{(s)} + \frac{1}{2} r_0^{(s)} p^2 + \ldots - ip\right]}$$

$$= -\frac{4\pi}{M} a^{(s)} \left\{ 1 - i a^{(s)} p + \left(\frac{a^{(s)} r_0^{(s)}}{2} - a^{(s)2} \right) p^2 + \ldots \right\}. \tag{3}$$

If $a^{(s)}$ was its natural size (i.e., $a^{(s)} \sim 1/\Lambda$) this would be the appropriate expansion to perform. However, in nature the S-wave NN scattering lengths are very large and the expansion above is only valid in the small region of momentum $p \lesssim |1/a^{(s)}| \ll 1/\Lambda$. Since the underlying physics is set by m_π and Λ_{QCD} there should be an expansion in p/Λ that is valid even when $p \gg |1/a^{(s)}|$. It is not difficult to deduce what this expansion is. In Eq. (3) keep $-1/a^{(s)} - ip$ in the denominator and expand in the remaining terms. This yields

$$\mathcal{A}^{(s)} = -\frac{4\pi}{M} \frac{1}{(1/a^{(s)} + ip)} \left[1 + \frac{r_0^{(s)} p^2/2}{(1/a^{(s)} + ip)} + \ldots \right]. \tag{4}$$

Now $\mathcal{A}^{(s)} = \sum_{n=-1}^{\infty} \mathcal{A}_n^{(s)}$, where $\mathcal{A}_n^{(s)} \sim \mathcal{O}(p^n)$. This is the appropriate expansion in the case where the scattering lengths are large. It has the unusual property that the leading term is order p^{-1}.

3 Effective Field Theory Without Pions

The effective field theory with the pions integrated out contains only nucleon fields, $N = \binom{p}{n}$, and we expect that the lowest dimension operators will be the most important ones. The Lagrange density is written as, $\mathcal{L} = \mathcal{L}_1 + \mathcal{L}_2 + \ldots$, where $\mathcal{L}_n$ contains n-body operators. The one and two body terms are:

$$\mathcal{L}_1 = N^\dagger \left[i\partial_t + \frac{\vec{\nabla}^2}{2M} \right] N + \ldots, \tag{5}$$

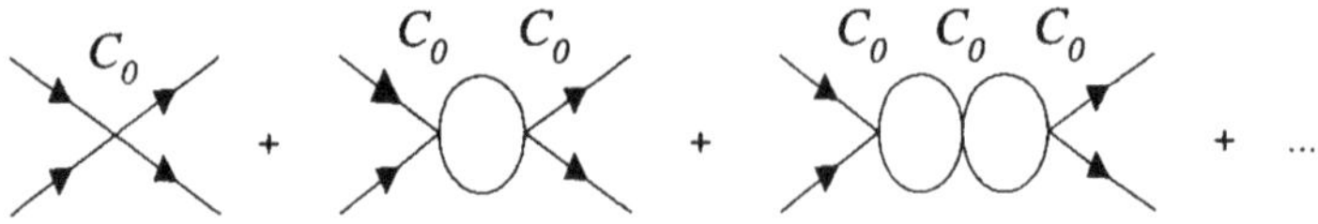

Figure 1: The leading contribution to NN scattering.

$$\mathcal{L}_2 = -\sum_s C_0^{(s)} (N^T P_i^{(s)} N)^\dagger (N^T P_i^{(s)} N) + \dots . \tag{6}$$

Here $s = {}^1S_0$ or ${}^3S_1$, the ellipses denote higher dimension operators and $P_i^{(s)}$ are the spin-isospin projectors

$$P_i^{({}^1S_0)} = \left(\frac{\sigma_2 \tau_2 \tau_i}{\sqrt{8}}\right), P_i^{({}^3S_1)} = \left(\frac{\sigma_2 \sigma_i \tau_2}{\sqrt{8}}\right), \tag{7}$$

where the Pauli matrices σ_i act in spin space and the Pauli matrices τ_i act in isospin space.

Neglecting higher dimension operators the scattering amplitudes in the ${}^1S_0$ and ${}^3S_1$ channels come from the sum of bubble-type Feynman diagrams shown in Fig. 1. Each bubble is linearly divergent in the ultraviolet so the coefficients $C_0^{(s)}$ depend on the regulator and subtraction scheme adopted. We use dimensional regularization and start with minimal subtraction (we will switch to a different subtraction scheme momentarily). Since the divergences are linear the Feynman diagrams have poles at $D = 3$ but not at $D = 4$. In MS (minimal subtraction) the coefficients of the operators explicitly displayed in Eq. (6), are subtraction point independent and we denote them by $\bar{C}_0^{(s)}$. In this scheme the sum of bubble-type Feynman diagrams gives

$$\mathcal{A}^{(s)} = -\frac{\bar{C}_0^{(s)}}{1 + iMp\bar{C}_0^{(s)}/4\pi}. \tag{8}$$

Comparing Eq. (8) with eqs. (1) and (2) it is evident that this corresponds to keeping only the scattering length term in the expansion of $p \cot \delta^{(s)}$, (i.e., the first term of Eq. (4)) and that

$$\bar{C}_0^{(s)} = \frac{4\pi a^{(s)}}{M}. \tag{9}$$

So in this subtraction scheme the coefficients $\bar{C}_0^{(s)}$ are very large and also very different in the two channels. However as $a^{(s)} \to \infty, \mathcal{A}^{(s)} \to 4\pi i/Mp$ which is the same in both channels. This form for the scattering amplitudes is consistent with Wigner spin-isospin $SU(4)$ symmetry, and also with scale invariance.

In MS when $p > 1/a^{(s)}$ the terms in the perturbative series for the scattering amplitude get larger and larger. We would like to use a subtraction scheme where the various Feynman diagrams in Fig. 1 are the same size as their sum and where the symmetries that arise as $a^{(s)} \to \infty$ are manifest at the level of the Lagrangian. Examples of such subtraction schemes are [2] PDS where poles at $D = 3$ are also subtracted and the OS momentum space subtraction scheme. [3,4] In these schemes the coefficients are subtraction point dependent, $C_0^{(s)} = C_0^{(s)}(\mu)$, and the sum of bubble diagrams gives

$$\mathcal{A}^{(s)} = -\frac{C_0^{(s)}(\mu)}{1 + M(\mu + ip)C_0^{(s)}(\mu)/4\pi}. \tag{10}$$

This still corresponds to keeping just the scattering length, and is the leading term in Eq. (4). But now

$$C_0^{(s)}(\mu) = -\frac{4\pi}{M}\frac{1}{\mu - 1/a^{(s)}}, \tag{11}$$

which as $a^{(s)} \to \infty$ becomes $C_0^{(s)}(\mu) = -4\pi/M\mu$. In this limit, the coefficients are the same in both channels and with $\mu \sim p$ each term in sum of bubble type Feynman diagrams in Fig. 1 is the same size as the sum itself.

The operators with coefficients $C_0^{(s)}$ are nonrenormalizable dimension six operators. Naively they are irrelevant operators and at low momentum can be treated in perturbation theory. However as $a^{(s)} \to \infty$ the coefficients $C_0^{(s)}(\mu)$ flow to a nontrivial fixed point [2,3] where $\mu d[\mu C_0^{(s)}(\mu)]/d\mu = 0$. For large $a^{(s)}$ the power counting is controlled by this fixed point and the leading contribution to the NN scattering amplitude comes from treating $C_0^{(s)}$ nonperturbatively. It is straightforward to show that in PDS or OS the coefficients of S-wave operators with $2n$ spatial derivatives scale as, [2]

$$C_{2n}^{(s)}(\mu) \sim \frac{4\pi}{M\Lambda^n \mu^{n+1}}, \tag{12}$$

for $\mu \gg |1/a^{(s)}|$. With $\mu \sim p, C_{2n}^{(s)}(\mu)p^{2n} \sim p^{n-1}$ and the two body operators with derivatives can be treated perturbatively. In a non-relativistic theory a loop integration $\int d^4q = \int dq^0 d^3q \sim \mathcal{O}(p^5)$ (since the dq^0 integration is of order p^2 and the d^3q is order p^3) and the nucleon propagator $i/(p^0 - p^2/2M + i\epsilon) \sim \mathcal{O}(p^{-2})$. Consequently each loop gives a factor p plus whatever factors of p are associated with the vertices. The power counting [2,5] is now evident. The leading order (LO) contribution $\mathcal{A}_{-1}^{(s)}$ comes from $C_0^{(s)}$ treated nonperturbatively, the next to leading order (NLO) contribution $\mathcal{A}_0^{(s)}$ comes from $C_0^{(s)}$ treated nonperturbatively and $C_2^{(s)}$ inserted once, the next-to-next to leading order (N^2LO) contribution comes from $C_0^{(s)}$ treated non-perturbatively, $C_2^{(s)}$ inserted twice or $C_4^{(s)}$ inserted once, etc.

With the pions integrated out the effective field theory expansion applied to NN scattering reproduces Eq. (4) and has no more content than the momentum expansion of $p\cot\delta^{(s)}$. However, even with the pions integrated out one can couple photons or W and Z gauge bosons to the nucleons. The relative importance of operators containing these fields depends on their renormalization group scaling near the fixed point.

In the two nucleon sector predictions based on the effective field theory without pions are similar to those made by effective range theory.[6] However the effective field theory approach has a number of advantages. Predictions based on effective range theory are only valid to a given order in the p/Λ expansion. In the effective field theory new two-body operators containing the gauge fields arise which spoil the predictions of effective range theory. For the thermal neutron capture cross section, $\sigma(n+p\to d+\gamma)$, this occurs at NLO while for the deuteron matter (charge) radius $< r_m >$ this doesn't occur until the next-to-next-to-next to leading order (N^3LO). This explains why the effective range theory prediction for $\sigma(n+p\to d+\gamma)$ is off by 10% while the effective range theory prediction for $< r_m >$ is accurate to better than a percent.[7] For these static deuteron properties the relevant momentum in the p/Λ expansion is set by the deuteron binding energy, i.e., $p\sim 40$ MeV. Another useful aspect of the effective field theory formalism is that it is straightforward to include relativistic corrections.

As $a^{(s)}\to\infty, \mathcal{L}_2\to -(2\pi/M\mu)(N^\dagger N)^2+\ldots$, where the ellipses denote two body operators with derivatives. In this limit the leading one and two body terms are invariant under the following symmetries:[8]

(i.) Wigner Symmetry[9]

Under infinitesimal Wigner symmetry $SU(4)$ transformations

$$\delta N = i\alpha_{\mu\nu}\sigma^\mu\tau^\nu N, \tag{13}$$

where $\sigma^\mu=(1,\boldsymbol{\sigma})$ and $\tau^\mu=(1,\boldsymbol{\tau})$ with $\mu=0,1,2,3$ and repeated indices summed. The symmetry group corresponding to Eq. (13) is actually $SU(4)\times U(1)$, with α_{00} the group parameter for the additional baryon-number $U(1)$. Associated with this symmetry are the conserved charges,

$$Q^{\mu\nu} = \int d^3x N^\dagger\sigma^\mu\tau^\nu N. \tag{14}$$

The two body terms with derivatives are not invariant under Wigner symmetry even if $a^{(s)} \to \infty$. Hence in the two body sector the violations of Wigner symmetry go as, $(1/[a^{({}^1S_0)}p] - 1/[a^{({}^3S_1)}p])$ and p/Λ. Wigner symmetry will not be a good approximation if the momentum p is too low or if it is too large.

Wigner symmetry is relevant for nuclei with many nucleons. [10] It is not difficult to see that the higher body terms with no derivatives are automatically invariant under Wigner symmetry. Since these contact terms are antisymmetric in the nucleon fields N and in the hermitian conjugates $N^\dagger$, contact terms without derivatives cannot occur for five body operators and higher. The nucleons N are in the $\mathbf{4}$ of $SU(4)$ and the $N^\dagger$'s are in the $\bar{\mathbf{4}}$. Four nucleons combined anti-symmetrically are an $SU(4)$ singlet and so the four-body terms are invariant under $SU(4)$. The three body terms transform as $\bar{\mathbf{4}} \otimes \mathbf{4} = \mathbf{1} \otimes \mathbf{15}$. However the operators in the $\mathbf{15}$ are not invariant under the total spin or isospin $SU(2)$ subgroups of $SU(4)$. Hence the allowed three body terms are also invariant under $SU(4)$ Wigner symmetry.

A complete extension of the general fixed point power counting to the higher body terms has not been made. However there has been considerable recent progress. [5] This work indicates that the 3-body term with no derivatives is leading order (i.e., as important as effects coming from $C_0^{(s)}$).

(ii.) Scale Invariance
The leading one and two body terms are invariant under the scale transformation $N(t, \mathbf{x}) \to N'(t, \mathbf{x})$ and $\mu \to \mu'$ where

$$N'(t, \mathbf{x}) = \lambda^{-3/2} N(t/\lambda^2, \mathbf{x}/\lambda), \tag{15}$$

$$\mu' = \mu/\lambda. \tag{16}$$

Note that Eq. (15) corresponds to $N'(t', \mathbf{x}') = \lambda^{-3/2} N(t, \mathbf{x})$ with $\mathbf{x}' = \lambda \mathbf{x}$ and $t' = \lambda^2 t$. The different scaling of space and time coordinates is dictated by invariance of the leading one-body terms in the Lagrange density.

4 Including Pions

With pions included the power counting is taken to be in powers of Q/Λ_{NN} where $p \sim m_\pi \sim Q$. A subscript NN has been put on Λ as a reminder that

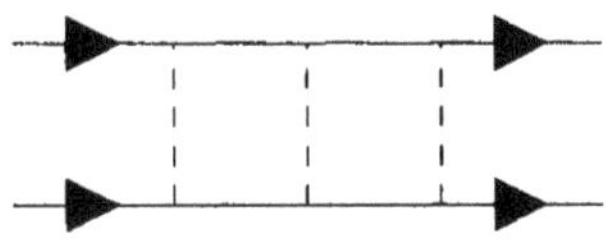

Figure 2: Contribution to $^3S_1$ scattering from three potential pion exchange. The dashed lines denote potential pion exchange.

the expansion should work better if the pions are included as explicit fields, i.e., we expect that $\Lambda_{NN} > \Lambda$. Potential pion exchange arises from the term

$$\mathcal{L}_{\rm int} = -\frac{g_A}{\sqrt{2} f_\pi} \nabla^i \pi^j N^\dagger \sigma^i \tau^j N, \tag{17}$$

where $g_A \simeq 1.25$ is the axial coupling and $f_\pi \simeq 131$ MeV is the pion decay constant. Exchange of a potential pion between nucleons is order Q^0 (the two factors of Q from the vertices cancel the $1/Q^2$ from the pion propagator). This is the same size as the two body contact terms with two derivatives and consequently pion exchange can be treated perturbatively. Including pion exchanges without the two derivative two body contact terms is not a systematic improvement and is no better (from a power counting perspective) than just including the effects of $C_0^{(s)}$. Note that this power counting is very different from the one originally proposed by Weinberg [3,12] where the leading contribution came from treating both potential pion exchange and $C_0^{(s)}$ nonperturbatively. The effects of two body terms with derivatives and insertions of the light quark mass matrix were considered subdominant.

Weinberg's power counting treats the nucleon mass M as large and $MQ \sim \mathcal{O}(1)$. It assumes that factors of M only arise from the loop integrations. In a toy model where perturbative matching between the full relativistic theory and the nonrelativistic effective theory can be explicitly performed Luke and Manohar [13] found that the two body local operators in the effective nonrelativistic theory have coefficients that contain factors of M. This is the origin of the problem with Weinberg's power counting.

In the $^3S_1$ channel the Feynman diagram shown in Fig. 2 with three potential pion exchanges is logarithmically divergent. Neglecting the pion mass

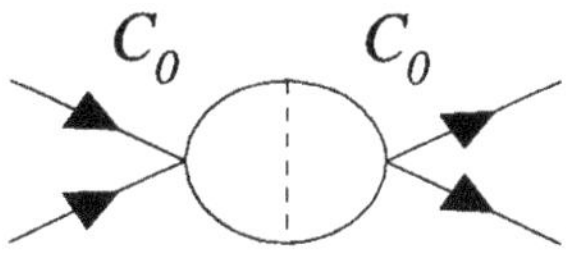

Figure 3: Contribution to NN scattering that renormalizes $D_2^{(s)}$.

it gives a contribution to $\mathcal{A}^{(^3S_1)}$ of order

$$\left(\frac{4\pi}{M}\right)\left(\frac{Mg_A^2}{8\pi f_\pi^2}\right)^3 p^2[\ln\mu^2 + K], \tag{18}$$

where K is a constant. The μ-dependence above is canceled by the μ-dependence of $C_2^{(^3S_1)}$. There is no point to including this Feynman diagram without including the effects of the two body $^3S_1$ operator with 2-derivatives. Eq. (18) is an N^3LO contribution. With the pions included a single insertion of $C_2^{(s)}$ is not just NLO it contributes at higher levels at the Q expansion as well. For that reason $C_2^{(s)}$ and the other contact term coefficients are sometimes written as a sum $C_2^{(s)} = \sum_{a=1}^{\infty} C_{2,a}^{(s)}$ where $C_{2,1}^{(s)}$ gives the NLO contribution, etc. When this is done predictions for physical quantities are exactly μ independent, at each order in the Q expansion. If $C_2^{(s)}$ is not expanded in this way then predictions at a given order on the Q expansion have some subtraction point dependence, which is higher order in the Q expansion.

There are two body S-wave contact terms with no derivatives but with an insertion of the light quark mass matrix,

$$m_q = \begin{pmatrix} m_u & 0 \\ 0 & m_d \end{pmatrix}. \tag{19}$$

Since $m_\pi^2 \propto (m_u + m_d)$ an insertion of m_q counts as two powers of Q and the coefficients of these operators $D_2^{(s)}$ scale with μ in the same way as the coefficients $C_2^{(s)}$. At NLO they must also be included. The Feynman diagram in Fig. 3 is logarithmically divergent and it gives a contribution to the $^1S_0$ and $^3S_1$ scattering amplitudes of order

$$\left(\frac{4\pi}{M}\right)\left(\frac{g_A^2 M}{8\pi f_\pi^2}\right)\left(\frac{C_0^{(s)}M}{4\pi}\right)^2 m_\pi^2[\ln\mu^2 + K], \tag{20}$$

where K is a constant. The μ dependence here is canceled by that of the coefficients $D_2^{(s)}$. Including one pion exchange without the effects of the two body terms with one insertion of the quark mass matrix does not systematically improve the theoretical prediction for the NN scattering amplitude.

If a momentum cutoff regulator is used instead of dimensional regularization then including pion exchange without the two body contact operators that have an insertion of the quark mass matrix results in a cutoff dependent amplitude $\mathcal{A}^{(s)}$. It is possible in the $^1S_0$ channel to sum to all orders potential pion exchange and when this is done the cutoff dependence does not become subdominant [14] (compared with the finite cut off independent parts of pion exchange). The effects of local four nucleon (i.e., two body) operators with an insertion of the quark mass matrix cannot be viewed as less important than the effects of pion exchange.

The conventional explanation for the discrepancy between the prediction of effective range theory for the thermal neutron capture cross section $\sigma(n + p \to d + \gamma)$ and its experimental value is meson exchange currents, [15] which roughly speaking are the contribution of Feynman diagrams where the photon couples to a potential pion. In the effective field theory approach with the pions included this discrepancy is made up [16] (at least partly) by the NLO contribution which involves both meson exchange current Feynman diagrams and the contribution of a local two body operator involving the magnetic field.

5 An Application of Wigner Symmetry

Potential pions have $k^0 \sim \mathbf{k}^2/M$ while radiation pions have $k^0 \sim \sqrt{\mathbf{k}^2 + m_\pi^2}$. The coupling of the radiation pions to the nucleons is done by performing a multipole expansion on Eq. (17). At leading order this amounts to evaluating the pion field in Eq. (17) at the space time point $(t, \mathbf{x}) = (t, \mathbf{0})$. Hence, for radiation pions the term in the action corresponding to Eq. (17) is

$$S_{int} = -\frac{g_A}{\sqrt{2}f} \int dt (\nabla^i \pi^j)|_{\mathbf{x}=0} Q^{ij}, \tag{21}$$

where Q^{ij} are the charges of Wigner symmetry in Eq. (14). In the limit $a^{(s)} \to \infty$ these charges are conserved and the Q^{ij}'s are time independent. Hence, as $a^{(s)} \to \infty$ only the $k^0 = 0$ mode of the pion couples in Eq. (21). This is incompatible with the radiation pion condition, $k^0 \sim \sqrt{\mathbf{k}^2 + m_\pi^2}$. Hence the leading contribution from radiation pions is suppressed by $1/a^{(^1S_0)} - 1/a^{(^3S_1)}$. In a recent paper [17] Mehen and Stewart found this suppression by an explicit calculation of the leading radiation pion contribution to the NN scattering amplitudes, $\mathcal{A}^{(s)}$. It involved a cancellation between different Feynman diagrams.

6 Outlook

Effective field theory methods are a viable model independent approach to the physics of the two nucleon sector. The power counting is slightly unusual due to the large S-wave NN scattering lengths. This approach is useful up to a center of mass momentum around 200 MeV, however, the expansion parameter at such a momentum is probably not much smaller than $\frac{1}{2}$. It seems likely that for many quantities calculations at N^2LO will reach the same precision as conventional potential model approaches, however, with such a large expansion parameter there are likely to be some failures.

Extension of the effective field theory approach to the three nucleon sector is underway. Several theoretical issues remain to be resolved before there is a complete power counting, but recent progress in this area is very encouraging.

The holy grail of this field is the application of these effective field theory methods to nuclear matter. We are still a long way from having the theoretical tools to tackle this problem and even with these tools the Fermi momentum associated with nuclear density may be too large for a Q expansion to be useful. However, given the importance of understanding the properties of nuclear matter continuing to develop the effective field theory approach is very worthwhile.

This work was supported in part by the Department of Energy under grant number DE-FG03-92-ER 40701.

References

1. W.E. Burcham, *Elements of Nuclear Physics* (John Wiley and Sons Inc., 1979).
2. D.B. Kaplan, M.J. Savage and M.B. Wise, *Phys. Lett.* B **424**, 390 (1998); *Nucl. Phys.* B **534**, 329 (1998).
3. S. Weinberg, *Nucl. Phys.* B **363**, 3 (1991).
4. T. Mehen and I.W. Stewart, *Phys. Lett.* B **445**, 378 (1999); T. Mehen and I.W. Stewart, `nucl-th/9809095`; J. Gegelia, `nucl-th/9802038`.
5. U. van Kolck, `hep-ph/971222`; *Nucl. Phys.* A **645**, 273 (1999).
6. H.A. Bethe, *Phys. Rev.* **76**, 38 (1949); H.A. Bethe and C. Longmire, *Phys. Rev.* **77**, 647 (1950).
7. J.-W. Chen, G. Rupak and M.J. Savage, `nucl-th/9962056`.
8. T. Mehen, I.W. Stewart and M.B. Wise, `hep-ph/9902370`.
9. E. Wigner, *Phys. Rev.* **51**, 106 (1937); *ibid.* **56**, 519 (1939).

10. *Group Symmetries in Nuclear Structure*, J.C. Parikh (Plenum Press, 1978); T.W. Donnelly and G.E. Walker, *Ann. of Phys.* **60**, 209 (1970); J.D. Walecka, *Theoretical Nuclear Physics and Subnuclear Physics* (Oxford University Press, 1995); J.P. Elliott, *Isospin in Nuclear Physics*, ed. D.H. Wilkinson (North Holland Pub., 1969); P. Vogel and M.R. Zirnbauer, *Phys. Rev. Lett.* **57**, 3148 (1986); P. Vogel and W.E. Ormand, *Phys. Rev.* C **47**, 623 (1993); Y.V. Gaponov, N.B. Shulgina, and D.M. Vladimirov, *Nucl. Phys.* A **391**, 93 (1982).
11. P.F. Bedaque and U. van Kolck, *Phys. Lett.* B **428**, 221 (1998); P.F. Bedaque, H.W. Hammer and U. van Kolck, *Phys. Rev.* C **58**, R641 (1998); *Phys. Rev. Lett.* **82**, 463 (1999); *Nucl. Phys.* A **646**, 444 (1999).
12. S. Weinberg, *Phys. Lett.* B **25**, 288 (1990).
13. M. Luke and A. V. Manohar, *Phys. Rev.* D **55**, 4129 (1997).
14. D.B. Kaplan, M.J. Savage and M.B. Wise, *Nucl. Phys.* B **478**, 629 (1996).
15. D.O. Riska and G.E. Brown, *Phys. Lett.* B **38**, 193 (1972).
16. M.J. Savage, K.A. Scaldeferri and M.B. Wise, `nucl-th/9811029`.
17. T. Mehen and I. Stewart, `nucl-th/9901064`.

Discussion during M. Wise's presentation

Chair: *B. Holstein (University of Massachusetts)*
Notes taken by: *S. Fleming (University of Toronto)*

M. Rho asked if Wigner symmetry is preserved when electroweak interactions are coupled to NN EFT. M. Wise pointed out that it is straightforward to couple external currents. Then all one needs to do is classify the transformation properties of the generators, and it should be clear if Wigner symmetry is broken. D. Kaplan pointed out that the NN wavefunction is $SO(4)$ symmetric so if any given operator is a generator of Wigner symmetry one would know.

W. Haxton raised the question whether pions should be iterated or not. He pointed out that many-body nuclear physics seems to indicate that iterating pions is important. M. Wise responded that you just can not tell if perturbative pions work until the calculations are done and theory is compared to experiment. Moreover, if it turns out that iterating pions is important one needs to be able to justify this. D. Kaplan commented that C_0 can compensate for the short range part of the pion interaction, and thus it may not be necessary to iterate pions. W. Haxton pressed his point further by saying that spin frustration is huge, and has to be accounted for. D. Kaplan replied by pointing out that perturbative pions get the phase shift right even though Nijmegen potential models lead one to believe that about half of the contribution to the phase shift comes from pions. T. Cohen then asked if there are any observables that probe the pion contribution. M. Wise said that at the level of the deuteron it is within our grasp and should be done. He once again emphasized that this issue will only be resolved by computing to higher orders. J. Gegelia asked why not just treat the pion nonperturbatively. Since the Weinberg power counting is in renormalizable operators this should not present a problem. M. Wise responded by saying that in this approach it is difficult to get an estimate of the difference between the operators and the counter terms.

THE TWO-NUCLEON SECTOR WITH EFFECTIVE FIELD THEORY

M. J. SAVAGE
Department of Physics, University of Washington, Seattle, WA 98195, USA
E-mail: savage@phys.washington.edu

I present the results obtained for several observables in the two-nucleon sector using effective field theory with Kaplan-Savage-Wise (KSW) power-counting and dimensional regularization. In addition to the phase shifts for nucleon-nucleon scattering, several deuteron observables are discussed, including electromagnetic form factors, polarizabilities, $\gamma d \to \gamma d$ Compton scattering, and $np \to d\gamma$. A detailed comparison between the effective field theory with pions, the theory without pions, and effective range theory is made.

1 Introduction

During the last year there has been a tremendous effort to understand the role of effective field theory (EFT) in the two- and three-nucleon sectors[1–32]. Since the previous *Nuclear Physics with Effective Field Theory* workshop held twelve months ago at Caltech I have been primarily focused on exploring the implications of Kaplan-Savage-Wise (KSW) power counting[10] in the two-nucleon sector. In addition to the nucleon-nucleon scattering phase shifts in the spin-singlet $^1S_0$ and spin-triplet $^3S_1$ channels, various properties of the deuteron and processes involving the deuteron have been considered.

If one is interested in the static properties of the deuteron or very low-energy ($|\mathbf{p}| \ll m_\pi$) nucleon-nucleon scattering the pion can be "integrated" out of the theory, leaving a very simple EFT involving only nucleons and external currents (we will denote this theory as EFT($\not\pi$)). The simplicity of the theory allows for many observables to be computed up to next-to-next-to-leading order (NNLO) with ease[32], and enables a direct comparison with effective range theory[33,34] to be made. The expansion parameter in EFT($\not\pi$) is $Q \sim |\mathbf{p}|/m_\pi$, which for static properties of the deuteron corresponds to $Q \sim \frac{1}{3}$. A direct comparison between EFT($\not\pi$) and effective range theory shows that effective range theory reproduces EFT up to the order at which multi-nucleon-external-current operators enter. At that order and beyond, effective range theory fails to reproduce effective field theory, and thus is an incomplete description of the strong interactions.

For processes involving higher momenta, such as $|\mathbf{p}| \gtrsim m_\pi$, the pions must be included as a dynamical field. In KSW power-counting, the exchange of potential pions is a subleading interaction compared to the momentum-independent, quark mass independent four-nucleon operator, and is treated in perturbation theory. Some have questioned the convergence of perturbative pions[23], and we will see that for the deuteron observables we have considered the subleading contribution from potential pion exchange is smaller than the contribution from higher derivative four-nucleon operators. While this does not directly answer the questions raised by Cohen and Hansen [23], it is an important observation. For the static properties of the deuteron, one recovers the results of the EFT($\not\pi$) with corrections suppressed by terms higher order in EFT($\not\pi$), as expected.

2 Effective Field Theory with Pions

2.1 The Nucleon-Nucleon Interaction

The two-nucleon sector contains length scales that are much larger than one would naively expect from QCD. Scattering lengths in the s-wave channels are $a^{(^1S_0)} = -23.7$ fm in the $^1S_0$ channel and $a^{(^3S_1)} = 5.4$ fm in the $^3S_1$ channel, much greater than both $1/\Lambda_\chi \sim 0.2$ fm and $1/f_\pi \sim 1.5$ fm, typical hadronic scales. The solid curve in fig. (1) shows the phase shift in the $^1S_0$ channel, $\delta^{(^1S_0)}$,

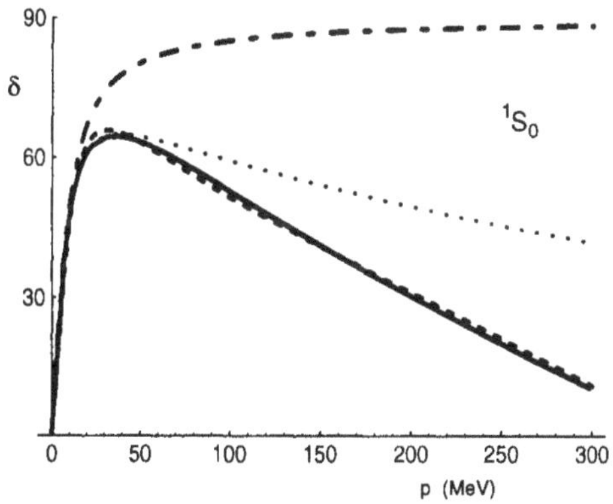

Figure 1: The phase shift $\delta^{(^1S_0)}$ verses the momentum of each nucleon in the center-of-mass. The solid line denotes the Nijmegen partial wave analysis of the experimental data.

from the Nijmegen partial wave analysis[35] of the experimental data. $\delta^{(^1S_0)}$ rises very fast at low momentum and then turns over at about $|\mathbf{p}| \sim 40$ MeV.

The lagrange density that describes the interaction between two S-wave nucleons has the form

$$\mathcal{L} = -C_0^{(^1S_0)}(N^T P^a N)^\dagger (N^T P^a N) - D_2^{(^1S_0)}\omega\ Tr[m_q](N^T P^a N)^\dagger (N^T P^a N)$$
$$+\frac{C_2^{(^1S_0)}}{8}\left[(N^T P^a N)^\dagger \left(N^T P^a (\overleftarrow{\nabla} - \overrightarrow{\nabla})^2 N\right) + h.c.\right]$$
$$-g_A N^\dagger \sigma \cdot \mathbf{A} N \ + \ (^3S_1) \ + \ \ldots \tag{1}$$

where the ellipses denote terms with more insertions of the small expansion parameters $Q \sim m_q$ and $Q \sim \nabla$. The reason for writing such an expansion is that one hopes or expects that QCD will give rise to S-matrix elements for multiple nucleon processes that have a systematic expansion in the light quark masses and in external momenta. Only interactions in the $^1S_0$ channel are explicitly shown in Eq. (1), while interactions in the $^3S_1$ channel are denoted by "($^3S_1$)". In addition, terms from D_2 involving pion field operators (required by chiral symmetry), and terms involving the photon field are not explicitly shown in Eq. (1). $\mathbf{A}$ is the axial vector meson field. The P^a is a spin-isospin projector, $P^a = \frac{1}{\sqrt{8}}\sigma_2\tau_2\tau^a$, projects onto the $^1S_0$ channel.

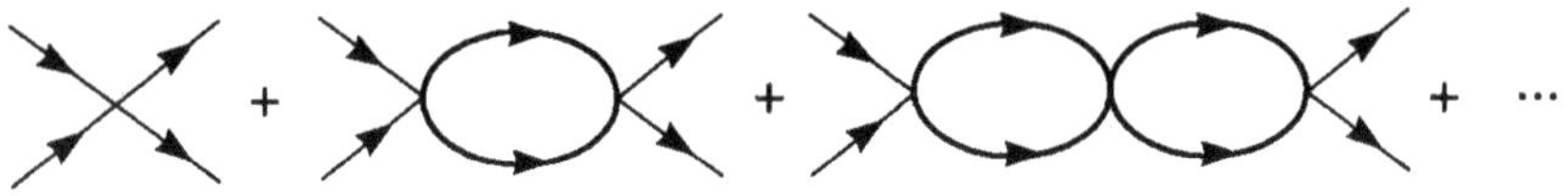

Figure 2: The leading contribution to nucleon-nucleon scattering arising from local operators.

NN scattering in the $^1S_0$ channel receives contributions at leading order from the graphs shown in fig. (2), i.e the bubble-chain of contact operators. Choosing the coefficient $C_0^{(^1S_0)}$ to reproduce the scattering length gives the dot-dashed curve in fig (1). At subleading order, there are contributions from the C_2, D_2 operators, in addition to the exchange of a potential pion, as shown in fig (3). The best fit to the phase shift for momenta less than $\sim$ 250 MeV is shown by the dashed curve in fig (1), while the fit that reproduces both the scattering length and effective range is shown as the dotted curve in fig (1).

Analysis of scattering in the $^3S_1 - {}^3D_1$ channel is a straightforward extension of the analysis in the $^1S_0$ channel. The important difference is that the nucleons in the initial and final states with total angular momentum $J = 1$ can be in an orbital angular momentum state of either $L = 0$ or $L = 2$. The

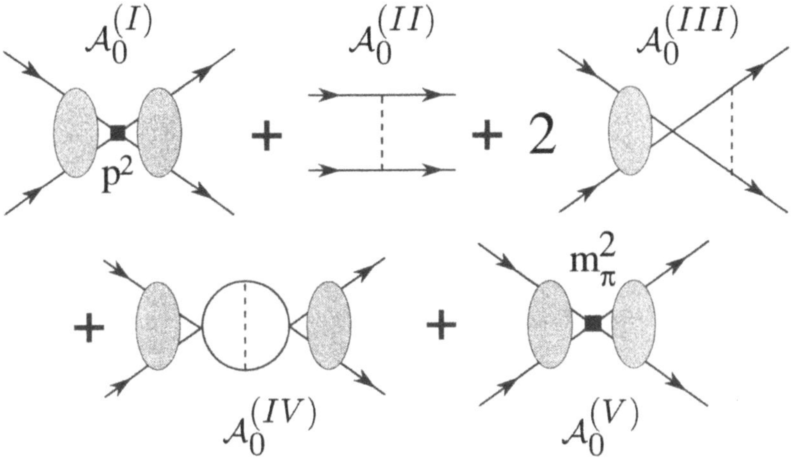

Figure 3: Graphs contributing to the subleading amplitude $\mathcal{A}_0$. The shaded ovals correspond to the bubble chain in Fig. (2) and free propagation.

S-matrix describing scattering in the coupled channel $J = 1$ system is written as

$$S = \begin{pmatrix} e^{i2\delta_0}\cos 2\bar{\varepsilon}_1 & ie^{i(\delta_0+\delta_2)}\sin 2\bar{\varepsilon}_1 \\ ie^{i(\delta_0+\delta_2)}\sin 2\bar{\varepsilon}_1 & e^{i2\delta_2}\cos 2\bar{\varepsilon}_1 \end{pmatrix} \quad , \tag{2}$$

where we use the "barred" parameterization of [36], also used in [37]. The power counting for amplitudes that take the nucleons from a $^3S_1$-state to a $^3S_1$-state is identical to the analysis in the $^1S_0$-channel. Operators between two $^3D_1$ states are not directly renormalized by the leading operators, which project out only $^3S_1$ states. However, they are renormalized by operators that mix the $^3S_1$ and $^3D_1$ states, which in turn are renormalized by the leading interactions. Further, they involve a total of four spatial derivatives, two on the incoming nucleons, and two on the out-going nucleons. Therefore, such operators contribute at order Q^3, and can be neglected in the present computation. Consequently, amplitudes for scattering from an $^3D_1$ state into an $^3D_1$ state are dominated by single potential pion exchange at order Q^0. Operators connecting $^3D_1$ and $^3S_1$ states are renormalized by the leading operators, but only on the $L = 0$ "side" of the operator. Therefore the coefficient of this operator, $C_2^{(sd)} \sim Q^{-1}$, indicates that this interaction contributes at order Q^1 and can be neglected at order Q^0. Thus, mixing between $^3D_1$ and $^3S_1$ states is dominated by single potential pion exchange dressed by a bubble chain of $C_0^{(^3S_1)}$ operators. A

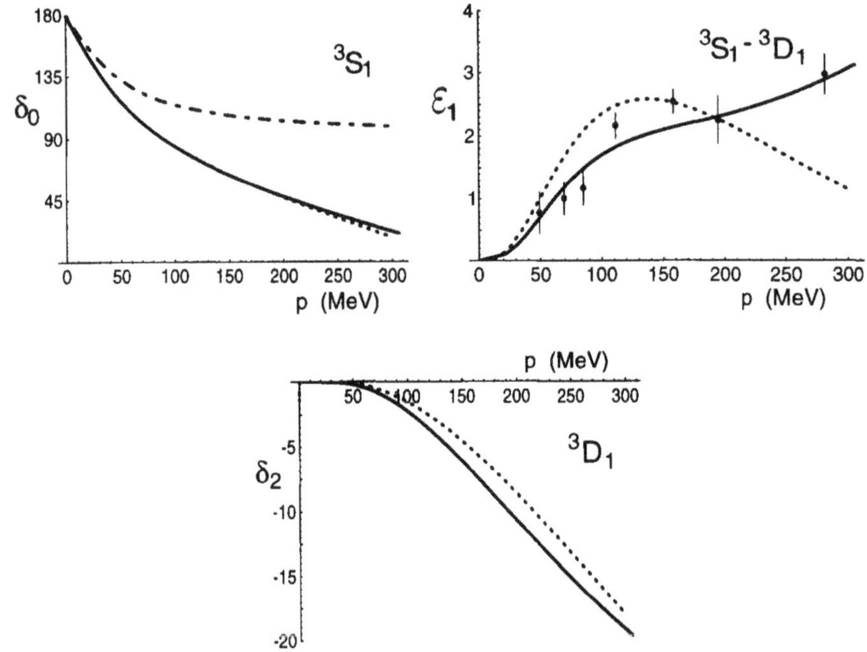

Figure 4: The phase shifts δ_0, δ_2 and mixing parameter $\bar{\varepsilon}_1$ for the $^3S_1 - {}^3D_1$ channel. The solid line denotes the results of the Nijmegen partial wave analysis. The dot-dashed curve is the fit at order Q^{-1} for δ_0, while $\delta_2 = \bar{\varepsilon}_1 = 0$ at this order. The dashed curves are the results of the order Q^0 fit of δ_0 to the partial wave analysis over the momentum range $p \leq 200$ MeV.

parameter free prediction for this mixing exists at order Q^0. The dashed curves in Fig. (4) show the phase shifts δ_0, δ_2 and mixing parameter $\bar{\varepsilon}_1$ compared to the Nijmegen partial wave analysis[35] for this set of coefficients. There are no free parameters at this order in either $\bar{\varepsilon}_1$ or δ_2.

2.2 The Deuteron

Electromagnetic Form Factors and moments

Once the Lagrange density in the nucleon sector has been established the standard tools of field theory can be used to determine the properties of the deuteron[12]. To compute the electromagnetic form factors of the deuteron one first computes the three point correlation function between a source that cre-

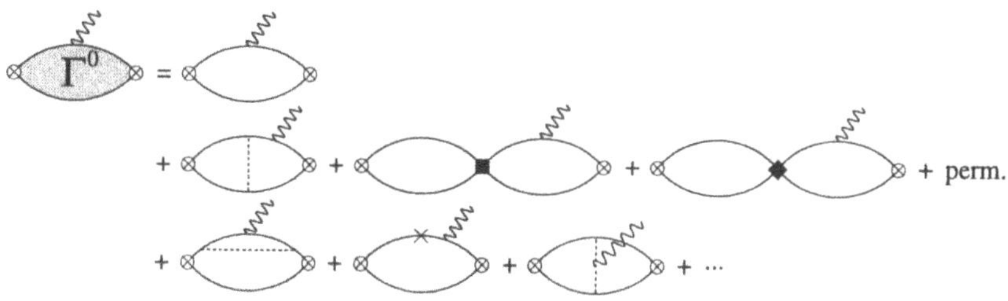

Figure 5: The diagrams contributing to the electric form factors of the deuteron.

ates a nucleon pair in a $^3S_1$ state, a source that destroys a nucleon pair in a $^3S_1$ state and a source that creates a photon. After LSZ reduction and wavefunction renormalization one obtains the electromagnetic form factors. Leading order (LO), next-to-leading (NLO) and next-to-next-to-leading order (NNLO) graphs contributing to the electric form factors of the deuteron are shown in Fig. (5). The resulting form factors $A(q^2)$ and $B(q^2)$ that appear in the differential cross section for electron deuteron scattering are shown in Fig. (6). $A(q^2)$ is dominated by the charge form factor and $B(q^2)$ depends only upon the magnetic form factor. One sees that the form factors computed at subleading order agree well with the data. The charge radius of the deuteron is found to be

$$\begin{aligned}\sqrt{< r^2 >_{\rm th}} &= \frac{1}{2\sqrt{2}\gamma} + C_2(\mu)\frac{M_N(\mu-\gamma)^2}{8\sqrt{2}\pi} + \frac{g_A^2 M_N m_\pi^2(3m_\pi+10\gamma)}{48\sqrt{2}\pi f_\pi^2(m_\pi+2\gamma)^3} + \ldots \\ &= 1.53 + 0.36 + \ldots = 1.89 \text{ fm} + \ldots \\ \sqrt{< r^2 >_{\rm expt}} &= 2.11 \text{ fm} \quad , \end{aligned} \tag{3}$$

where $\gamma = \sqrt{M_N B}$ is the binding momentum of the deuteron. In computing the numerical value we have used the numerical value of C_2 extracted from a fit to data over the range of momenta $|\mathbf{p}| < 250$ MeV. Given the expansion parameter for the theory is $\sim \frac{1}{3}$, we expect $\sqrt{< r^2 >_{\rm th}}$ given in Eq. (3) is within $\sim 10\%$ of the actual value, which it is.

In effective range theory the electromagnetic form factors are assumed to be dominated by the asymptotic S-wave deuteron wave function,

$$\psi^{(\rm ER)}(\mathbf{r}) = \sqrt{\frac{\gamma}{2\pi(1-\gamma r_0)}}\frac{e^{-\gamma r}}{r} \quad . \tag{4}$$

Assuming the small r part of the deuteron wave function is only important for establishing the normalization condition, $F_C(0) = 1$, the prediction of effective

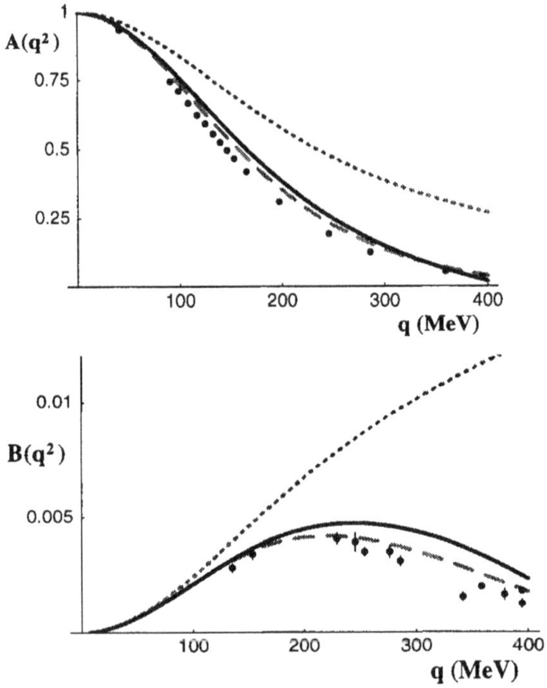

Figure 6: The form factors $A(q^2)$ and $B(q^2)$ measured in elastic electron-deuteron scattering. The dotted curve is the LO prediction, while the solid curve NLO prediction. The dashed curve is the prediction of effective range theory. There is one counterterm contributing to $B(q^2)$ at this order which is fixed by the deuteron magnetic moment.

range theory for the form factor $F_C(q^2)$ follows from the Fourier transform of $|\psi^{(ER)}(\mathbf{r})|^2$,

$$F_C^{(ER)}(q^2) = \left(\frac{1}{1-\gamma\rho_d}\right)\left(\left(\frac{4\gamma}{\sqrt{-q^2}}\right)\tan^{-1}\left(\frac{\sqrt{-q^2}}{4\gamma}\right) - \gamma\rho_d\right) \quad . \tag{5}$$

This yields a deuteron charge radius,

$$\begin{aligned}\sqrt{\langle r_d^2\rangle^{\rm ER}} &= \frac{1}{2\sqrt{2}\gamma}\frac{1}{\sqrt{1-\gamma\rho_d}} \\ &= \frac{1}{2\sqrt{2}\gamma}\left[1+\frac{1}{2}\gamma\rho_d + \frac{3}{8}\gamma^2\rho_d^2 + \ldots\right] \quad ,\end{aligned} \tag{6}$$

which gives a numerical value of $\sqrt{\langle r_d^2\rangle^{\rm ER}} = 1.98$ fm, very close to the quoted value of the matter radius of the deuteron[48,49] of $r_m = 1.967 \pm 0.002$ fm. Conventionally, the charge radius is obtained by combining the nucleon charge radius in quadrature with the matter radius, which agrees very well with the experimental value.

Expanding the EFT expression for $\sqrt{< r^2 >_{\rm th}}$ in Eq. (3) in powers of γ/m_π, we have

$$\gamma \ll m_\pi \rightarrow \frac{1}{2\sqrt{2}\gamma}\left[1 + \frac{1}{2}\gamma r_0 + \frac{g_A^2 M_N \gamma}{4\pi f_\pi^2}\frac{\gamma^2}{m_\pi^2} + \ldots\right] , \tag{7}$$

where r_0 is the effective range. It is clear that the NLO calculation is the same order as a single insertion of r_0, but there is also a contribution arising from pion exchange that is beyond effective range theory.

The quadrupole moment vanishes at leading order in the expansion but receives a contribution at subleading order from the exchange of one potential pion, giving

$$\begin{aligned} \mu_{Q,th} &= \frac{g_A^2 M_N (6\gamma^2 + 9m_\pi\gamma + 4m_\pi^2)}{30\pi f_\pi^2 (m_\pi + 2\gamma)^3} + \ldots. \\ &= 0.40 \text{ fm}^2 + \ldots \\ \mu_{Q,expt} &= 0.2859 \text{ fm}^2 \quad , \end{aligned} \tag{8}$$

which is approximately 30% larger than the experimental value. Clearly, a NNLO calculation[38] is needed to ensure that the EFT value is indeed converging to the experimental value.

In contrast, there is a local counterterm contributing to the magnetic moment at NLO,

$$\begin{aligned} \mu_M &= \mu_p + \mu_n + L_2(\mu)\frac{\gamma}{2\pi}(\mu - \gamma)^2 + \ldots \\ &= 0.88 - 0.02 \quad (fit) \quad , \end{aligned} \tag{9}$$

which determines the counterterm L_2 at the scale μ. Once this counterterm has been determined from the deuteron magnetic moment, the entire form factor $B(q^2)$ is determined to NLO.

Polarizabilities

The scalar α_{E0} and tensor α_{E2} electric polarizabilities are found to be[13]

$$\alpha_{E0} = \frac{\alpha M_N}{32\gamma^4} + \frac{\alpha M_N^2}{64\pi\gamma^3}C_2(\mu)(\mu - \gamma)^2$$

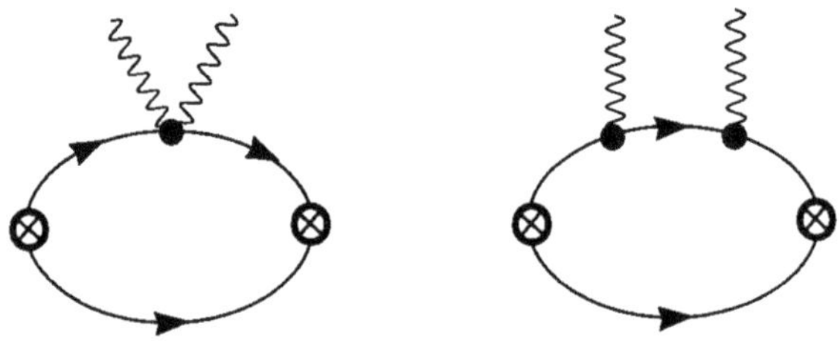

Figure 7: The leading diagrams contributing to the electric polarizability of the deuteron. Crossed graphs are not shown.

$$+ \frac{\alpha g_A^2 M_N^2}{384\pi f^2} \frac{m_\pi^2(3m_\pi^2 + 16m_\pi\gamma + 24\gamma^2)}{\gamma^3(m_\pi + 2\gamma)^4} + \ldots$$

$$\alpha_{E2} = -\frac{\alpha g_A^2 M_N^2}{80\pi f^2} \frac{2m_\pi^3 + 11m_\pi^2\gamma + 16m_\pi\gamma^2 + 8\gamma^3}{\gamma^2(m_\pi + 2\gamma)^4}$$

$$\alpha_{E0} = 0.595 \text{ fm}^3 + \ldots , \quad \alpha_{E2} = -0.062 \text{ fm}^3 + \ldots \quad . \tag{10}$$

Numerically, the value of α_{E0} in Eq. (10) is expected to be within $\sim$ 10% of the actual value. Potential models, which find a value of[89] $\alpha_{E0} = 0.632 \pm 0.003$ fm^3, are expected to reproduce this observable to high precision as the short-distance contributions are expected to be small.

It is informative to perform a momentum expansion of α_{E0} to make contact with effective range theory,

$$\alpha_{E0} \to \frac{\alpha M_N}{32\gamma^4} \left[1 + C_2(\mu) \frac{M_N\gamma(\mu-\gamma)^2}{2\pi} + \frac{g_A^2 M_N \gamma}{4\pi f^2} \left(1 - \frac{8}{3}\frac{\gamma}{m_\pi} + \frac{16}{3}\frac{\gamma^2}{m_\pi^2} + \ldots \right) \right]$$

$$= \frac{\alpha M_N}{32\gamma^4} \left[1 + \gamma r_0 + \frac{g_A^2 M_N \gamma}{4\pi f^2} \frac{10}{3} \frac{\gamma^2}{m_\pi^2} + \ldots \right] \quad . \tag{11}$$

The momentum expansion of Eq. (10) agrees with effective range theory up to order γ^2 at linear order in r_0. The deviations from effective range theory arise from the electromagnetic interactions of the pions, which are beyond effective range theory.

γ-Deuteron Compton Scattering

A process that may allow for a determination of the neutron polarizabilities (given the proton polarizabilities) is γ-deuteron Compton scattering. In

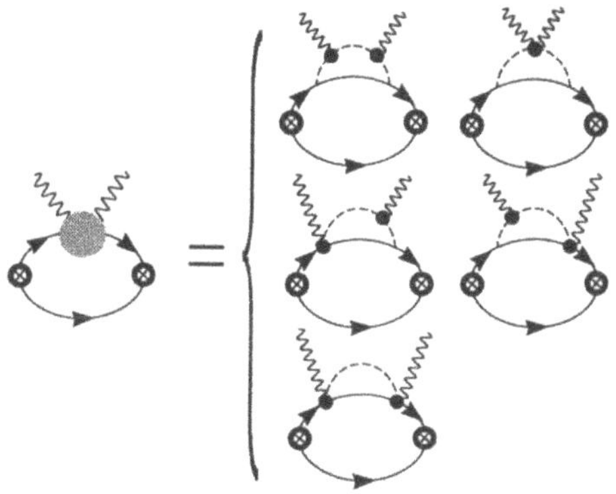

Figure 8: *Pion graphs that contribute to the nucleon polarizabilities and to γ-deuteron Compton scattering at NLO. The crossed circles denote operators that create or annihilate two nucleons with the quantum numbers of the deuteron. The dark solid circles correspond to the photon coupling via the nucleon or pion kinetic energy operator or via the gauged axial pion-nucleon interaction. The solid lines are nucleons and the dashed lines are pions. The photon crossed graphs are not shown.*

power counting the graphs that contribute to the deuteron polarizability, it was assumed[13] that $E_\gamma \sim |\mathbf{p}_\gamma| \sim Q^2$. However, for Compton scattering at photon energies of $\sim$ 100 MeV, the appropriate scaling is $E_\gamma \sim |\mathbf{p}_\gamma| \sim Q$. With this power counting (called Regime II counting[13]), the nucleon isoscalar polarizabilities contribute at NLO through the graphs shown in fig (8).

The cross section for γ-deuteron Compton scattering[13] up to NLO is shown in fig (9) for incident photon energies of E_γ = 49 MeV and 69 MeV. There are no published data for this cross section, but there is unpublished data in the PhD Thesis of M. Lucas[40], consisting of four data points at $E_\gamma = 49$ MeV and two data points at $E_\gamma = 69$ MeV. In fig (9) the dashed curve is the LO calculation which is consistently below the data. The dotted curve is the NLO calculation with the nucleon polarizabilities resulting from the graphs in fig (8) set equal to zero. This overestimates the cross section, with the exception of the back angle data point at $E_\gamma = 49$ MeV. The solid curve is the complete NLO calculation which agrees reasonably well with the data.

There are two published potential model calculations of this cross section[41,42]. Their results disagree with each other at the $\sim 10 - 15\%$ level, and do not reproduce the data at both energies. A more sophisticated potential model calculation has recently been completed by Karakowski and Miller[43]. Further, a

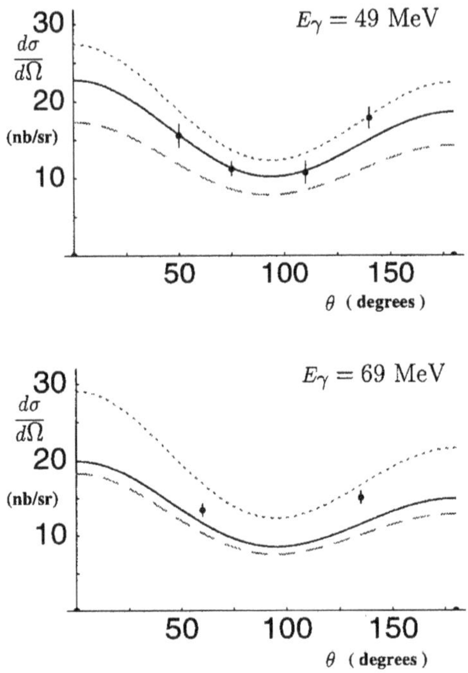

Figure 9: *The differential cross section for γ-deuteron Compton scattering at incident photon energies of* $E_\gamma = 49$ MeV *and* 69 MeV. *The dashed curves correspond to the LO result. The dotted curves correspond to the NLO result without the graphs that contribute to the polarizability of the nucleon. The solid curves correspond to the complete NLO result with no free parameters, as described in the text. Systematic and statistical errors associated with each data point have been added in quadrature.*

calculation of this process is being performed using Weinberg's power counting by Beane, Philips and van Kolck[44]. In the near future more data is expected to become available which will allow for a clearer picture of our understanding of this process.

$np \rightarrow d\gamma$

One of the classic nuclear physics processes is the radiative capture $np \rightarrow d\gamma$. It is a clear demonstration of the existence of "meson exchange currents" in nuclei. The cross section for this process has been measured long ago, and for neutrons incident in the lab with speed $|\mathbf{v}| = 2200$ m/s, the cross section is

$\sigma^{\rm expt} = 334.2 \pm 0.5$ mb[45]. The theoretical cross section for this process has been calculated with effective range theory[33,34], and also more recently using a potential model motivated by Weinberg's power counting[4].

The amplitude for the radiative capture of extremely low momentum neutrons $np \to d\gamma$ has contributions from both the $^1S_0$ and $^3S_1$ NN channels. It can be written as

$$i\mathcal{A} = e\ X\ N^T \tau_2\, \sigma_2\ [\sigma \cdot \mathbf{k}\ \ \epsilon(d)^* \cdot \epsilon(\gamma)^*\ -\ \sigma \cdot \epsilon(\gamma)^*\ \ \mathbf{k} \cdot \epsilon(d)^*]\, N \qquad (12)$$
$$+\, ie\ Y\ \epsilon^{ijk}\ \epsilon(d)^{i*}\ k^j\ \epsilon(\gamma)^{k*}\ (N^T \tau_2 \tau_3 \sigma_2 N) \quad ,$$

where $e = |e|$ is the magnitude of the electron charge, N is the doublet of nucleon spinors, $\epsilon(\gamma)$ is the polarization vector for the photon, $\epsilon(d)$ is the polarization vector for the deuteron and $\mathbf{k}$ is the outgoing photon momentum. The term with coefficient X corresponds to capture from the $^3S_1$ channel while the term with coefficient Y corresponds to capture from the $^1S_0$ channel. For convenience, we define dimensionless variables $\tilde{X}$ and $\tilde{Y}$, by

$$X = i\frac{2}{M_N}\sqrt{\frac{\pi}{\gamma^3}}\ \tilde{X} \quad , \quad Y = i\frac{2}{M_N}\sqrt{\frac{\pi}{\gamma^3}}\ \tilde{Y} \quad . \qquad (13)$$

Both $\tilde{X}$ and $\tilde{Y}$ have the Q expansions, $\tilde{X} = \tilde{X}_0 + \tilde{X}_1 + \ldots$, and $\tilde{Y} = \tilde{Y}_0 + \tilde{Y}_1 + \ldots$, where a subscript n denotes a contribution of order Q^n.

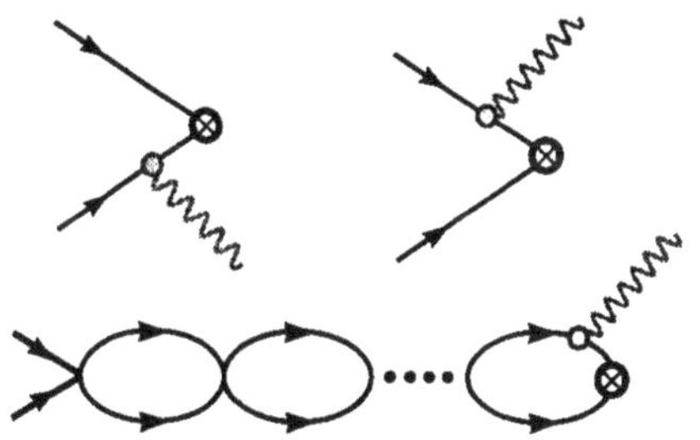

Figure 10: *Graphs contributing to the amplitude for $np \to d\gamma$ at leading order in the effective field theory expansion. The solid lines denote nucleons and the wavy lines denote photons. The light solid circles correspond to the nucleon magnetic moment coupling to the electromagnetic field. The crossed circle represents an insertion of the deuteron interpolating field which is taken to have $^3S_1$ quantum numbers.*

At leading order in the EFT[17] the graphs shown in fig (10) give a contribution to the cross section of

$$\sigma^{LO} = \frac{8\pi\alpha\gamma^5\kappa_1^2(a^{(^1S_0)})^2}{|\mathbf{v}|M_N^5}\left(1-\frac{1}{\gamma a^{(^1S_0)}}\right)^2 = 297.2 \text{ mb} \quad . \tag{14}$$

This agrees with the effective range theory calculation of Bethe and Longmire[33] and Noyes [34] when terms in their expression involving the effective range are neglected. Eq. (14) is about 10% less than the experimental value, $\sigma^{\rm expt} = 334.2 \pm 0.5$ mb[45].

At NLO there are contributions from insertions of the C_2, D_2 operators and from potential pion exchange with the isovector nucleon magnetic moment.

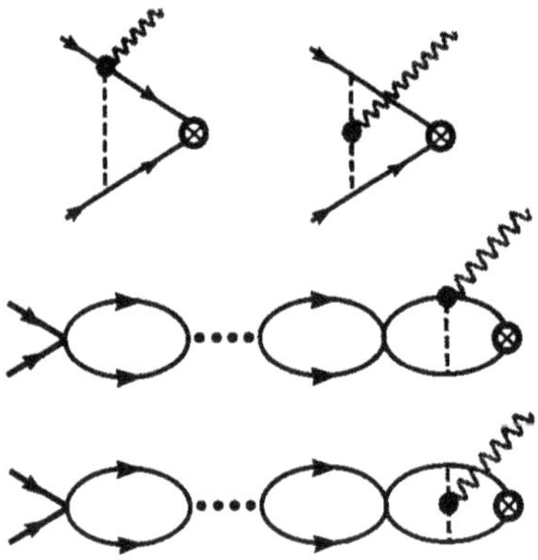

Figure 11: *Pion exchange current contributions to the amplitude for $np \to d\gamma$ arising at NLO. The solid lines denote nucleons and the wavy lines denote photons. The dashed line denotes a pion. The dark solid circles correspond to minimal coupling of the photon. The crossed circle represents an insertion of the deuteron interpolating field.*

In addition, there are contributions from meson exchange where the photon is minimally coupled either to the axial current or to the pion itself. These diagrams, shown in fig (11), are called "meson exchange currents". Also, at this order there is a contribution from a four-nucleon-one-photon counterterm, shown in fig (12). The contributions from the meson exchange currents and the counterterm[17] are

$$\tilde{Y}^{(\pi,{\rm E})} = \frac{g_A^2 M_N\gamma^2}{12\pi f^2}\left(\frac{m_\pi-\gamma}{(m_\pi+\gamma)^2}\right.$$
$$\left. + \frac{M_N}{4\pi}\mathcal{A}_{-1}^{(^1S_0)}(0)\left(\frac{3m_\pi-\gamma}{2(m_\pi+\gamma)} + \ln\left(\frac{m_\pi+\gamma}{\mu}\right) - \frac{1}{6} + \delta\right)\right)$$

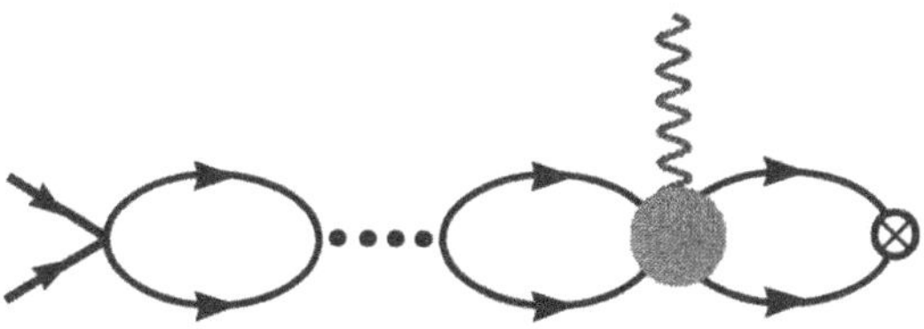

Figure 12: *Local counterterm contribution to the amplitude for $np \to d\gamma$ at NLO. The solid lines denote nucleons and the wavy lines denote photons. The solid circle corresponds to an insertion of the L_1 operator. The crossed circle represents an insertion of the deuteron interpolating field. The graphs corresponding to final state interactions are not shown.*

$$\tilde{Y}^{(L_1)} = L_1 \, \gamma^2 \, \frac{\mathcal{A}_{-1}^{(^1S_0)}(0)}{\bar{C}_0^{(^1S_0)}\bar{C}_0^{(^3S_1)}} \quad , \tag{15}$$

where δ is a subtraction constant. The important point to note here is that the contribution from the meson exchange currents is UV divergent, as is clear from the presense of the $\log(\mu)$ contribution in eq. (15), and the subtraction constant δ. This divergence is absorbed by the local counterterm $L_1(\mu)$, and therefore it makes no sense to discuss the contribution from meson exchange currents alone.

The two four-nucleon-one-photon counterterms that arise at NLO are renormalized very differently. In contrast to the counterterm $L_1(\mu)$, the counterterm $L_2(\mu)$ is not driven by meson exchange currents and satisfies a simple renormalization group equation,

$$\mu\frac{d}{d\mu}\left[\frac{L_1 - \frac{1}{2}\kappa_1\left(C_2^{(^1S_0)} + C_2^{(^3S_1)}\right)}{\bar{C}_0^{(^1S_0)}\bar{C}_0^{(^3S_1)}}\right] = \frac{g_A^2 M^2}{48\pi^2 f^2}$$
$$\mu\frac{d}{d\mu}\left[\frac{L_2}{(\bar{C}_0^{(^3S_1)})^2}\right] = 0 \quad . \tag{16}$$

3 Effective Field without Pions

For processes involving external momenta much less than the pion mass, it is much simpler to work with an EFT where the pions do not appear. In this pionless theory[32] (EFT($\not{\pi}$)), one can examine each of the observables we have just discussed in the theory with dynamical pions. In many of the cases the calculation can be pushed to higher order with little effort, which makes the pionless theory particularly attractive for static properties. Without pions the only expansion parameter is the external momentum, $Q \sim \mathbf{p}$, where we will neglect the electromagnetic interaction and isospin breaking[27].

3.1 *The Nucleon-Nucleon Interaction*

In EFT($\not\pi$), the mixing parameter $\bar{\varepsilon}_1$ is supressed by Q^2 compared with $\delta_0^{(0)}$, and therefore, up to N^{4}LO, we can isolate the S-wave from the D-wave in the deuteron channel, leaving

$$S_{00} = e^{i2\delta_0} = 1 + \frac{2i}{\cot\delta_0 - i} \quad . \tag{17}$$

The phase shift δ_0 has an expansion in powers of Q, $\delta_0 = \delta_0^{(0)} + \delta_0^{(1)} + \delta_0^{(2)} + \ldots$, where the superscript denotes the order in the Q expansion. By forming the logarithm of both sides of eq. (17) and expanding in powers of Q, it is straightforward to obtain

$$\begin{aligned}
\delta_0^{(0)}(|\mathbf{k}|) &= \pi - \tan^{-1}\left(\frac{|\mathbf{k}|}{\gamma}\right) \\
\delta_0^{(1)}(|\mathbf{k}|) &= -\frac{\rho_d}{2}|\mathbf{k}| \\
\delta_0^{(2)}(|\mathbf{k}|) &= -\left[\frac{\rho_d^2\gamma}{4} + \frac{\gamma^3}{8M_N^2(\gamma^2 + |\mathbf{k}|^2)}\right]|\mathbf{k}| \quad ,
\end{aligned} \tag{18}$$

which are shown in fig. (13). Up to NNLO, the shape parameter term, w_2 does not contribute to the S-wave phase shift and therefore, in addition to being numerically small (about a factor of 5 smaller than ρ_d), w_2 enters only at higher orders. Formally, the first deviations from linear $|\mathbf{k}|$ dependence at large momenta arises from relativistic corrections (the second term in $\delta_0^{(2)}$ in eq. (18). The Q expansion is clearly demonstrated in eq. (18), by simply counting powers of $|\mathbf{k}|$ and γ, which both scale like Q. We expect this perturbative expansion of the phase shifts to converge up to momenta of order $\sim m_\pi/2 \sim 70$ MeV, at which point one encounters the t-channel cut from potential pion exchange.

Scattering between the S-wave and D-wave is induced by local operators first arising at Q^1 in the power counting. At order Q^1 and Q^2, the lagrange density describing such interaction is

$$\begin{aligned}
\mathcal{L}_2^{(sd)} = &\frac{1}{4}{}^{\not\pi}C_0^{(sd)}\left(N^T P^i N\right)^\dagger\left(N^T \mathcal{O}^{(sd),xyj} N\right) \mathcal{T}^{ijxy} + \text{h.c.} \\
&- \frac{1}{16}{}^{\not\pi}C_2^{(sd)}\left(N^T\left[P_i \overrightarrow{\mathbf{D}}^2 + \overleftarrow{\mathbf{D}}^2 P_i - 2\overleftarrow{\mathbf{D}} P_i \overrightarrow{\mathbf{D}}\right] N\right)^\dagger \left(N^T \mathcal{O}^{(sd),xyj} N\right) \mathcal{T}^{ijxy} \\
&+ \text{h.c.}
\end{aligned} \tag{19}$$

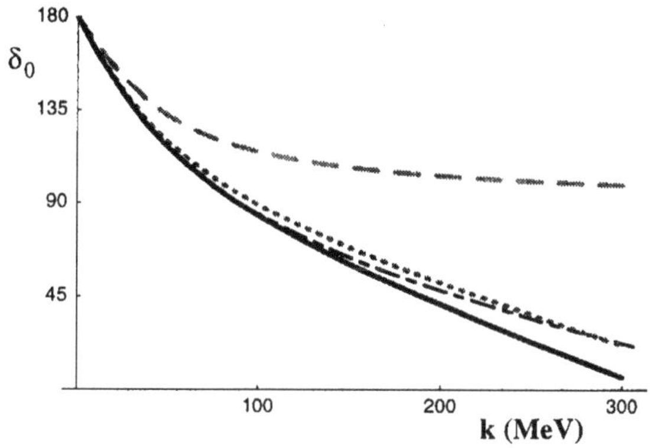

Figure 13: *The phase shift in the $^3S_1$ channel, δ_0, as a function of the center of mass momentum $|\mathbf{k}|$. The dashed curve corresponds to $\delta_0^{(0)}$, the dotted curve corresponds to $\delta_0^{(0)} + \delta_0^{(1)}$, the solid curve corresponds to $\delta_0^{(0)} + \delta_0^{(1)} + \delta_0^{(2)}$, and the dot-dashed curve is the Nijmegen partial wave analysis*[35].

where

$$\mathcal{O}^{(sd),xyj} = \overleftarrow{\mathbf{D}}^x \overleftarrow{\mathbf{D}}^y P^j + P^j \overrightarrow{\mathbf{D}}^x \overrightarrow{\mathbf{D}}^y - \overleftarrow{\mathbf{D}}^x P^j \overrightarrow{\mathbf{D}}^y - \overleftarrow{\mathbf{D}}^y P^j \overrightarrow{\mathbf{D}}^x$$
$$\mathcal{T}^{ijxy} = \left(\delta^{ix}\delta^{jy} - \frac{1}{n-1}\delta^{ij}\delta^{xy}\right) \quad , \tag{20}$$

and where n is the number of spacetime dimensions. The lagrange density in eq. (20) appears somewhat complex. However, when the electromagnetic field is ignored, the operator collapses to

$$\mathcal{O}^{(sd),xyj} \rightarrow P^j \left(\overleftarrow{\nabla} - \overrightarrow{\nabla}\right)^x \left(\overleftarrow{\nabla} - \overrightarrow{\nabla}\right)^y \quad , \tag{21}$$

explicitly Galilean invariant.

The coefficients that appear in eq. (20) themselves have an expansion in powers of Q, e.g. ${}^{\not\pi}C_0^{(sd)} = {}^{\not\pi}C_{0,-1}^{(sd)} + {}^{\not\pi}C_{0,0}^{(sd)} + \ldots$. Performing a Q expansion on the mixing parameter $\bar{\varepsilon}_1 = \bar{\varepsilon}_1^{(2)} + \bar{\varepsilon}_1^{(3)} + \ldots$ it is straightforward to demonstrate that

$$\bar{\varepsilon}_1^{(2)}(|\mathbf{k}|) = \frac{\sqrt{2}}{3}\left(\frac{{}^{\not\pi}C_{0,-1}^{(sd)}}{{}^{\not\pi}C_{0,-1}^{({}^3S_1)}}\right)\frac{|\mathbf{k}|^3}{\sqrt{\gamma^2 + |\mathbf{k}|^2}}$$
$$= -\frac{M_N}{4\pi}(\mu - \gamma)\frac{\sqrt{2}}{3}\, {}^{\not\pi}C_{0,-1}^{(sd)} \frac{|\mathbf{k}|^3}{\sqrt{\gamma^2 + |\mathbf{k}|^2}} \quad . \tag{22}$$

Renormalization group invariance of this leading order contribution to $\overline{\varepsilon}_1$ indicates that ${}^{\not\pi}C^{(sd)}_{0,-1} \propto (\mu-\gamma)^{-1}$. Renormalizing at the scale $\mu = m_\pi$, and fitting to the Nijmegen phase shift analysis[35], we find ${}^{\not\pi}C^{(sd)}_{0,-1} = -4.57\ \mathrm{fm}^4$.

At the next order, Q^3, the contribution to $\overline{\varepsilon}_1$ is

$$\overline{\varepsilon}_1^{(3)}(|\mathbf{k}|) = -\frac{\sqrt{2}}{3}\frac{M_N}{4\pi}\left[\left(\frac{\mu\gamma\rho_d}{2}\ {}^{\not\pi}C^{(sd)}_{0,-1} \ + \ (\mu-\gamma)\ {}^{\not\pi}C^{(sd)}_{0,0}\right)\frac{|\mathbf{k}|^3}{\sqrt{\gamma^2+|\mathbf{k}|^2}}\right.$$
$$\left. + \ \left(\frac{\rho_d}{2}\ {}^{\not\pi}C^{(sd)}_{0,-1} + (\mu-\gamma)\ {}^{\not\pi}C^{(sd)}_{2,-2}\right)\frac{|\mathbf{k}|^5}{\sqrt{\gamma^2+|\mathbf{k}|^2}}\right] \quad , \tag{23}$$

where we have expanded ${}^{\not\pi}C^{(sd)}_2 = {}^{\not\pi}C^{(sd)}_{2,-2} + ...$ The first term in eq. (23) has the same momentum dependence as, $\overline{\varepsilon}_1^{(2)}$ the leading contribution in eq. (22). In the same way that we have required the position of the deuteron pole is not modified by higher orders in perturbation theory, we may require that the coefficient contributing to $\overline{\varepsilon}_1^{(2)}$ is not modified by higher order contributions. This constraint leads to

$${}^{\not\pi}C^{(sd)}_{0,0} = -\frac{\mu\gamma\rho_d}{2(\mu-\gamma)}\ {}^{\not\pi}C^{(sd)}_{0,-1} \quad . \tag{24}$$

Further, as the second term in eq. (23) is an observable, we obtain the RG equation

$$\mu\frac{d}{d\mu}\left(\frac{\rho_d}{2}\ {}^{\not\pi}C^{(sd)}_{0,-1} + (\mu-\gamma)\ {}^{\not\pi}C^{(sd)}_{2,-2}\right) = 0 \quad , \tag{25}$$

and hence the term in the bracket is of order Q^0. Therefore, $\overline{\varepsilon}_1^{(3)} = 0$. A fit to the Nijmegen phase shift analysis[35] up to order Q^4, is shown in fig. (14).

Electric Polarizability of the Deuteron

With the EFT($\not\pi$) we have computed α_{E0} up to NNLO, including relativistic corrections. It has a perturbative expansion in Q, and we write $\alpha_{E0} = \alpha_{E0}^{(-4)} + \alpha_{E0}^{(-3)} + \alpha_{E0}^{(-2)} + ...$ Explicit calculation gives

$$\begin{aligned}\alpha_{E0}^{(-4)} + \alpha_{E0}^{(-3)} + \alpha_{E0}^{(-2)} &= \frac{\alpha M_N}{32\gamma^4}\left[1 \ + \ \gamma\rho_d \ + \ \gamma^2\rho_d^2 \ + \ \frac{2\gamma^2}{3M_N^2}\right] \\ &= \quad 0.377 \ + \ 0.153 \ + \ 0.062 \ + \ 0.0006 \\ &= 0.592\ \mathrm{fm}^3 \quad .\end{aligned} \tag{26}$$

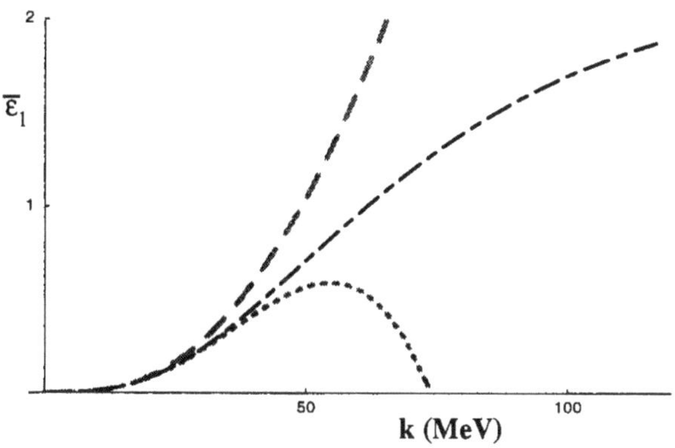

Figure 14: *The S-D mixing parameter $\overline{\varepsilon}_1$ as a function of the center of mass momentum $|\mathbf{k}|$. The dashed curve corresponds to $\overline{\varepsilon}_1^{(2)}$, the dotted curve corresponds to $\overline{\varepsilon}_1^{(2)} + \overline{\varepsilon}_1^{(4)}$, and the dot-dashed curve is the Nijmegen partial wave analysis.*

Numerically, the relativistic corrections are very small, two orders of magnitude smaller than the NNLO corrections from the four-nucleon interactions. The value of α_{E0} in Eq. (26) is within $\sim 5\%$ of that computed with potential models and with effective range theory. Despite being numerically small, relativistic corrections can be calculated easily with the EFT. The S-wave-D-wave mixing operators (corresponding to the D-wave component of the deuteron in potential model language) make contributions to $\alpha_E^{(-2)}$. They do not contribute to the scalar polarizability, $\alpha_{E0}^{(-2)}$, but do contribute to the tensor polarizability, $\alpha_{E2}^{(-2)}$. Such operators will contribute to α_{E0} at higher orders in the expansion.

Electromagnetic Form Factors of the Deuteron

The charge radius of the deuteron at NNLO one finds in EFT($\not\pi$) is

$$\begin{aligned} \langle r_d^2 \rangle^{\rm EFT} &= \langle r_{N,0}^2 \rangle + \frac{1}{8\gamma^2}\left[1 + \gamma\rho_d + \gamma^2\rho_d^2\right] + \frac{1}{32 M_N^2} \\ &= 0.62 + 2.33 + 0.95 + 0.39 + 0.0014 \\ &= 4.30 \text{ fm}^2 \quad , \end{aligned} \tag{27}$$

where the last term is the relativistic correction. Taking the square root of this value gives $\sqrt{\langle r_d^2 \rangle} = 2.07$ fm, which is within a few percent of the measured value of $\sqrt{\langle r_d^2 \rangle} = 2.1303$ fm[46,48,49,50].

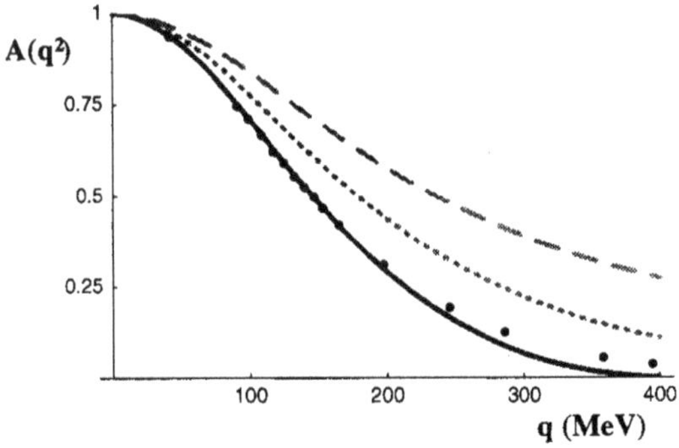

Figure 15: *The form factor $A(q^2)$ as a function of $|\mathbf{q}| = \sqrt{-q^2}$. The dashed curve corresponds to the LO prediction, the dotted curve corresponds to the NLO prediction, and the solid curve corresponds to the NNLO prediction, in* EFT($\not\pi$).

At NLO the deuteron magnetic moment is

$$\mu_d = \frac{e}{2M_N}\left(\kappa_p + \kappa_n + {}^{\not\pi}L_2 \frac{2M_N\gamma(\mu-\gamma)^2}{\pi}\right) \quad , \tag{28}$$

the same expression as is found in the theory with pions. Comparing the numerical value of this expression with the measured value of μ_d gives (at NLO)

$${}^{\not\pi}L_2(m_\pi) = -0.149 \text{ fm}^4 \quad , \tag{29}$$

at the renormalization scale $\mu = m_\pi$. The evolution of the ${}^{\not\pi}L_2(\mu)$ operator as the renormalization scale is changed is determined by the RG equation

$$\mu\frac{d}{d\mu}\left[\frac{{}^{\not\pi}L_2}{\left({}^{\not\pi}C_{0,-1}^{({}^3S_1)}\right)^2}\right] = 0 \quad . \tag{30}$$

It is combinations of the electric, magnetic and quadrupole form factors that are measured in elastic electron-deuteron scattering. The differential cross section for unpolarized elastic electron-deuteron scattering is given by

$$\frac{d\sigma}{d\Omega} = \left.\frac{d\sigma}{d\Omega}\right|_{\text{Mott}}\left[A(q^2) + B(q^2)\tan^2\left(\frac{\theta}{2}\right)\right] \quad , \tag{31}$$

where A and B are related to the form factors[46] by

$$A = F_C^2 + \frac{2}{3}\eta F_M^2 + \frac{8}{9}\eta^2 F_Q^2 \quad ,$$
$$B = \frac{4}{3}\eta(1+\eta) F_M^2 \ , \qquad (32)$$

with $\eta = -q^2/4M_d^2$. In order to compare with data, we take our analytic results for the form factors and expand the expression eq. (32) in powers of Q. At the order we are working, A is sensitive both the electric and magnetic form factors, while B depends only on the magnetic form factor. The predictions for $A(q^2)$ and $B(q^2)$ along with data are shown in fig. (15) and fig. (16), respectively.

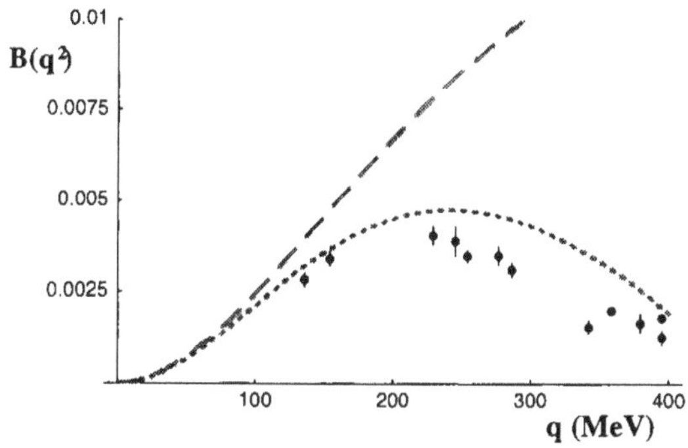

Figure 16: *The form factor $B(q^2)$ as a function of $|q| = \sqrt{-q^2}$. The dashed curve corresponds to the LO prediction, and the dotted curve corresponds to the NLO prediction, in* EFT($\not\pi$).

At LO and NLO the deuteron quadrupole moment, μ_Q, is found to be

$$\mu_Q^{(0)} = -\frac{M_N \, {}^{\not\pi}C_{0,-1}^{(sd)}(\mu-\gamma)}{12\pi}$$
$$\mu_Q^{(1)} = \frac{1}{2}\rho_d\gamma\ \mu_Q^{(0)} \ - \ {}^{\not\pi}C_Q \ \frac{(\mu-\gamma)^2\gamma}{\pi} \quad . \qquad (33)$$

Numerically, one finds that the quadrupole moment is given by the sum of $\mu_Q^{(0)} = 0.273\ \mathrm{fm}^2$ and $\mu_Q^{(1)} = \left(0.056 \ - \ 0.0165\ {}^{\not\pi}C_Q\right)\ \mathrm{fm}^2$ giving, at NLO,

$$\mu_Q = \left(0.329 \ - \ 0.0165\ {}^{\not\pi}C_Q\right)\ \mathrm{fm}^2 \quad , \qquad (34)$$

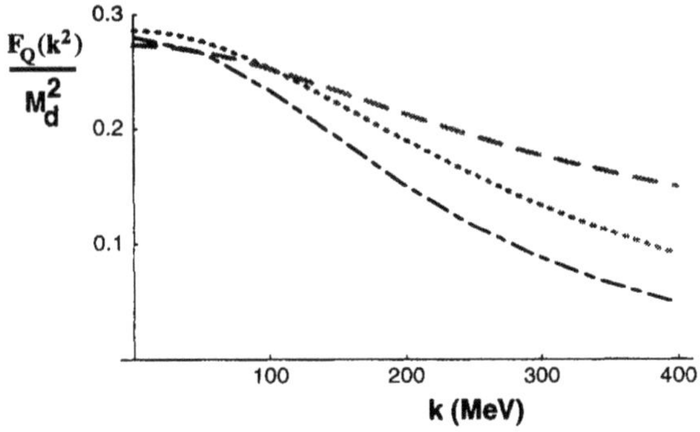

Figure 17: *The quadrupole form factor $F_Q(|\mathbf{k}|^2)$ as a function of $|\mathbf{k}|$. The dashed curve corresponds to the LO prediction, and the dotted curve corresponds to the NLO prediction, in* EFT($\not\pi$). *The dot-dashed curve corresponds to a calculation with the Bonn-B potential in the formulation of*[51] .

where ${}^{\not\pi}C_Q$ is measured in fm^5. Setting ${}^{\not\pi}C_Q = 0$, we find a quadrupole moment of 0.329 fm^2, which is to be compared with the measured value of 0.286 fm^2. In order to reproduce the measured value of the quadrupole moment ${}^{\not\pi}C_Q = +2.60$ fm^5, at $\mu = m_\pi$. The quadrupole form factor resulting from this fit is shown in fig. (17).

$np \to d\gamma$

In EFT($\not\pi$) we find a cross section for $np \to d\gamma$, at NLO and at $|\mathbf{v}| = 2200$ m/s incident neutron speed, of

$$\sigma_{\not\pi} = \left(287.1 \ + \ 6.51 \ {}^{\not\pi}L_1 \right) \text{ mb} \quad , \tag{35}$$

where ${}^{\not\pi}L_1$ is the coefficient of a four-nucleon-one-magnetic-photon operator, with units of fm^4 and is renormalized $\mu = m_\pi$. Requiring $\sigma_{\not\pi}$ to reproduce the measured cross section σ^{expt} fixes ${}^{\not\pi}L_1 = 7.24$ fm^4.

We see that even in the theory without dynamical pions, one is able to recover the cross section for radiative neutron capture at higher orders. It is clear that in this theory the four-nucleon-one-photon operators play a central role in reproducing the low energy observables. In the theory with pions, one can see by examining the contributing Feynman diagrams[17], that in the limit that the momentum transferred to the photon is small the pion propagators can be replaced by $1/m_\pi^2$, while keeping the derivative structure in the numerator.

This contribution, as well as the contribution from all hadronic exchanges, is reproduced order by order in the momentum expansion by the contributions from local multi-nucleon-photon interactions. ¿From the calculations in the theory with dynamical pions, the value of ${}^{\not\pi}L_1$ is not saturated by pion exchange currents as these contributions are divergent, and require the presense of the L_1 operator[17]. Therefore, estimates of ${}^{\not\pi}L_1$ based on meson exchanges alone are model dependent.

The effective range calculation of $np \to d\gamma$ was first performed by Bethe and Longmire[33] and revisited by Noyes[34]. After correcting the typographical errors in the expression for σ that appears in the Noyes article, the expressions in the two papers[33,34] are identical,

$$\sigma^{(\mathrm{ER})} = \frac{2\pi\alpha\ \kappa_1^2\ \gamma^6\ (a^{({}^1S_0)})^2\ a^{({}^3S_1)}}{|\mathbf{v}|M_N^5\ (2\ -\ \gamma a^{({}^3S_1)})} \left(1\ +\ \frac{1}{\gamma a^{({}^3S_1)}}\ -\ \frac{2}{\gamma a^{({}^1S_0)}} - \frac{\gamma r_0^{({}^1S_0)}}{2}\right)^2 \tag{36}$$

which when expanded in powers of Q is

$$\sigma^{(\mathrm{ER})} = \frac{8\pi\alpha\kappa_1^2\gamma^3}{|\mathbf{v}|M_N^5}(1-\gamma a^{({}^1S_0)})\Big[(1-\gamma a^{({}^1S_0)}) + \frac{1}{2}\gamma\left(\rho_d - r_0^{({}^1S_0)}\right)(1-\gamma a^{({}^1S_0)})\ +\ \frac{1}{2}\gamma\left(\rho_d + r_0^{({}^1S_0)}\right)\ +\ \ldots\Big] \tag{37}$$

At LO in the EFT($\not\pi$) expansion the cross sections agree, however, at NLO the expressions are very different. In addition to the counterterm that appears at this order in the EFT($\not\pi$), the contributions from the effective range parameters are found to disagree, and more importantly be renormalization scale dependent. In the EFT($\not\pi$) the local counterterm is renormalized by the short-distance behavior of graphs involving the C_2 operators and hence the effective range parameters in both channels. Given, this behavior it is no surprise that the effective range contributions differ between the two calculations. It is amusing to ask if there is a scale for which the expressions are identical, with ${}^{\not\pi}L_1 = 0$. Indeed such a scale exits,

$$\mu^{\mathrm{ER}} = \frac{\gamma r_0^{({}^1S_0)} a^{({}^1S_0)} - \rho_d}{a^{({}^1S_0)}\left(r_0^{({}^1S_0)} - \rho_d\right)}\quad , \tag{38}$$

which, by inserting the appropriate values, gives scale $\mu^{\mathrm{ER}} \sim 144$ MeV, coincidentally close to $\mu = m_\pi$.

4 Conclusions

We have performed a systematic analysis of the two-nucleon sector with effective field theory using KSW power-counting and dimensional regularization. It is clear that for each observable considered in this work the perturbative expansion appears to be converging, and the expansion parameter(s) is $\sim \frac{1}{3}$, both in the theory with pions and for the static properties of the deuteron in the pionless theory. I have deliberately avoided a discussion of the scale at which the theory with dynamical pions breaks down, as this has been discussed in other talks at this workshop[52,53,54,55]. In the theory with pions, the NLO contributions from potential pion exchange are found to be $\sim \frac{1}{3}$ the size of the NLO contributions from the C_2 operators. ¿From these calculations, there is nothing to indicate that treating pions in perturbation theory is not converging. In order to be more confident that this is generally true, higher order calculations in this theory must be performed.

As the pionless theory is very simple, computations beyond NNLO will be performed, thereby giving calculations with better than $\sim 1\%$ precision. As relativistic contributions to the deuteron static properties are easy to calculate and are found to be very small (suppressed by additional factors of m_π^2/M_N^2) such high precision calculations are possible in the near future.

5 Acknowledgements

I would like to thank the *Institute for Nuclear Theory* for sponsoring this workshop and particularly Wick Haxton who enabled us (the organizers) to hold the type of workshop we envisaged. I would also like to thank Maria Francom, Linda Vilet and Shizue Shikuma for making the organization of the workshop pain-free. Finally, I would like to thank my co-organizers, Paulo Bedaque, Ryoichi Seki and Bira van Kolck, for a superb job.

References

1. S. Weinberg, *Phys. Lett.* B **251**, 288 (1990); *Nucl. Phys.* B **363**, 3 (1991); *Phys. Lett.* B **295**, 114 (1992).
2. C. Ordonez and U. van Kolck, *Phys. Lett.* B **291**, 459 (1992); C. Ordonez, L. Ray and U. van Kolck, *Phys. Rev. Lett.* **72**, 1982 (1994); *Phys. Rev.* C **53**, 2086 (1996); U. van Kolck, *Phys. Rev.* C **49**, 2932 (1994).
3. J. Friar, `nucl-th/9601012`; `nucl-th/9601013`; *Few Body Systems Suppl.* **99**, 1 (1996); `nucl-th/9804010`; J.L. Friar, D. Huber, and U. van Kolck, *Phys. Rev.* C **59**, 53 (1999).

4. T.S. Park, D.P. Min and M. Rho, *Phys. Rev. Lett.* **74**, 4153 (1995) ; *Nucl. Phys.* A **596**, 515 (1996); nucl-th/9807054.
5. T.D. Cohen, *Phys. Rev.* C **55**, 67 (1997); D.R. Phillips and T.D. Cohen, *Phys. Lett.* B **390**, 7 (1997); K.A. Scaldeferri, D.R. Phillips, C.W. Kao and T.D. Cohen, *Phys. Rev.* C **56**, 679 (1997); S.R. Beane, T.D. Cohen and D.R. Phillips, nucl-th/9709062; D.R. Phillips, S.R. Beane and T.D. Cohen, *Annals Phys.* **263**, 255 (1998).
6. M.J. Savage, *Phys. Rev.* C **55**, 2185 (1997).
7. G.P. Lepage, nucl-th/9706029, Lectures at 9th Jorge Andre Swieca Summer School: Particles and Fields, Sao Paulo, Brazil, Feb 1997.
8. M. Luke and A.V. Manohar, *Phys. Rev.* D **55**, 4129 (1997).
9. D.B. Kaplan, M.J. Savage, and M.B. Wise, *Nucl. Phys.* B **478**, 629 (1996).
10. D.B. Kaplan, M.J. Savage and M.B. Wise, *Phys. Lett.* B **424**, 390 (1998); *Nucl. Phys.* B **534**, 329 (1998).
11. U. van Kolck, *Nucl. Phys.* A **645**, 273 (1999).
12. D.B. Kaplan, M.J. Savage and M.B. Wise, *Phys. Rev.* C **59**, 617 (1999).
13. J.W. Chen, H.W. Griesshammer, M.J. Savage and R.P. Springer, *Nucl. Phys.* A **644**, 221 (1999);*Nucl. Phys.* A **644**, 245 (1999).
14. J.W. Chen, nucl-th/9810021.
15. M.J. Savage and R.P. Springer, *Nucl. Phys.* A **644**, 235 (1999).
16. D.B. Kaplan, M.J. Savage, R.P. Springer and M.B. Wise, *Phys. Lett.* B **449**, 1 (1999).
17. M.J. Savage, K.A. Scaldeferri, M.B.Wise, nucl-th/9811029.
18. T. Mehen and I.W. Stewart, nucl-th/9901064; nucl-th/9809095; *Phys. Lett.* B **445**, 378 (1999).
19. G. Rupak and N. Shoresh, private communication.
20. J. Gegelia, *Phys. Lett.* B **429**, 227 (1998); nucl-th/9806028; nucl-th/9805008; nucl-th/9802038.
21. A.K. Rajantie, *Nucl. Phys.* B **480**, 729 (1996).
22. J.V. Steele and R.J. Furnstahl, *Nucl. Phys.* A **637**, 46 (1998); *Nucl. Phys.* A **645**, 439 (1999).
23. T.D. Cohen and J.M. Hansen, *Phys. Rev.* C **59**, 13 (1999); nucl-th/9901065.
24. T.-S. Park, K. Kubodera, D.-P. Min, and M. Rho, *Nucl. Phys.* A **646**, 83 (1999); astro-ph/9804144; *Phys. Rev.* C **58**, 637 (1998).
25. X. Kong and F. Ravndal, nucl-th/9803046; *Phys. Lett.* B **450**, 320 (1999).
26. X. Kong and F. Ravndal, nucl-th/9902064; hep-ph/9903523.

27. E. Epelbaoum and U.G. Meissner, nucl-th/9902042; E. Epelbaoum, W. Glockle, A. Kruger and U.G. Meissner, *Nucl. Phys.* A **645**, 413 (1999); E. Epelbaoum, W. Glockle and U.G. Meissner, *Phys. Lett.* B **439**, 1 (1998); E. Epelbaoum, W. Glockle and U.G. Meissner, *Nucl. Phys.* A **637**, 107 (1998); E. Epelbaoum and U.G. Meissner, nucl-th/9903046, *this volume.*
28. T. Mehen, I.W. Stewart and M.B. Wise, hep-ph/9902370.
29. D.R. Phillips, S.R. Beane and M.C. Birse, hep-ph/9810049.
30. M.N. Butler and J.W. Chen, private communication.
31. P.F. Bedaque, H.W. Hammer and U. van Kolck, *Phys. Rev. Lett.* **82**, 463 (1999); *Phys. Rev.* C **58**, R641 (1998); P.F. Bedaque and U. van Kolck, *Phys. Lett.* B **428**, 221 (1998).
32. J.W. Chen, G. Rupak and M.J. Savage, nucl-th/9902056.
33. H.A. Bethe, *Phys. Rev.* **76**, 38 (1949); H.A. Bethe and C. Longmire, *Phys. Rev.* **77**, 647 (1950).
34. H.P. Noyes, *Nucl. Phys.* **74**, 508 (1965).
35. V.G.J. Stoks, R.A.M. Klomp, C.P.F. Terheggen and J.J. de Swart, *Phys. Rev.* C **49**, 2950 (1994); nucl-th/9406039.
36. H.P. Stapp, T.J. Ypsilantis and N. Metropolis, *Phys. Rev.* **105**, 302 (1957).
37. J.J. de Swart, C.P.F Terheggen and V.G.J. Stoks, nucl-th/9509032.
38. M. Binger, nucl-th/9901012.
39. J.L. Friar and G.L. Payne, *Phys. Rev.* C **55**, 2764 (1997).
40. M.A. Lucas, Ph. D. thesis, University of Illinois at Urbana-Champaign (1994)
41. M.I. Levchuk and A.I. L'vov, *Few Body Systems Suppl.* **9**, 439 (1995).
42. T. Wilbois, P. Wilhelm and H. Arenhovel, *Few Body Systems Suppl.* **9**, 263 (1995).
43. J. Karakowski and G. Miller, private communication.
44. S. Beane, D. Philips and U. van Kolck, private communication.
45. A.E. Cox, S.A.R. Wynchank and C.H. Collie, *Nucl. Phys.* **74**, 497 (1965).
46. C.W. Wong, *Int. J. of. Mod. Phys.* E **3**, 821 (1994).
47. J.L. Friar, *Ann. of Phys.* **81**, 332 (1973).
48. A.J. Buchmann, H. Henning and P.U. Sauer, *Few-Body Systems,* **21**, 149 (1996).
49. J.L. Friar, J. Martorell and D.W.L. Sprung, *Phys. Rev.* A **56**, 4579 (1997).
50. T. Ericson and W. Weise, *Pions and Nuclei,* (Oxford University Press, 1988).

51. J. Adam Jr. , H. Göller and H. Arenhövel, *Phys. Rev.* C **48**, 370 (1993).
52. J.V. Steele, `nucl-th/9904023`, *this volume.*
53. D.B. Kaplan, *this volume.*
54. G. Rupak, *this volume.*
55. T. Mehen, *this volume.*

Discussion during M. Savage's presentation

Chair: *B. Holstein (University of Massachusetts)*
Notes taken by: *S. Fleming (University of Toronto)*

T. Cohen remarked that the EFT does not do as well as the Effective Range Expansion (ERE) because the ERE keeps r_0 to all orders and therefore includes more of the physics. For example the asymptotic part of the wave function is well described by $1/a$ and r_0. M. Savage responded that such a resummation does not imply that the calculation will be more accurate. The critical issue is at what order the higher dimension operators coupling to the electroweak fields enter.

B. Holstein asked if it is possible to determine the neutron polarizability from Compton scattering off the deuteron. M. Savage said that at present it is not possible to get a constraint with error bars less than 100%. He added that it may be possible to carry out such a calculation in the theory without pions, since pions are not the dominant contribution at NLO.

SESSION 3

Session Chair: *R. J. Furnstahl (Ohio State University)*
Notes taken by: *R. P. Springer (Duke University)*

OPEN DISCUSSION

R. J. FURNSTAHL
Department of Physics, Ohio State University
Columbus, OH 43210
E-mail: furnstahl.1@osu.edu

R. P. SPRINGER
Department of Physics, Duke University
Durham, NC 27708
E-mail: rps@phy.duke.edu

Summary of the open discussion closing the first day of the workshop.

1 Overview

A year ago, the speakers on day one of the first "Nuclear Physics with Effective Field Theory" workshop presented a broad spectrum of positive and negative opinions on the prospects for a useful and systematic effective field theory (EFT) for two-nucleon physics.[1] At the end of a long day of presentations and debate, an open discussion revealed wide (though not complete) consensus that nucleon-nucleon EFT's worked and observables were regularization scheme independent.[2] There was particular enthusiasm for the regularization and power counting scheme proposed by Kaplan, Savage, and Wise (KSW). A sizable laundry list of calculations was proposed for the coming year.

One year later, much has been accomplished. The first day of the second Nuclear Physics EFT workshop concentrated on issues in the two-nucleon sector.[3–10] It was gratifying to see how much progress has been made on the laundry list from the first workshop, particularly in the KSW scheme. In the afternoon, results were summarized for a wide range of calculations in next-to-leading-order (NLO) for NN scattering, deuteron properties, Compton scattering, and more. Progress on pushing calculations to the next order (NNLO) were encouraging. The morning session focused more generally on power counting in NN EFT's and contrasted the two candidate schemes, referred to as KSW and Weinberg (or Λ) counting.

The open discussion session at the end of the day provided an opportunity to supplement discussions from previous sessions, introduce new topics, sharpen debate on the most important unresolved issues, and seek consensus on the highest priorities for the nuclear physics EFT community. This is a summary of the discussion, which includes many verbatim excerpts (with some artistic license taken to make the order of discussion more coherent).

2 Final Comments from Session 2

The discussion session opened with some comments to supplement the discussion from the afternoon session on two-body systems:

Grießhammer: In γ-deuteron scattering we are sensitive to only the scalar polarizabilities of the nucleon. We have the electric scalar nuclear polarizability fixed, not as a free parameter, but as calculated in chiral perturbation theory by Meißner, *et al.* If we treat it as a free parameter, the back angle skewness of the data and the large error bars prevent an accurate determination of it through fitting. We can use the magnetic susceptibility to get skewness, but numerically the magnetic scalar nuclear polarizability is very small and looks like NNLO although it is NLO in the power counting. In extracting it one therefore gets huge error bars. The higher the energy we go to, the better such a determination will be, but it will require an NNLO calculation.
Hemmert: One must go to one order higher to be sensitive to n-p splitting. Only that way can neutron polarizabilities be extracted.
Phillips: One needs to do an order Q^4 calculation to fit polarizabilities.

3 Power Counting Schemes

While discussion at the first workshop a year ago focused heavily on regularization schemes (cut-off vs. dimensional regularization), those issues are largely settled and debate at this workshop shifted to power counting schemes. Thus, the argument has shifted from the question of "Is there any EFT for nuclear physics?" to "Which power counting scheme is most consistent?" "Which converges best?" and "Which is easiest to apply?" The major issues discussed during the day for the KSW and Weinberg power counting schemes included the role of perturbative (KSW) vs. nonperturbative (Weinberg) pions, the consistency of power counting in the Weinberg scheme, the breakdown scale for each, and how to fit parameters. The following excerpts typified the debate. The ultimate resolution to many of these questions await definitive NNLO calculations (expected soon) and further calculations with Weinberg counting.

3.1 KSW and Perturbative Pions

While the virtues of an EFT with pions entering only perturbatively are clear, there was considerable skepticism about whether it worked.

Gegelia: The whole method breaks down at 80 MeV. Including nonleading orders nonperturbatively is necessary. Otherwise, you encounter very large

higher order corrections. There are cancellations between orders, and a fine tuning, not a power counting. NNLO is not necessary and instead a nonperturbative treatment is needed.
Mehen: David Kaplan addressed this in his talk. The uncertainty in the NLO calculation can be large. We see that dramatic oscillations quickly settle down to correct predictions. Just summing C_2 nonperturbatively does not help.
Lutz: You always have large cancellations.
Cohen: Certain observables do much better when higher order terms are iterated to all orders. For example, when the shape term is small and the effective range is included to all orders. Is this an accident or is it generic?
Savage: It is not true for $np \to d\gamma$. Effective range theory fails at NLO.

3.2 Convergence in the KSW Power Counting Scheme

Previous theoretical arguments by KSW suggested a breakdown scale of $\Lambda_{NN} \approx$ 300 MeV in their approach. An important issue is whether this scale is, in fact, a barrier.

Savage: In the last few months, Mehen and Stewart have increased Λ_{NN}. This should be emphasized as nuclear matter is central to this program.
Furnstahl: What is the status on that?
Mehen: We have theoretical arguments for why the 300 MeV limit is wrong. We know that in two nucleon scattering 500 MeV is O.K., but what about other observables? We are working on it. $^1S_0$ is done. $^3S_1$ should be done in the next few months. Then there is $S-D$ mixing, deuteron properties, quadrupole moments, etc. The Binger calculation may be helpful for this.
Phillips: Take the Binger calculation and plug in C_2^{S-D}.
Stewart: There is too much freedom in the phase shift fits for this to be definitive. We need to use these to predict other observables and see how it does.
Phillips: I would like to see a comparison of A_{-1}/A_0 versus A_0/A_1, or some other measure of convergence, rather than just reproduction of the data.
Lutz: I have done the problem analytically and the convergence is great.
Phillips: I'd like to see that calculation.

3.3 Weinberg (or Λ) Counting

Wise: Without pions, Weinberg is fine. Kaplan has shown that for power law divergent things there is an issue. But are logarithms in the theory with pions an issue? Nobody knows.
Furnstahl: How do we resolve this?

Wise: Try summing the pions to all orders. We need a naive dimensional analysis argument.
Furnstahl: Tom Cohen said in his talk that with KSW there is an expansion with error estimates, which Weinberg counting does not have. But when does Weinberg not work?
Cohen: We have six parameters, and do a fit to 6th order. How well do we do? KSW is an expansion in amplitude, Weinberg an expansion in potential. In the latter case, what do you expect for the amplitude? I don't know. Kaplan has worked some of this out. Weinberg rules are not rules for amplitudes.
Furnstahl: Steele sees Weinberg fails at 300 MeV, just like everything else.
van Kolck: We don't know if it is better than KSW or not.
Cohen: Add parameters in the potential and it gets better.
Furnstahl: I have seen higher order potentials which do not get better.
Phillips: This may depend on how the fit is done.
Cohen: With a calculation on the phase shifts done, we can see how the rest of the physics does based on this as the momentum increases. If you have a potential model or Weinberg without pions and a cutoff going to infinity, you can get a number. But you cannot get an error. With KSW, you know the error.
van Kolck: Yes, we do have an error. The error is in the potential.
Mehen: We do know the error in Weinberg because we can identify when a counterterm enters and know from its order what the error will be.
van Kolck: We have C_2, plus some other junk that KSW does not have, but not C_4, and we know the error.
Cohen: Then you have an explicit cutoff.
Mehen: The cutoff should not enter into the error estimate.
Cohen: It *shouldn't*, but how does it actually feed through the analysis?

4 New Topic: Wilson Renormalization Group

The open discussion session also provided an opportunity for new ideas not addressed by the scheduled talks. Michael Birse took advantage of the forum to present insights from a Wilsonian renormalization group approach. He identified the two fixed points in nonrelativistic nuclear two-body scattering and discussed the conditions for which it is appropriate to expand about one or the other. [See Ref. 11 for more details.] This discussion followed:

Furnstahl: So what do we learn from this analysis?
Birse: We can decide which power counting to use based on this.
van Kolck: If you are suggesting that Weinberg is wrong to iterate the potential to all orders and use a cutoff, then I disagree with you.

Birse: But the RG flow is clearly controlled by the fixed points, not by Weinberg's power counting.
Perry: In particular, if the power counting is incorrect, operators which should be irrelevant turn out to be relevant.
Wise: I think what Birse has shown is that iterating C_2 to all orders is not more important than including C_4.
Cohen: So we have two fixed points, and we can expand around these. If I am near a bound state which is near zero, KSW is sensible. But is this new? We knew that last year.
Perry: Wilsonian RG treatments have been known in the condensed matter community for a long time and seem to have been ignored by particle and nuclear types. We now know about non-Gaussian fixed points and know that linearizing around them is better than going into some complicated equation and looking for points. The Wilsonian method will also make these problems interesting to condensed matter people, as well as cleaning up some issues.
Gegelia: If you re-expand around a nontrivial fixed point, you get that Weinberg's counting is O.K.
Birse: But then you have large coefficients.
Gegelia: There is nothing wrong with Weinberg's power counting.
Birse: The question is whether to expand in K or in 1/K, corresponding to the two types of fixed points. Which converges most quickly in powers of energy? Here we have a way of making a practical choice and establishing a range of validity. For low momentum, it doesn't matter, but for higher momentum, and large a, the flow is to the other point.

5 What Should Be Calculated?

The discussion next turned to more global issues, such as how to convince the community of nuclear physicists that effective field theory approaches are important, and what should be the priority calculations.

The participants were asked to consider the advantages of an EFT approach. The *a priori* estimate of errors and systematic improvement of results was considered of primary importance. Despite the tremendous phenomenological success of potential models, the physics is often obscured or misleading. An EFT can identify what is cleanly known (long-distance physics, constrained by symmetries) and what is generic short-distance physics, to be parametrized. Thus one obtains a better understanding of the "knob turning" in potential models. The ability to fit using one process and then to make predictions for other processes, particularly inelastic processes, was also stressed. Finally, only an EFT approach offers the hope of direct connection to QCD through matching.

The question of what calculation(s) would be most convincing focused on areas where potential models are not already highly successful. Some suggestions were (see the excerpts below):

- gamma deuteron scattering,
- pion photoproduction on deuteron,
- spin observables,
- neutron gas,
- do higher order everywhere in Mehen/Stewart picture to see if the 500 MeV convergence scale survives.

The importance of a global set of calculations was also stressed.

To spur the discussion, Furnstahl showed a graph with error plots of ${}^1S_0$ nucleon-nucleon scattering as a "reality check." It compared the errors with phase shift data from state-of-the-art potential models to those from state-of-the-art EFT results (namely NLO calculations in the KSW scheme). The potential models, for example the updated Reid potential, do much better at describing the data than the EFT to NLO. The following discussion ensued:

Shoresh: But how many parameters does this take?
Steele: Two in this channel with the Reid (RSC) potential.

Holstein: Arthur Kerman would say "Thirty years ago I did this point interaction analysis and we have learned a lot since then. There is no reason to turn back the clock." How do we answer this?
Perry: This would be a foolish statement to make. He did not invent EFT and did not have something systematic.
Holstein: So, what we have gained is power counting?
Perry: Yes, if indeed we do have it.
Savage: We now have an a priori estimate of errors. We have actual predictions. Potential models did not agree with themselves or the data until after the data arrived and tuning was done.
Cohen: So we are doing badly but we know just how badly? Potential modelers just don't understand why they do as well as they do. We need to be systematic and useful.
Savage: The quadrupole moment potential model prediction is off by 6%. That is not great by potential model standards. We need to make predictions which can be measured: Observables in photon-deuteron scattering, the many precise measurements which will come out of the TUNL-FELL project at Duke.

Lepage: I think we need to concentrate on what is new which can be predicted.
van Kolck: Pion photoproduction was predicted before the data. We were a factor of two lower than conventional predictions. New data appeared, and we were right. But this is just a corner of nuclear physics.
Wise: The trouble is that this is an old subject, with much data, and much time to fine-tune models and prediction.
Kaplan: We can go to three and more body systems.
Meißner: It is misleading to say that the data is all there. In the pion-nucleon scattering sector, Nijmegen was based on neutral pion measurements which were really very crude. But a chiral perturbation theory prediction, which was substantially different from the Nijmegen one, turned out to be correct when the process was re-measured. The data is not as good as you think. Probably half of it is terrible.

Lepage: We need a list of important calculations to do.
Savage: Higher order static predictions. NNLO with pions. Get form factors also. N^3LO without pions. This will give 1% precision.
Phillips: The BLAST program will be doing electromagnetic probes of light nuclei to high precision. We might look for observables there. In particular, is parity violating electron scattering on the deuteron not well predicted by potential models?
Cohen: If something is not well predicted by potential models, will it be in EFT?
Phillips: If radiative pions and relativistic effects are small, what will EFT do for you? The answer is, give you counterterms when external fields are included.
Cohen: If there are counterterms, what is the difference between this and a parameter fit?
Savage: Counterterms found from one observable can be used in the prediction of another. For example, if you fix the counterterm in parity violation electron scattering, it can be used in $\nu d \to \nu d$.
Someone: I think that pion production is where we should be going. Chiral perturbation theory is good, but here there is the Λ_{NN} bound. That scale needs to be extended so that we have predictions in this regime.
V. Kolck: There is the A_y puzzle.
Phillips: The three body system cannot be satisfied by tweaking the two nucleon forces according to the LANL people.

Kaplan: How about looking at a neutron gas, where a is huge and the density is controllable. [Reference was made to George Bertsch's challenge problem.]

Or how about atomic systems with low density where the magnetic field can be used to tune a. Then there would be the hierarchy of scales with a much greater than the interatomic spacing which is much greater than the effective range.
Cohen: There are already experiments on this.
Wise: The momentum is too low. $ap << 1$ still.
Kaplan: They can try to dial into the nonperturbative regime.

Furnstahl: Ultimately, I think we need to make a connection to QCD and have a hope of finding the EFT constants (say through lattice calculations) before the method will have its greatest impact.

6 The Future

Keister: From a funding perspective, it is important to know under what conditions this community would leave these problems and do something else.
Cohen: If we do NNLO and it goes nowhere.
Keister: There does not seem to be agreement about the convergence even at this stage.
Meißner: This may well depend upon the observables. What we really need to see is if we can establish a consistent picture. For instance, in chiral perturbation theory there are observables for which the expansion does not converge, and then you try to understand the physics from which this comes. You do not trash the whole program because of one observable. We need to think globally. We should not stop because Nijmegen does NN scattering better.
Keister: We just don't want this to take on a life of its own for its own sake.
Savage: That would be true if in thirty years we are still looking at NNLO.
Grießhammer: I think that the higher order, higher precision predictions on the NN system are not as important as going to three and more bodies. After all, except for many body terms, the power counting must be expected to be the same as in the two body system. If two-bodies does work then three and many-bodies should work.
Steele: A lot of weight is given to the Weinberg statements about understanding three and four bodies. The physical understanding, the Phillips line, etc., has impact, even if the calculations are not more precise than potential models.

7 Adjournment

With many issues still active, the workshop was adjourned. Heated discussions continued at dinner and throughout the second day!

Acknowledgments

We thank the organizers of the workshop for a stimulating environment and the participants for lively and fruitful discussions.

References

1. "Nuclear Physics with Effective Field Theory," R. Seki, U. van Kolck, and M. J. Savage, eds. (World Scientific, Singapore, 1998.
2. "Discussion Session," R. J. Furnstahl and H.-W. Hammer, in ref. 1.
3. U. van Kolck, "Aims of the Workshop," *this volume.*
4. D. Kaplan, "EFT Expansion for Nonrelativistic Scattering," *this volume.*
5. T. D. Cohen, "Perturbative Pions and the Effective Range Expansion," *this volume.*
6. J. Steele, "Convergence Issues in NN EFT," *this volume.*
7. G. Rupak, "NN EFT at NNLO," *this volume.*
8. T. Mehen, "NN EFT at Higher Order," *this volume.*
9. M. Wise, "2-Body Systems in EFT," *this volume.*
10. M. Savage, "2-Body Systems in EFT," *this volume.*
11. M. C. Birse, J. A. McGovern, and K. G. Richardson, "A renormalization-group treatment of two-body scattering," `hep-ph/9807302`.

SESSION 4

Session Chair: *F. Ravndal (University of Oslo)*
Notes taken by: *J. McGovern (University of Manchester)*

GAMMA-DEUTERON SCATTERING

S.R. BEANE
Department of Physics, University of Maryland,
College Park, MD 20742-4111
E-mail: sbeane@physics.umd.edu

We discuss a recent computation of Compton scattering on the deuteron at photon energies of order the pion mass. An interaction kernel is computed in baryon chiral perturbation theory and sewn to phenomenological deuteron wave functions. As an appetizer, we consider a computation of the pion-deuteron scattering length using this method.

1 Introduction

Precise calculations of hadron processes are possible only where a small dimensionless expansion parameter is identified. This is the main motivation behind the ongoing intense effort to develop a perturbative theory of nuclear interactions [1]. The dimensionless parameters relevant to low energy QCD and therefore to nuclear physics consist of ratios of external momenta to various characteristic energy scales, like the nucleon mass. Effective field theory is the technology which develops a hierarchy of scales into a perturbative expansion of physical observables. In a system with broken symmetries this technology is especially powerful. When a continuous symmetry is spontaneously broken there are massless Goldstone modes which dominate the low-energy dynamics and couple only derivatively. Hence at energies small relative to the characteristic symmetry breaking scale, the interactions of the Goldstone bosons can be ordered in an effective Lagrangian which is constrained by chiral symmetry and in which each operator contains a nonvanishing number of derivatives acting on the pion fields. Observables computed from the effective Lagrangian are therefore power series in momenta, with the non-analyticities required by perturbative unitarity.

In this paper we describe several recent effective field theory calculations whose objective is to extract nucleon properties from nuclear scattering processes in a systematic way. We first discuss the paradigmatic problem of the pion-deuteron scattering length and its dependence on nucleon parameters. We then describe a recent calculation of Compton scattering on the deuteron at photon energies of order the pion mass. Here the ultimate objective is to learn about neutron polarizabilities from nuclear Compton scattering. The basic power-counting scheme is reviewed in Sec. 2. In Sec. 3 we discuss the problem of pion-deuteron scattering at threshold as a heuristic tool. Sec. 4 is dedicated to Compton scattering on the deuteron. We conclude in Sec. 5.

2 Weinberg power-counting

At energies well below the chiral symmetry breaking scale, $\Lambda_\chi \sim 4\pi f_\pi \sim M \sim m_\rho$, the interactions of pions, photons and nucleons can be described systematically using an effective field theory. This effective field theory, known as chiral perturbation theory (χPT), reflects the observed QCD pattern of symmetry breaking. In QCD the chiral $SU(2)_L \times SU(2)_R$ symmetry is spontaneously broken. Here we are interested in processes where the typical momenta of all external particles is $p \ll \Lambda_\chi$, so we identify our expansion parameter as p/Λ_χ. In QCD $SU(2)_L \times SU(2)_R$ is softly broken by the small quark masses. This explicit breaking implies that the pion has a small mass in the low-energy theory. Since m_π/Λ_χ is then also a small parameter, we have a dual expansion in p/Λ_χ and m_π/Λ_χ. We take Q to represent either a small momentum *or* a pion mass.

In few-nucleon systems, a complication arises in χPT due to the existence of shallow nuclear bound states and related infrared singularities in A-nucleon reducible Feynman diagrams evaluated in the static approximation [2]. The fundamental problem is that nuclear physics introduces a new mass scale, the nuclear binding energy, which is very small compared to a typical hadronic scale like Λ_χ. One way to overcome this difficulty is to adopt a modified power-counting scheme in which χPT is used to calculate an effective potential which generally consists of all A-nucleon irreducible graphs. The S-matrix, which includes all reducible graphs as well, is then obtained through iteration by solving a Lippmann-Schwinger equation [2]. This version of nuclear effective theory is known as the Weinberg formulation. To date the Weinberg formulation can be carried through rigorously only using finite cutoff regularization [3,4,5,6]. This limitation has spawned an intense theoretical effort geared at formulating an effective field theory for low-lying bound states which is verifiably (analytically) consistent in the sense of renormalization [1]. One result of this effort is a new power-counting scheme in which all nonperturbative physics responsible for the presence of low-lying bound states arises from the iteration of a single operator in the effective theory, while all other effects, including all higher dimensional operators *and* pion exchange, are treated perturbatively [7,8]. This version of the effective theory is known as the Kaplan-Savage-Wise (KSW) formulation. This is relevant here because Compton scattering on the deuteron has been computed to next-to-leading order in the KSW formulation [9]. We will discuss this result and its relation to our calculation. A comprehensive and up-to-date review of nuclear applications of effective field theories can be found in Ref. [1].

In the interactions of the deuteron with pionic and electromagnetic probes, Weinberg power-counting is encoded in the following simple diagrammatic rules:

- A nucleon propagator contributes Q^{-1};
- A pion propagator contributes Q^{-2};
- Each derivative or power of the pion mass at a vertex contributes Q. Therefore, an operator insertion from the π-N Lagrangian with n derivatives or powers of m_π, $\mathcal{L}_{\pi N}^{(n)}$, contributes Q^n;
- Each loop integral contributes Q^4;
- Each deuteron wavefunction, $\triangle$ or $\bigtriangledown$, contributes $Q^{-\frac{1}{2}}$.

It should be noted that typical nucleon momenta inside the deuteron are small—on the order of $\sqrt{MB}$ or m_π, with B the deuteron binding energy—and consequently, *a priori* we expect no convergence problems in the χPT expansion of any low-momentum electromagnetic or pionic probe of the deuteron. Although in principle we could use wavefunctions computed in χPT, we will consider wavefunctions generated using modern nucleon-nucleon potentials. Generally we find that any wavefunction with the correct binding energy gives equivalent results to within the theoretical error expected from neglected higher orders in the chiral expansion. Presumably we are insensitive to short distance components of the wavefunction because we are working at low energies and the deuteron is a large object.

Weinberg power-counting has led to fruitful computation of many pionic and photonic probes of the two-nucleon system [1], including the pion-deuteron scattering length [10,11], neutral pion photoproduction on the deuteron at threshold [12], Compton scattering on the deuteron [13], as well as pn radiative capture [14] and the solar burning process $pp \to de^+\nu$ [15].

3 Appetizer: threshold π-deuteron scattering

Effective field theory relates scattering processes involving a single nucleon to nuclear scattering processes. For instance, one can relate π-N scattering to π-nucleus scattering [10,11]. The non-perturbative effects responsible for nuclear binding are accounted for using phenomenological nuclear wavefunctions, as noted above. One can compute matrix elements using a variety of wavefunctions in order to ascertain the theoretical error induced by the off-shell behavior of different wavefunctions.

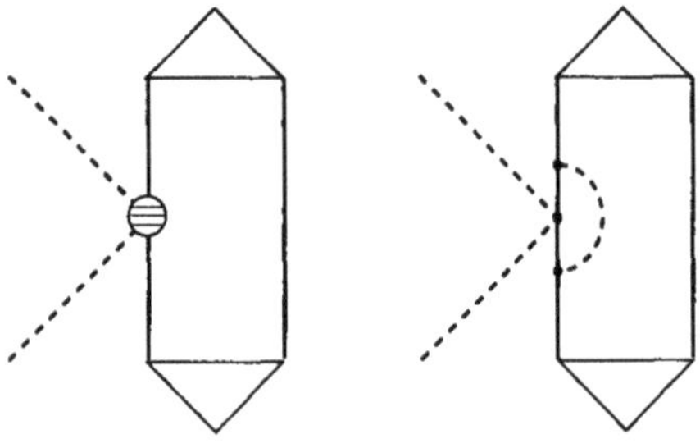

Figure 1: Contributions to the π-deuteron scattering length at order Q^2 (a) and Q^3 (b). The dots are vertices from $\mathcal{L}_{\pi N}^{(1)}$ and the sliced blob is from $\mathcal{L}_{\pi N}^{(2)}$. All topologies are not shown.

To $O(Q^3)$ in χPT the π-deuteron scattering length can be written as [10]

$$a_{\pi d} = \frac{(1+\mu)}{(1+\mu/2)}(a_{\pi n} + a_{\pi p}) + a^{(2a)} + a^{(2b,2c)}, \tag{1}$$

where $\mu \equiv m_\pi/M$ is the ratio of the pion and the nucleon mass and the π-N scattering lengths have the decomposition

$$a_{\pi n} + a_{\pi p} = 2a^+ = a^{(1a)} + a^{(1b)} \tag{2}$$

where a^+ is the isoscalar S-wave scattering length and the superscripts refer to the figures. The various diagrammatic contributions to $a_{\pi d}$ are illustrated in Fig. 1 and Fig. 2. The leading contribution, Fig. 1(a), has three nucleon propagators (Q^{-3}), a vertex from $\mathcal{L}_{\pi N}^{(2)}$ (Q^2), one loop (Q^4) and two deuteron wavefunctions, (Q^{-1}) giving a total of $Q^{-3+2+4-1} = Q^2$. Fig. 1(b) has five nucleon propagators (Q^{-5}), one pion propagator (Q^{-2}), three vertices from $\mathcal{L}_{\pi N}^{(1)}$ (Q^3), two loops (Q^8) and two deuteron wavenfunctions, (Q^{-1}) giving a total of $Q^{-5-2+3+8-1} = Q^3$. One can further verify that the graphs of Fig. 2 are $O(Q^3)$. Together, Figs. 1 and 2 are all that contribute at $O(Q^3)$ at threshold.

The contributions to $a_{\pi d}$ from the graphs of Fig. 2 are:

$$a^{(2a)} = -\frac{m_\pi^2}{32\pi^4 f_\pi^4(1+\mu/2)}\langle\frac{1}{\vec{q}^{\,2}}\rangle_{wf} \tag{3}$$

$$a^{(2b,2c)} = \frac{g_A^2 m_\pi^2}{128\pi^4 f_\pi^4(1+\mu/2)}\langle\frac{\vec{q}\cdot\vec{\sigma}_1\vec{q}\cdot\vec{\sigma}_2}{(\vec{q}^{\,2}+m_\pi^2)^2}\rangle_{wf} \tag{4}$$

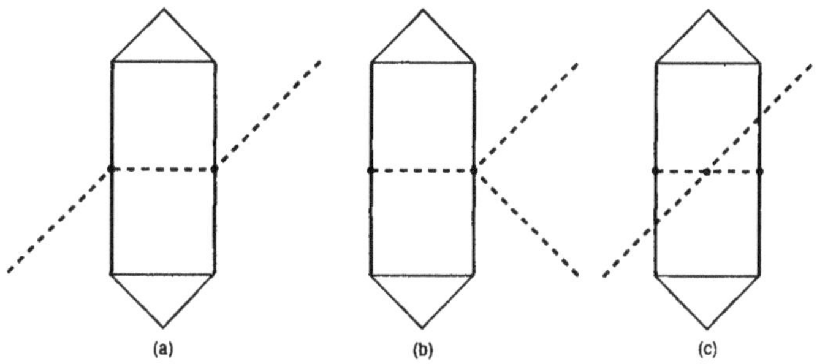

Figure 2: Two loop contributions (with an exchanged pion) to the π-deuteron scattering length at Q^3.

where $\langle\vartheta\rangle_{wf}$ indicates that ϑ is sandwiched between deuteron wavefunctions. These matrix elements have been evaluated using a cornucopia of wavefunctions; results are displayed in table 1. Clearly $a^{(2a)}$ dominates. This is the result of the shorter range nature of $a^{(2b,2c)}$. It is important to stress that the dominant contribution from these graphs is quite independent of the wavefunction used. This implies that the χPT approach, which relies on the dominance of the pion-exchange, is useful in this context.

To $O(Q^3)$ in χPT[16]:

$$4\pi(1+\mu)a^+ = \frac{m_\pi^2}{f_\pi^2}\left(-4c_1+2c_2-\frac{g_A^2}{4M}+2c_3\right)+\frac{3g_A^2 m_\pi^3}{64\pi f_\pi^4}, \tag{5}$$

where the c_i are low-energy constants from $\mathcal{L}_{\pi N}^{(2)}$[a]. The sole undetermined parameter entering the $O(Q^3)$ computation of $a_{\pi d}$ is therefore a combination of c_1, c_2 and c_3:

$$\Delta \equiv -4c_1 + 2(c_2+c_3). \tag{6}$$

There is recent experimental information about both the π-N and π-deuteron scattering lengths[17,18]. Since a^+ involves constants that are not fixed by chiral symmetry we can use experimental information about π-deuteron scattering to predict a^+; the recent Neuchatel-PSI-ETHZ (NPE) pionic deuterium measurement[17] gives

$$a_{\pi d} = -0.0259 \pm 0.0011\, m_\pi^{-1}. \tag{7}$$

[a]It should be stressed that to this order there appear large cancellations between the individual terms[16] which lead one to suspect that a calculation at $O(Q^4)$ should be performed to obtain a more precise prediction for this anomalously small observable.

Table 1: Contributions of Fig. 2 for various deuteron wavefunctions in units of m_π^{-1}. We use $f_\pi = 92.4\,\mathrm{MeV}$, $g_A = 1.32$ and $m_{\pi^+} = 139.6\,\mathrm{MeV}$.

wf	$a^{(2a)}$	$a^{(2b,2c)}$
Bonn [19]	−0.02021	−0.0005754
ANL-V18 [20]	−0.01960	−0.0007919
Reid-SC [21]	−0.01941	−0.0008499
SSC [22]	−0.01920	−0.0006987

For the contributions of Fig. 2 we take the average of the $a^{(2a)}$ and $a^{(2b,2c)}$ values in table 1:

$$a^{(2a+2b+2c)} = -0.0203\, m_\pi^{-1}. \tag{8}$$

We then find from Eq. (1):

$$a^+ = -(2.6 \pm 0.5) \cdot 10^{-3} m_\pi^{-1}. \tag{9}$$

Note that although $a^{(2b,2c)}$ is small, there is a strong cancellation between $a^{(2a)}$ and $a_{\pi d}$ which leads to a sensitivity to $a^{(2b,2c)}$. Our value for a^+ is not consistent with the Karlsruhe-Helsinki value [23],

$$a^+ = -(8.3 \pm 3.8) \cdot 10^{-3} m_\pi^{-1}, \tag{10}$$

or the new NPE value deduced from the strong interaction shifts in pionic hydrogen and deuterium, which is small and positive [18]

$$a^+ = (0...5) \cdot 10^{-3} m_\pi^{-1}. \tag{11}$$

The result Eq. (9) agrees, however, with the value obtained in the SM95 partial-wave analysis [24], $a^+ = -3.0 \cdot 10^{-3} m_\pi^{-1}$.

Given the ambiguous experimental situation regarding a^+, it seems most profitable to turn our formula around and use the π-deuteron scattering data to constrain Δ. We can write

$$\Delta = \frac{2\pi f_\pi^2}{m_\pi^2}(1+\mu/2)\{a_{\pi d} - (a^{(2a)} + a^{(2b,2c)})\} + \frac{g_A^2}{4M}(1 - \frac{3Mm_\pi}{16\pi f_\pi^2}). \tag{12}$$

Using Eqs. (3), (4) and (7) we find

$$\Delta = -(0.08 \pm 0.02)\,\mathrm{GeV}^{-1}, \tag{13}$$

where we have taken into account the error in the determination of $a_{\pi d}$.

Table 2: Values of the LECs c_i in GeV^{-1} for $i = 1, \ldots, 3$. Also given are the central values (cv) and the ranges for the c_i from resonance exchange. The * denotes an input quantity.

i	c_i	c_i^{Res} cv	c_i^{Res} ranges
1	-0.93 ± 0.10	-0.9^*	–
2	3.34 ± 0.20	3.9	$2 \ldots 4$
3	-5.29 ± 0.25	-5.3	$-4.5 \ldots -5.3$
Δ	-0.18 ± 0.75	0.8	$-3.0 \ldots +2.6$

In table 2 we give values of the relevant c_i obtained from a realistic fit to low-energy pion-nucleon scattering data and subthreshold parameters [25]. Central values lead to $\sigma(0) = 47.6\,\text{MeV}$ and $a^+ = -4.7 \cdot 10^{-3} m_\pi^{-1}$. These values of the c_i give the conservative determination:

$$\Delta = -(0.18 \pm 0.75)\,\text{GeV}^{-1}. \tag{14}$$

Also shown in table 2 are values of the c_i deduced from resonance saturation. It is worth mentioning that an independent fit to pion-nucleon scattering including also low-energy constants related to dimension three operators finds results consistent with the fit values of table 2 [26].

To summarize, we have shown that recent precise data on the π-deuteron scattering length can be used to constrain a combination of dimension two low-energy constants of the pion-nucleon chiral Lagrangian. This constraint can be improved by going to $O(Q^4)$ in the chiral expansion. Thus this simple calculation provides an example of how using effective field theory one can extract nucleon properties from a nuclear process in a systematic way.

4 Entrée: γ-deuteron scattering

4.1 Motivation

Nucleon Compton scattering has been studied in χPT in Ref. [27], where the following results for the polarizabilities were obtained to order Q^3:

$$\begin{aligned} \alpha_p = \alpha_n &= \frac{5e^2 g_A^2}{384\pi^2 f_\pi^2 m_\pi} = 12.2 \times 10^{-4}\,\text{fm}^3; \\ \beta_p = \beta_n &= \frac{e^2 g_A^2}{768\pi^2 f_\pi^2 m_\pi} = 1.2 \times 10^{-4}\,\text{fm}^3. \end{aligned} \tag{15}$$

Here we have used $g_A = 1.26$ for the axial coupling of the nucleon, and $f_\pi = 93$ MeV as the pion decay constant. Note that the polarizabilities are *predictions* of χPT at this order. The $O(Q^3)$ χPT predictions diverge in the chiral limit because they arise from pion loop effects. In less precise language, the power-counting of χPT implies that polarizabilities are dominated by the dynamics of the long-ranged pion cloud surrounding the nucleon, rather than by short-range dynamics. The polarizabilities should thus provide a sensitive test of chiral dynamics. At the next order in the chiral expansion, Q^4, there are contributions to the polarizabilities from undetermined parameters which must be fixed independently [28]. These counterterms account for short-range contributions to the nucleon structure.

Recent experimental values for the proton polarizabilities are [29] [b]

$$\begin{aligned} \alpha_p + \beta_p &= 13.23 \pm 0.86^{+0.20}_{-0.49} \times 10^{-4}\,\text{fm}^3, \\ \alpha_p - \beta_p &= 10.11 \pm 1.74^{+1.22}_{-0.86} \times 10^{-4}\,\text{fm}^3, \end{aligned} \tag{16}$$

where the first error is a combined statistical and systematic error, and the second set of errors comes from the theoretical model employed. These values are in good agreement with the χPT predictions.

On the other hand, the neutron polarizabilities are difficult to obtain experimentally due to the absence of suitable neutron targets and so the corresponding χPT prediction is not well tested. One way to extract neutron polarizabilities is to consider Compton scattering on nuclear targets. Consider coherent photon scattering on the deuteron. The cross section in the forward direction naively goes as:

$$\left.\frac{d\sigma}{d\Omega}\right|_{\theta=0} \sim (f_{Th} - (\alpha_p + \alpha_n)\omega^2)^2. \tag{17}$$

The sum $\alpha_p + \alpha_n$ may then be accessible via its interference with the dominant Thomson term for the proton, f_{Th} [30]. This means that with experimental knowledge of the proton polarizabilities it may be possible to extract those for the neutron. Coherent Compton scattering on a deuteron target has been measured at $E_\gamma = 49$ and 69 MeV by the Illinois group [31]. An experiment with tagged photons in the energy range $E_\gamma = 84.2 - 104.5$ MeV is under analysis at Saskatoon [32], while data for E_γ of about 60 MeV is being analyzed at Lund [33].

Clearly the amplitude for Compton scattering on the deuteron involves mechanisms other than Compton scattering on the individual constituent nucleons. Hence, extraction of nucleon polarizabilities requires a theoretical calculation of Compton scattering on the deuteron that is under control in the

[b] These are the result of a model-dependent fit to data from Compton scattering on the proton at several angles and at energies ranging from 33 to 309 MeV.

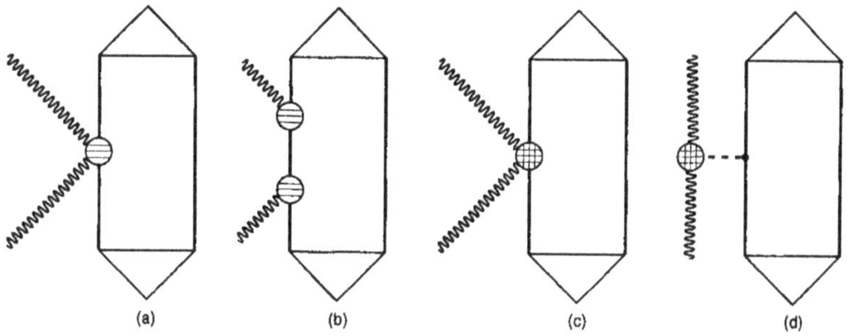

Figure 3: One loop graphs which contribute to Compton scattering on the deuteron at order Q^2 (a) and at order Q^3 (b-d) (in the Coulomb gauge). The sliced and diced blobs are from $\mathcal{L}^{(3)}_{\pi N}$ (c) and $\mathcal{L}^{(4)}_{\pi\gamma}$ (d). Crossed graphs are not shown.

sense that it accounts for *all* mechanisms to a given order in a systematic expansion in a small parameter. There exist a few calculations of this reaction in the framework of conventional potential models [34,35,36]. These calculations yield similar results if similar input is supplied, but typically mechanisms for nucleon polarizabilities and two-nucleon contributions are not treated consistently. We will see that χPT provides an alternative framework where this drawback can be eliminated.

In the remainder of this paper I will review a recent computation of Compton scattering on the deuteron for incoming photon energies of order 100 MeV in the Weinberg formulation. As in the computation of the pion-deuteron scattering length, baryon χPT is used to compute an irreducible scattering kernel to order Q^3, which is then sewn to external deuteron wavefunctions.

4.2 Compton scattering to $O(Q^3)$

The Compton amplitude we wish to evaluate is (in the γd center-of-mass frame):

$$T^{\gamma d}_{M'\lambda' M\lambda}(\vec{k}\,',\vec{k}) = \int \frac{d^3p}{(2\pi)^3}\,\psi_{M'}\left(\vec{p}+\frac{\vec{k}-\vec{k}\,'}{2}\right)\,T^{\gamma d\ c.m.}_{\gamma N_{\lambda'\lambda}}(\vec{k}\,',\vec{k})\,\psi_M(\vec{p})$$
$$+\int \frac{d^3p\,d^3p'}{(2\pi)^6}\,\psi_{M'}(\vec{p}\,')\,T^{2N}_{\gamma NN_{\lambda'\lambda}}(\vec{k}\,',\vec{k})\,\psi_M(\vec{p}) \qquad (18)$$

where M (M') is the initial (final) deuteron spin state, and λ (λ') is the initial (final) photon polarization state, and $\vec{k}$ ($\vec{k}\,'$) the initial (final) photon three-

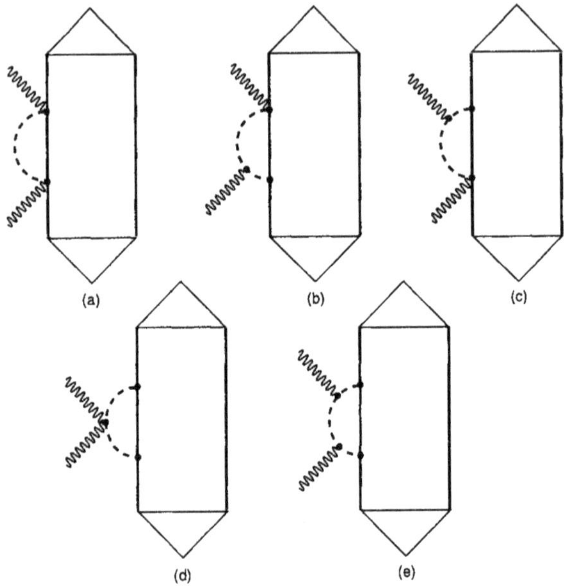

Figure 4: Two loop graphs which contribute to Compton scattering on the deuteron at order Q^3. Crossed graphs are not shown.

momentum, which are constrained to $|\vec{k}| = |\vec{k}\,'| = \omega$. The amplitude $T_{\gamma N}^{\gamma d\ c.m.}$ represents the graphs of figures 3 and 4 where the photon interacts with only one nucleon. Of course this amplitude must be evaluated in the γd center-of-mass frame. The amplitude $T_{\gamma NN}^{2N}$ represents the graphs of Fig. 5 where there is an exchanged pion between the two nucleons.

The leading contribution to Compton scattering on the deuteron is shown in Fig. 3(a). This graph has three nucleon propagators (Q^{-3}), a vertex from the second order pion-nucleon Lagrangian (Q^2), one loop (Q^4) and two deuteron wavefunctions, (Q^{-1}) giving a total of $Q^{-3+2+4-1} = Q^2$. This contribution is simply the Thomson term for scattering on the proton. At next order, $O(Q^3)$, there are several more one loop graphs Fig. 3(b,c,d) and two loop graphs without (Fig. 4) and with (Fig. 5) a pion exchanged between the nucleons. The full amplitudes are given in Ref. [13].

For the wave function ψ we use the energy-independent Bonn OBEPQ wave function parameterization which is found in Ref. [37]. The photon-deuteron T-matrix (18) is then calculated and the laboratory differential cross section evaluated directly from it:

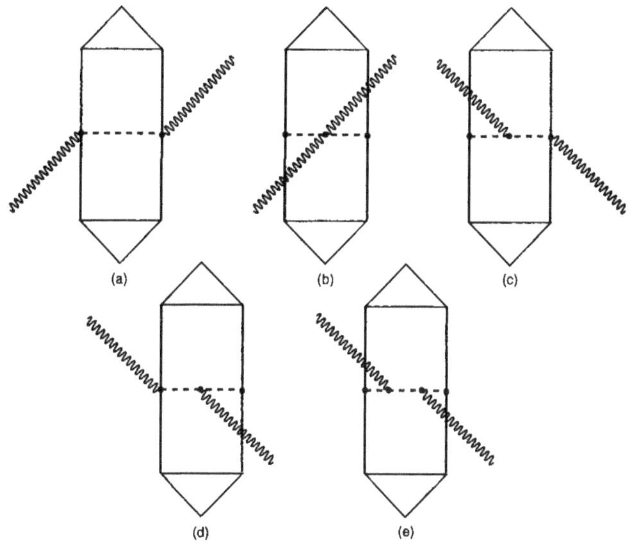

Figure 5: Two loop graphs which contribute to Compton scattering on the deuteron at order Q^3. Crossed graphs are not shown.

$$\frac{d\sigma}{d\Omega_L} = \frac{1}{16\pi^2}\left(\frac{E'_\gamma}{E_\gamma}\right)^2 \frac{1}{6} \sum_{M'\lambda' M\lambda} |T^{\gamma d}_{M'\lambda' M\lambda}|^2, \tag{19}$$

where E_γ is the initial photon energy in the laboratory frame and E'_γ is the final photon energy in the laboratory frame.

Convergence tests indicate that with the numbers of quadratures chosen the cross section evaluated in this fashion is numerically accurate at about the 1% level. Of course, this error does not include the theoretical error from uncertainties due to different deuteron wave functions[c], and the effect of higher-order terms in χPT. The errors due to omitted higher orders in χPT will be discussed further below.

In figures 6, 7 and 8 we display our results at 49, 69, and 95 MeV. For comparison we have included the calculation at $O(Q^2)$, where the second contribution in Eq. (18) is zero, and the γN T-matrix in the single-scattering contribution is given by the Thomson term on a single nucleon. It is remarkable that to $O(Q^3)$ no unknown counterterms appear. All contributions to the kernel are fixed in terms of known pion and nucleon parameters such as

[c]As discussed in detail in Ref. [13], wavefunction errors are minimal and well understood.

m_π, g_A, M, and f_π. Thus, to this order χPT makes *predictions* for Compton scattering.

The curves show that the correction from the $O(Q^3)$ terms gets larger as ω is increased, as was to be expected. Indeed, while at lower energies corrections are relatively small, in the 95 MeV results the correction to the differential cross section from the $O(Q^3)$ terms is of order 50%, although the contribution of these terms to the *amplitude* is of roughly the size one would expect from the power-counting: about 25%. Nevertheless, it is clear, even from these results, that this calculation must be performed to $O(Q^4)$ before conclusions can be drawn about polarizabilities from data at photon energies of order m_π. This is in accord with similar convergence properties for the analogous calculation for threshold pion photoproduction on the deuteron [12].

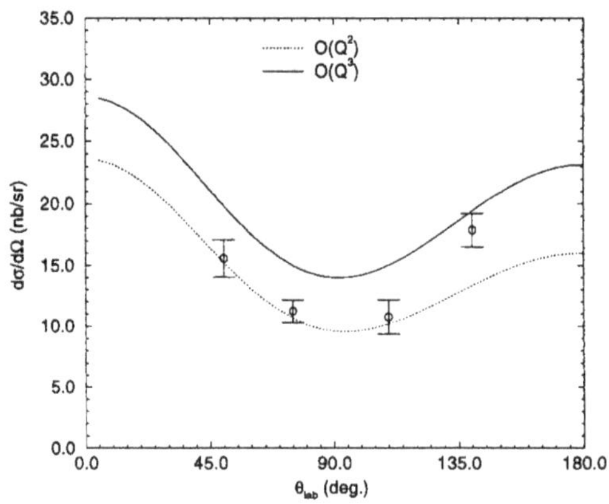

Figure 6: Results of the $O(Q^2)$ (dotted line) and $O(Q^3)$ (solid line) calculations at a photon laboratory energy of 49 MeV.

We have also shown the six Illinois data points at 49 and 69 MeV [31]. Statistical and systematic errors have been added in quadrature. It is quite remarkable how well the $O(Q^2)$ calculation reproduces the 49 MeV data. However, it is clear that the agreement at forward angles is somewhat fortuitous, as there are significant $O(Q^3)$ corrections. Meanwhile, the agreement of the $O(Q^3)$ calculation with the 69 MeV data is very good, although only limited conclusions can be drawn, given that there are only two data points, each with sizeable error bars.

Our results are qualitatively not very different from other existing calcu-

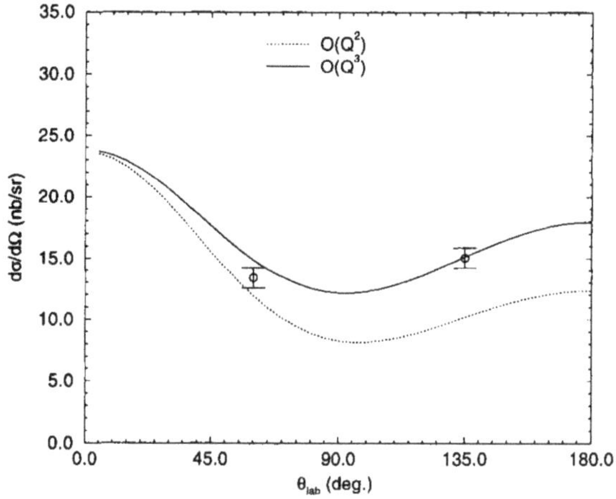

Figure 7: Results of the $O(Q^2)$ (dotted line) and $O(Q^3)$ (solid line) calculations at a photon laboratory energy of 69 MeV.

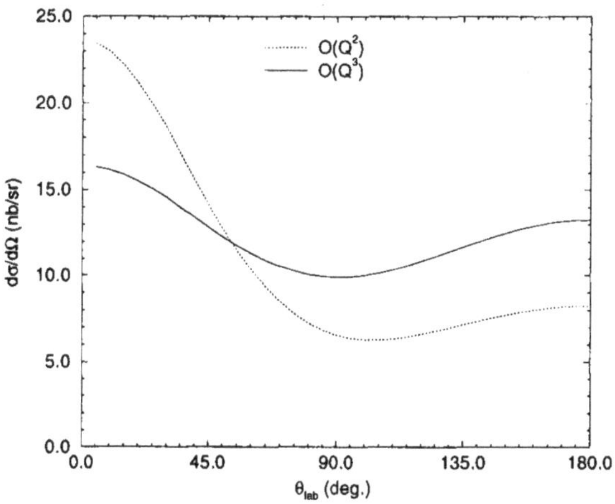

Figure 8: Results of the $O(Q^2)$ (dotted line) and $O(Q^3)$ (solid line) calculations at a photon laboratory energy of 95 MeV.

lations. At 49 and 69 MeV our $O(Q^3)$ results are very close to those in Ref. [34] and a few nb/sr higher, especially at back angles, than those of Refs. [35,36] (which are similar at these energies). At 95 MeV our $O(Q^3)$ result is close to that of Ref. [35], higher by several nb/sr at back angles than Ref. [36], and several nb/sr lower than the calculation with no polarizabilities of Ref. [34] [d]. Comparing to the calculations of deuteron Compton scattering in the KSW formulation of effective field theory [9], we see that the result of Ref. [9] is significantly lower than those presented here at both 49 and 69 MeV. At 49 MeV the agreement of Ref. [9]'s calculation with the data is better than ours. We shall show in the next section that this is partly because 49 MeV is at the lower end of the domain of applicability of the Weinberg formulation. At 69 MeV our calculation does a slightly better job of reproducing the (two) data points available. The qualitative agreement among these calculations is a reflection of the similarities of mechanisms involved. Ours is however the only calculation to incorporate the full single-nucleon amplitude instead of its polarizability approximation. As shown in Fig. 9 our tendency to higher relative cross sections in the backward directions is at least in part due to this feature.

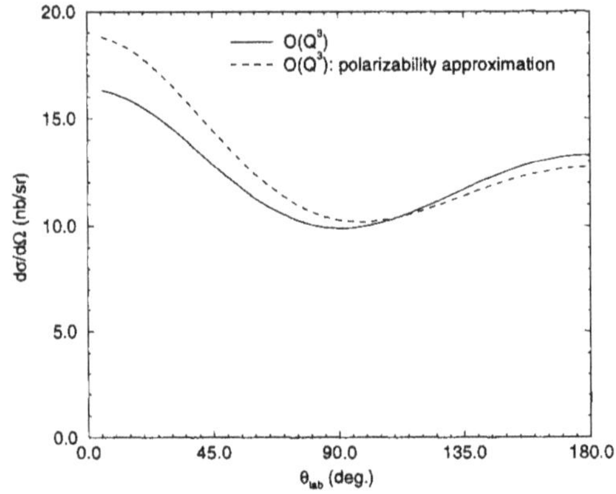

Figure 9: Unpolarized cross section for Compton scattering on the deuteron in deltaless χPT to $O(Q^3)$ at 95 MeV. The polarizability approximation to the single-nucleon amplitude (dashed line) is to be compared with the full $O(Q^3)$ calculation (solid line).

[d]At this energy Ref. [34] only presents results with $\alpha_p + \alpha_n = \beta_p + \beta_n = 0$, which in turn are considerably less forward peaked than the corresponding calculation of Ref. [35].

4.3 Effective theories of Compton scattering

Although nominally the domain of validity of the Weinberg formulation extends well beyond the threshold for pion production, the power-counting fails at low energies well before the Thompson limit is reached. Consider the $O(Q^4)$ contribution shown in Fig. 10. We can use this graph to illustrate the transition to the very-low energy regime $Q \sim m_\pi^2/M$. It is easy to see that this graph becomes comparable to the order Q^3 graph of Fig. 5(a) when

$$\frac{|\vec{p}\,|^2}{\omega M} \sim 1. \tag{20}$$

Here $\vec{p}$ is a typical nucleon momentum inside the deuteron and ω is the photon energy. Since our power-counting is predicated on the assumption that all momenta are of order m_π, we find that power-counting is valid in the region

$$\frac{m_\pi^2}{M} \ll Q \ll \Lambda_\chi. \tag{21}$$

Therefore, in the region $\omega \sim B$ the Weinberg power-counting is not valid, since the external probe momentum flowing through the nucleon lines is of order Q^2/M, rather than order Q. It is in this region that the Compton low-energy theorems are derived. Therefore our power-counting will not recover those low-energy theorems. Of course the upper bound on the validity of the effective theory should increase if the Δ-resonance is included as a fundamental degree of freedom [38].

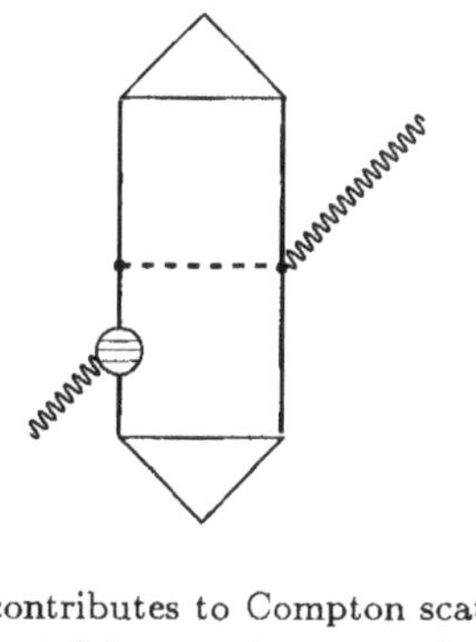

Figure 10: Interaction which contributes to Compton scattering on the deuteron at order Q^4. The sliced blob represents a $1/M$ correction vertex from $\mathcal{L}_{\pi N}^{(2)}$.

In Ref. [9] Compton scattering on the deuteron was computed to the same order discussed here, one order beyond leading non-vanishing order but in the KSW formulation of two-nucleon effective field theory. An advantage of

KSW power-counting is that the effective field theory moves smoothly between $Q < B$ and $Q > B$. KSW power-counting is valid for nucleon momenta $Q < \Lambda_{NN} \sim 300$ MeV. Thus in the KSW formulation deuteron polarizabilities and Compton scattering up to energies $\omega < \Lambda_{NN}^2/M \sim 90$ MeV can be discussed in the same framework. Here we are interested mostly in the region $\omega \sim m_\pi$, and so we regard ourselves as being firmly in the second regime. We stress that the value of Λ_{NN} is uncertain; it is conceivable that $\Lambda_{NN} \sim 500$ MeV in which case the range of the KSW formulation would extend well beyond pion production threshold.

4.4 Effects of higher order terms

In order to test the sensitivity of our calculation to higher-order effects we added a small piece of the $O(Q^4)$ amplitude for Compton scattering off a single nucleon. Specifically, we have modified the Compton amplitudes so that the polarizabilities in the calculation are changed to

$$\alpha = \alpha^{(Q^3)} + \Delta\alpha, \qquad \beta = \beta^{(Q^3)} + \Delta\beta, \tag{22}$$

where $\alpha^{(Q^3)}$ and $\beta^{(Q^3)}$ are the $O(Q^3)$ values of Eq. (15). We emphasize that this effect represents only a part of what will appear in the single-nucleon scattering amplitude $T_{\gamma N}$ at $O(Q^4)$. Furthermore, a number of additional mechanisms must be included in $T_{\gamma NN}^{2N}$ in any $O(Q^4)$ calculation of Compton scattering on the deuteron. Nevertheless, here we calculate the differential cross section with the modified polarizabilities in order to get a feel for the sensitivity of our result to the presence of such higher-order terms.

Two calculations were performed. In the first, $\Delta\alpha$ and $\Delta\beta$ were chosen so that the total polarizabilities (22) were equal to the "experimental" values[e]. The second calculation involved a more dramatic change in the polarizabilities: $\Delta\alpha$ and $\Delta\beta$ were chosen so that α and β were equal to the $O(Q^4)$ values of Ref. [28] where resonance-saturation has been used to estimate some of the $O(Q^4)$ χPT counterterms. In either case $\Delta\alpha_p + \Delta\alpha_n$ is relatively small, while $\Delta\beta_p + \Delta\beta_n$ is not large for "experimental" values, but is significant for the $O(Q^4)$ values. The results of these two calculations for the two photon energies $E_\gamma = 49$ MeV and $E_\gamma = 95$ MeV are shown in Fig. 11 and Fig. 12.

In both cases we see that, just as one would expect, the cross section at 95 MeV is much more sensitive to these $O(Q^4)$ terms than the cross section at 49 MeV. It is not surprising that the calculation with $O(Q^4)$ polarizabilities exhibits a larger change than that with "experimental" values. Continually

[e]Here we use the proton experimental values and neutron "experimental" values (see Ref. [13] for details).

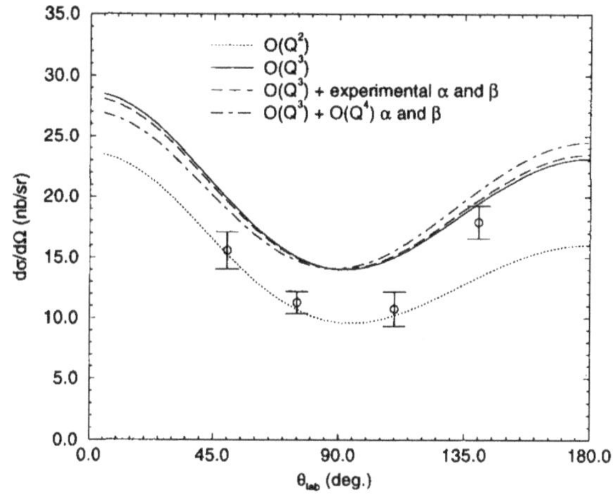

Figure 11: Results of calculations at 49 MeV using different values for the nucleon electromagnetic polarizabilities. The solid line is the result using the $O(Q^3)$ χPT value, the long-dashed line is the result using "experimental" polarizabilities, and the dot-dashed line represents a calculation with the $O(Q^4)$ polarizabilities.

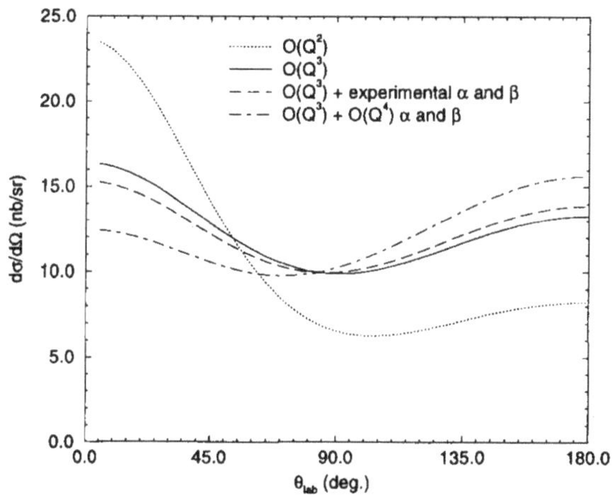

Figure 12: Results of calculations at 95 MeV using different values for the nucleon electromagnetic polarizabilities. Legend as in Fig. 11.

increasing $\beta_p + \beta_n$ at approximately constant $\alpha_p + \alpha_n$ decreases the cross section at forward angles and increases it at back angles. In fact, it seems that if $\beta_p + \beta_n$ is sufficiently large then the character of the cross section at 95 MeV can change completely from forward peaked to backward peaked.

Significant change in the 95 MeV results presented here from those obtained at $O(Q^3)$ mandates a cautious interpretation. At the same time that the cross section at 95 MeV is more sensitive to polarizabilities than at lower energies, it is also more sensitive to $O(Q^4)$ corrections. In our view, a full $O(Q^4)$ calculation in χPT is necessary if any attempt is to be made to extract the neutron polarizability from the Saskatoon data within this framework.

5 Dessert

Using effective field theory techniques, nucleon properties can be extracted systematically from experiments performed with nuclei. As a simple example, we have reviewed a χPT calculation which uses π-deuteron data to place constraints on low energy constants of the π-N chiral Lagrangian.

And, too, we have reviewed a recent computation of Compton scattering on the deuteron in χPT. We find reasonable agreement with the data at 49 MeV. At this energy $O(Q^3)$ corrections are not large compared to the leading $O(Q^2)$ result, and $O(Q^4)$ terms seem to be even smaller. However, as anticipated, the effective theory appears to break down as the Thomson limit is approached. We find good agreement with the data at 69 MeV. At this energy the convergence appears to be good. This suggests that χPT at $O(Q^3)$ is providing reasonable neutron and two-nucleon contributions. We find that the polarizability approximation should not be used in the calculation of the differential cross section at 95 MeV, since truncating the photon-nucleon amplitude at order ω^2 results in a significant change in the photon-deuteron differential cross section for forward angles. The wave function dependence is minor (on the order of 10% in the differential cross section). We make a prediction at 95 MeV which is, however, plagued by considerable uncertainties. Convergence is slow at this energy, as indicated by the relative size of both the full set of $O(Q^3)$ corrections and a partial set of $O(Q^4)$ corrections. The cross section tends to come out somewhat smaller than at lower energies, in particular in the backward directions, although the full $O(Q^4)$ amplitude is likely to be somewhat bigger at back angles. It seems that a more stringent test of χPT at these energies—including aspects of neutron structure beyond the $O(Q^3)$ "pion cloud" picture—will have to wait for a next-order calculation.

Acknowledgments

I thank Veronique Bernard, Harry Lee, Mané Malheiro, Ulf Meißner, Dan Phillips and Ubi van Kolck for enjoyable collaborations, Dan and Ubi for comments on the MS and the INT for a great workshop. This research was supported by DOE grant DE-FG02-93ER-40762 (DOE/ER/40762-183, UMPP#99-112).

References

1. *Nuclear Physics with Effective Field Theory*, eds. R. Seki, U. van Kolck and M. J. Savage (World Scientific, 1998); U. van Kolck, nucl-th/9902015.
2. S. Weinberg, *Phys. Lett.* B **251**, 288 (1990); *Nucl. Phys.* B **363**, 3 (1991).
3. C. Ordóñez and U. van Kolck, *Phys. Lett.* B **291**, 459 (1992); U. van Kolck, Ph. D. thesis, University of Texas (1993).
4. U. van Kolck, *Phys. Rev.* C **49**, 2932 (1994).
5. C. Ordóñez, L. Ray, and U. van Kolck, *Phys. Rev. Lett.* **72**, 1982 (1994); *Phys. Rev.* C **53**, 2086 (1996), hep-ph/9511380.
6. G. P. Lepage, lectures given at the 9th Jorge Andre Swieca Summer School: Particles and Fields, Sao Paulo, Brazil, nucl-th/9706029.
7. U. van Kolck, in *Mainz 1997, Chiral Dynamics: Theory and Experiment*, ed. A. Bernstein *et al*, (Springer-Verlag,1998), hep-ph/9711222; in Ref. 1; *Nucl. Phys.* A **645**, 273 (1999), nucl-th/9808007.
8. D. B. Kaplan, M. J. Savage, and M. B. Wise, *Phys. Lett.* B **424**, 390 (1998), nucl-th/9801034; *Nucl. Phys.* B **534**, 329 (1998), nucl-th/9802075.
9. J.-W. Chen, H. W. Griesshammer, M. J. Savage and R. P. Springer, *Nucl. Phys.* A **644**, 221 (1998); *Nucl. Phys.* A **644**, 245 (1998); J.-W. Chen, nucl-th/9810021.
10. S. Weinberg, *Phys. Lett.* B **295**, 114 (1992).
11. S. R. Beane, V. Bernard, T.-S. H. Lee and Ulf-G. Meißner, *Phys. Rev.* C **57**, 424 (1998), nucl-th/9708035.
12. S. R. Beane, C. Y. Lee, and U. van Kolck, *Phys. Rev.* C **52**, 2914 (1995), nucl-th/9506017; S.R. Beane, V. Bernard, T.S.H. Lee, Ulf-G. Meißner and U. van Kolck, *Nucl. Phys.* A **618**, 381 (1997), hep-ph/9702226; U. van Kolck, *this volume*.
13. S.R. Beane, M. Malheiro, D.R. Phillips and U. van Kolck, nucl-th/9905023.

14. T.-S. Park, K. Kubodera, D-P. Min, and M. Rho, *Phys. Rev.* C **58**, 637 (1998), hep-ph/9711463.
15. T.-S. Park, K. Kubodera, D-P. Min, and M. Rho, astro-ph/9804144; *Nucl. Phys.* A **646**, 83 (1999), nucl-th/9807054.
16. V. Bernard, N. Kaiser and Ulf-G. Meißner, *Phys. Lett.* B **309**, 421 (1993).
17. D. Chatellard et al., *Phys. Rev. Lett.* **74**, 4157 (1995); *Nucl. Phys.* A, in print.
18. D. Sigg et al., *Phys. Rev. Lett.* **75**, 3245 (1995); *Nucl. Phys.* A **609**, 269 (1996); (E) A617, 526 (1997).
19. R. Machleidt, ADV **19**, 189 (1989).
20. R.B. Wiringa, V.G. Stoks and R. Schiavilla, *Phys. Rev.* C **51**, 38 (1995).
21. R.V. Reid, *Ann. Phys.* **50**, 411 (1968).
22. R. de Tourrell and D.W. Sprung, *Nucl. Phys.* A **210**, 193 (1973).
23. R. Koch, *Nucl. Phys.* A **448**, 707 (1986).
24. R.A. Arndt et al., *Phys. Rev.* C **52**, 2120 (1995).
25. V. Bernard, N. Kaiser and Ulf-G. Meißner, *Nucl. Phys.* A **615**, 483 (1997).
26. M. Mojzis, hep-ph/9704415, Z.Phys.C, in print.
27. V. Bernard, N. Kaiser, and Ulf-G. Meißner, *Phys. Rev. Lett.* **67**, 1515 (1991); *Nucl. Phys.* B **383**, 442 (1992); V. Bernard, N. Kaiser, J. Kambor, and Ulf-G. Meißner, *Nucl. Phys.* B **388**, 315 (1992).
28. V. Bernard, N. Kaiser, A. Schmidt, and Ulf-G. Meißner, *Phys. Lett.* B **319**, 315 (1993), hep-ph/9309211; *Z. Phys.* A **348**, 317 (1994).
29. J. Tonnison, A. M. Sandorfi, S. Hoblit, and A. M. Nathan, *Phys. Rev. Lett.* **80**, 4382 (1998), nucl-th/9801008.
30. D. Drechsel, *et al.*, in *Mainz 1997, Chiral Dynamics: Theory and Experiment*, ed. A. Bernstein *et al*, (Springer-Verlag, 1998), p. 264, nucl-th/9712013.
31. M. Lucas, Ph. D. thesis, University of Illinois, unpublished (1994).
32. D. Hornidge, private communication; G. Feldman, private communication.
33. M. Lunding, private communication.
34. T. Wilbois, P. Wilhelm, and H. Arenhövel, *Few-Body Sys. Suppl.* **9**, 263 (1995).
35. M. I. Levchuk and A. I. L'vov, nucl-th/9809034, and references therein.
36. J. J. Karakowski and G. A. Miller, nucl-th/9901018; J. J. Karakowski, Ph. D. thesis, University of Washington (1999), nucl-th/9901011.
37. R. Machleidt, K. Holinde, and Ch. Elster, *Phys. Rep.* **149**, 1 (1987).
38. E. Jenkins and A. V. Manohar, Phys. Lett. **B255**, 558 (1991).

Discussion during S. Beane's presentation

Chair: *F. Ravndal (Oslo University)*
Notes taken by *J. McGovern (University of Manchester)*

Several members of the audience, including Cohen, Wise and Furnstahl, commented during and after the talk on the fact that the work presented used "off-the-peg" wavefunctions rather than ones calculated in the same effective field theory which provided the operators; Wise in particular pointed out that an operator which should vanish by the equations of motion would not vanish with an inconsistent wavefunction. Beane pointed out that the differences between the various wavefunctions, one of which was van Kolck's, were small, and that the success was related to the weak binding of the deuteron. In addition Phillips pointed out that the EFT wavefunctions were not yet very good, and van Kolck described the situation as a temporary expedient driven by a desire to get some results out quickly.

Meißner asked if the combination of HBCPT parameters that were termed Δ occurred elsewhere, and then answered his own question in the affirmative.

Cohen asked which diagrams appeared at $O(Q)$ in gauges other than Coulomb; Beane did not know.

NEAR–THRESHOLD PION (PHOTO)PRODUCTION: A SUCCESS AND AN OPPORTUNITY

U. van KOLCK
W. K. Kellogg Radiation Laboratory,
California Institute of Technology,
Pasadena, CA 91125, USA
E-mail: vankolck@krl.caltech.edu

I discuss neutral pion photoproduction on the deuteron and pion production in nucleon-nucleon collisions. The first reaction is *predicted* at threshold in the effective field theory (EFT) approach with *non*-perturbative pions. This is the first nuclear EFT prediction that differs significantly from conventional model predictions, and recent SAL data clearly favors the EFT result. The second reaction is currently not understood near threshold, and thus offers another opportunity to contrast EFT and *ad hoc* model approaches.

1 Introduction

There are two main motivations for studying scattering of photons and pions on light nuclei. First, simple nuclear targets can partially compensate for the lack of neutron targets. Theories and models of the nucleon are typically adjusted to existing proton data but differ in their predictions of neutron amplitudes. Under certain circumstances it is possible to extract information on the neutron from nuclear targets, the deuteron being a particular favorable case due to its low density. Second, probing nuclei can test theories and models of nuclear forces. The interactions between real pions (and photons) and nucleons are closely related to those between virtual pions (and photons) and nucleons that make up important pieces of the nuclear potential. After the potential is constrained to reproduce scattering and bound states in purely nucleonic systems, some predictive power usually results for scattering of probes on nuclear targets. Because —thanks to its small mass— the pion makes important contributions to nucleon structure and to the nuclear potential, processes involving pion production are of special interest.

Clearly, these motivations demand that scattering on nuclear targets be treated in a way that is simultaneously systematic and consistent with both the scattering on a single nucleon and the dynamics of the target. Otherwise, systematic errors will prevent much theoretical insight. Furthermore, for any insight to be relevant, we require consistency with QCD as well: the hadronic theory must contain all possible operators generated by the underlying theory of quarks and gluons; in particular, it has to exhibit the same pattern of symmetry breaking as QCD. Currently, the most general chiral Lagrangian ordered by power counting arguments is the only tool that satisfies these requirements when probes have typical momenta Q comparable to the pion mass m_π. In the meson and one-baryon sectors, this effective field theory (EFT) method goes by the name of chiral perturbation theory (ChPT); applications to systems of two or more nucleons involve at least some non-perturbative considerations, and these are the thrust of this workshop [1].

Here I will consider two near-threshold processes in this framework, pion photoproduction on the deuteron, which is paradigmatic, and pion production in nucleon-nucleon (NN) collisions, which is not. The former is work done in collaboration with Silas Beane, Chang-Yong Lee, Veronique Bernard, Harry Lee, and Ulf Meißner [2,3], the latter with Tom Cohen, Jim Friar, Jerry Miller, Dan Riska, and Carlos da Rocha [4,5,6]. Other reactions are discussed using the same methods by Beane [7] and Rho [8], and using a different power counting by Savage [9].

Pion photoproduction on the nucleon near threshold has been studied up to $O(Q^4)$ in ChPT with the delta integrated out in Ref. [10] (where references to experimental papers can be found). The differential cross-section for a photon of momentum k to produce a pion of momentum q comes at threshold from a pure S wave, $[(|\vec{k}|/|\vec{q}|)(d\sigma/d\Omega)]_{|\vec{q}|=0} = |E_{0+}|^2$. At leading order, $O(Q)$, there exist only contributions to charged channels stemming from minimal substitution on the pion-nucleon vertex ("Kroll-Ruderman term") and from the pion pole. We thus expect charged channels to be the largest. At $O(Q^2)$, minimal substitution on the recoil correction to the pion-nucleon vertex contributes to photoproduction on the proton. At $O(Q^3)$ there are tree-level contributions involving the magnetic moment of the nucleon. Together with the previous orders, this reproduces an old "low-energy theorem". However, still at $O(Q^3)$ there are loop corrections where the photon interacts with a virtual pion which then rescatters on the nucleon before flying away to the detector. This quantum effect has been missed in the old "theorem" and in most models. It is determined in terms of known parameters and thus constitutes a new, correct low-energy theorem; it is also numerically significant. At $O(Q^4)$ other loops and new counterterms appear. Results [10] for E_{0+} up to $O(Q^4)$ from a fit to data above threshold constrained by resonance saturation of parameters are shown in Table 1 [a].

[a] In this calculation, the dominant isospin breaking effect, namely the charged to neutral pion mass difference, which is almost entirely of electromagnetic origin, has been taken into account. A fully consistent calculation including all effects from virtual photons and the quark mass differences is not yet available. The result for $E_{0+}^{\pi^0 n}$ is based on the assumption that the counterterms entering at $O(Q^4)$ are the same for the proton and the neutron apart from trivial isospin factors.

Table 1: Values for E_{0+} in units of $10^{-3}/m_{\pi^+}$ at various orders ν in ChPT and in experiments, for the different channels.

$E_{0+}(10^{-3}/m_{\pi^+})$	$\nu=1$	$\nu=2$	$\nu=3$	$\nu=4$	Experiment
$\gamma p \to \pi^+ n$	34.0	26.4	28.9	28.2	27.9 ± 0.5
					28.8 ± 0.7
					27.6 ± 0.3
$\gamma n \to \pi^- p$	−34.0	−31.5	−32.9	−32.7	-31.4 ± 1.3
					-32.2 ± 1.2
					-31.5 ± 0.8
$\gamma p \to \pi^0 p$	0	−3.58	0.96	−1.16	-1.31 ± 0.08
					-1.32 ± 0.08
$\gamma n \to \pi^0 n$	0	0	3.7	2.13	

We observe that charged-pion channels are indeed larger and show pretty good convergence and agreement with experimental data. For neutral-pion channels convergence is less apparent, but the absolute values are much smaller. The $O(Q^4)$ result for $\gamma p \to \pi^0 p$ is in relatively good agreement with the recent results from Mainz and Saskatoon. To the same order there is a prediction for the near-threshold behavior of the $\gamma n \to \pi^0 n$ reaction; the cross-section is considerably larger than that obtained in models that omit the important one-loop $O(Q^3)$ diagram. Neutral pion photoproduction on the deuteron is the best candidate for a test of these discrepant predictions.

Pion production in NN collisions has also been an active field of research since good near-threshold data for $pp \to pp\pi^0$ [11], $np \to d\pi^0$ [12], $pp \to d\pi^+$ [13], and $pp \to pn\pi^+$ [14] became available. The $pp \to pp\pi^0$ reaction was observed to be a factor ~ 5 larger than previous theoretical estimates [15] based on an impulse term and on-shell pion rescattering. Soon this discrepancy was explained either by Z-graphs with σ and ω exchange [16] or by off-shell pion rescattering [17]. These explanations cannot be both correct, or otherwise the cross-section would be overpredicted. Moreover, both omit effects from the delta isobar. These effects are small because in order to produce an S-wave pion the delta has to decay in flight. However, competing terms are similarly small; for example, the impulse approximation is also a recoil effect. Since the pion threshold lies approximately mid-way between the elastic and delta production thresholds, it is not *a priori* obvious that delta contributions are smaller than the impulse approximation. Finally, the answer seems to depend on the off-shell extension of the pion-nucleon amplitude, that is, on the particular choice of pion field. This cannot be correct. For $pp \to d\pi^+$, too, theory was found to be in bad shape. There Z-graphs and off-shellness are relatively less important, and data are underpredicted[15] by a factor ~ 2. Such a spectacular failure of conventional models even when the best NN potentials are used underscores the need for a systematic approach.

In the following I will first give a brief account of our framework (Sect. 2) and then apply it to threshold neutral-pion photoproduction on the deuteron (Sect. 3). I will show that a *pre*diction was made in strong variance with the predictions of conventional models, and that recent data clearly supports the ChPT result. This is an unequivocal success of the EFT method. Next I consider pion production in NN collisions (Sect. 4). I will discuss why the framework has to be extended to account for heftier momenta, and show that the most naive extension suffers from the same problems of conventional models. Even this naive extension points, however, to the importance of ingredients that have so far been neglected in other calculations. This situation clearly affords an opportunity for further development of nuclear EFT. These conclusions are in Sect. 5.

2 Scattering on Few-Nucleon Systems

Suppose we probe light nuclei with photons and pions. Without any loss of generality we can decompose the full amplitude T into a kernel I, where the probes are attached, linked through nearly on-shell, free states to initial and final asymptotic nuclear wavefunctions Ψ_A and Ψ'_A,

$$T = < \Psi'_A | I | \Psi_A > . \tag{1}$$

This is represented in Fig. 1 for the particular case of coherent ($\Psi'_A = \Psi_A$) pion photoproduction.

It is well-known that nuclei are large objects when compared to the distance characteristic of the structure of nucleons that is not directly associated with pions: the typical size of light nuclei is $\langle r \rangle \sim 1/f_\pi$ (where f_π is the pion decay constant), much larger than the Compton wavelength of the nucleon, $1/m_N$. This implies that the typical momentum of nucleons in light nuclei is $p \sim f_\pi$. If we consider now probes with momenta $q \sim f_\pi$, all momenta that flow through I are $O(f_\pi)$. Because the probes carry energy $q^0 \sim f_\pi \sim m_\pi$, where m_π is the pion mass, energies are also $O(f_\pi)$. In fact, we *define* the kernel as those sub-diagrams where all intermediate energies are $O(f_\pi)$. By construction, the kernel is characterized by only one explicit scale $Q \sim f_\pi \sim m_\pi \sim 100$ MeV.

Because Q is much smaller than hadronic masses other than m_π and the delta-nucleon mass difference $m_\Delta - m_N \sim 3 f_\pi$, in the description of processes

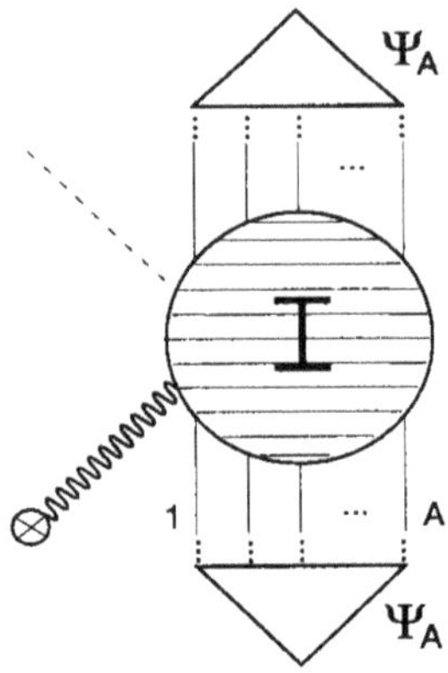

Figure 1: The scattering amplitude for coherent pion photoproduction. The Ψ_A's correspond to the A-nucleon wavefunctions and the blob I is the kernel to which external probes are attached.

of this type we need to consider only nucleons, photons, pions, and delta isobars as explicit degrees of freedom. They will generate the correct low-energy analytic structures in amplitudes, the effect of higher-energy states being well approximated by local interactions. The only constraint on interactions is that they respect the symmetries of the short-distance theory, QCD+QED: Lorentz invariance, approximate chiral symmetry, and, if we neglect higher-dimensional operators, parity and time-reversal invariances. Because $Q \ll m_N$, it suffices to treat nucleons and deltas as non-relativistic objects subject to Lorentz invariance under small-velocity boosts. Approximate chiral symmetry requires that pions couple to each other and to nucleons through either derivative interactions or quark masses. Photons couple via the electromagnetic field strength tensor and by minimal substitution, which has the simple effect of replacing derivatives by factors of the charge e. The symmetries still allow an infinite number of interactions and there is no reason to believe their strengths are small except where e appears. To proceed we need to make some assumption about the sizes of parameters, and the mildest one is that they are given by the characteristic mass scale of QCD, $M_{QCD} \sim 1$ GeV, to a power determined by the mass dimension of the corresponding field operator.

There are then two expansion parameters, Q/M_{QCD} and $\alpha_{\rm em} = e^2/4\pi$. Because the kernel involves only one low-energy scale, counting powers of Q in any kernel diagram can be done in the same way as in evaluating the superficial degree of divergence. Counting powers of $\alpha_{\rm em}$ is even simpler. Since in the following I will be working at a fixed order in the expansion in $\alpha_{\rm em}$ (in particular, leading order), it is convenient to count e as a derivative. A generic kernel diagram with A external nucleon lines, L loops, C separately connected pieces, and V_i vertices of type i, with f_i fermion lines, d_i derivatives and/or factors of the pion mass and/or the delta-nucleon mass difference and/or the charge, is of $O(Q^\nu)$ with [18]

$$\nu = 4 - A + 2(L - C) + \sum_i V_i \Delta_i, \tag{2}$$

where the so-called index of the interaction i is

$$\Delta_i = d_i + \frac{f_i}{2} - 2 \geq 0. \tag{3}$$

The last in/equality can be shown to follow from chiral symmetry. Contributions to the Lagrangian $\mathcal{L}$ that generates diagrams can be ordered according to the index,

$$\mathcal{L} = \sum_{\Delta=0}^{\infty} \mathcal{L}^{(\Delta)}, \tag{4}$$

where $\mathcal{L}^{(\Delta)}$ collects all interactions with index Δ that transform under chiral symmetry, parity and time-reversal in the same way as terms in the QCD Lagrangian.

This power counting immediately reveals a feature that has been found generically in models. The kernel, being a sub-amplitude, is not necessarily connected, but can have up to $C_{max} = A$ separately connected pieces. This maximally disconnected situation corresponds to diagrams where all the probes are attached to one nucleon, while the other $A-1$ nucleons are undisturbed. Eq. (2) implies that such diagrams will tend to contribute the smallest power of the expansion parameter. We then expect that, unless one-nucleon contributions are suppressed by choice of kinematics, the impulse approximation will be good up to corrections of $O((Q/M_{QCD})^2)$. Conversely, two-nucleon effects (meson exchange currents) will be particularly noticeable in kinematics where single-scattering contributions are non-vanishing only at subleading orders, that is, arise from maximally disconnected graphs that have $2L + \sum V_i \Delta_i \geq 1$.

To calculate the full amplitude we also need the asymptotic wavefunctions. These can be generated from the same Lagrangian used to derive the kernel. The power counting described above has to be adapted to the fact that, before and after interaction with probes, intermediate states exist in the amplitude that differ in energy from the initial (or final) state only by the kinetic energy of nucleons, $\Delta E \sim Q^2/M_{QCD}$. As is the case for the kernel, only sub-diagrams without such small denominators obey Eqs. (2) and (3). These sub-diagrams are A-nucleon irreducible; their sum is by definition the nuclear potential V. As noted in Ref. [18] and discussed in Ref. [1], if the potential is iterated to all orders it naturally generates bound states where the nucleons have momenta $p \sim f_\pi$. The decomposition of the full amplitude into "irreducible" parts that are free of $O(Q^2/M_{QCD})$ denominators is thus analogous to the decomposition of perturbative QCD amplitudes into infrared safe quantities.

The two- and three-nucleon potentials from the most general chiral Lagrangian (4) were derived to $O((Q/M_{QCD})^3)$ in Refs. [19,20]. The NN potential [19] is very complicated, including non-local terms and, in the basis used, energy-dependent terms as well. It was fitted [21] to the results of NN phase shift analyses. Unfortunately this fit preceded the determination of some parameters from one-nucleon processes. Quantitative agreement with data was found for deuteron properties and scattering up to about 100 MeV laboratory energies (see Ref. [1]), but the precision did not match that of conventional potential models. In principle consistency requires that we use wavefunctions computed from the NN potential with appropriate one-nucleon parameters. However, for simplicity and accuracy we can use a hybrid approach, with wavefunctions generated by modern phenomenological NN potentials that describe the NN system more precisely. Because all such potentials incorporate the appropriate long-range dynamics (one-pion exchange and a mock-up of two-pion exchange), they differ from each other and from the chiral potential mostly in the fit of short-range effects. The error introduced by this hybrid approach should be within the theoretical error expected from neglected higher orders in the chiral expansion. Below we will check this approximate wavefunction independence numerically by comparing results for different phenomenological potentials, and by using the chiral potential for calculation of the largest contribution.

More details of the EFT method applied to nuclear systems can be found in Ref. [22].

3 Pion Photoproduction

We now apply the above ideas to pion photoproduction [2]. Consider a process with A nucleons, and a single photon and a single pion in the initial and final state, respectively —see Fig. 1. We can order the chiral expansion of the kernel [b] by way of the counting index ν:

[b]Note that there are no irreducible graphs with $C = A$ where the photon and the pion are attached to different nucleons, because energy and momentum conservation require the exchange of a pion and somewhere along the nucleon lines there is an energy flow of the order of the pion mass.

- $\nu = \mathbf{3 - 3A}$: At leading order in small momenta, the matrix element is given by tree graphs with the maximum number of separately connected pieces ($L = 0$, $C = A$) constructed out of interactions with $\Delta = 0$. For example, at this order one finds the Kroll-Ruderman term (Fig. 2a).

- $\nu = \mathbf{4 - 3A}$: The first corrections to the leading terms are still tree graphs with $C = A$, but have one vertex with $\Delta = 1$. For example, we get a $1/m_N$ correction (Fig. 2b).

- $\nu = \mathbf{5 - 3A}$: There are three classes of corrections at this order:

  (i) One-loop ($L = 1$) graphs with $C = A$ having only $\Delta = 0$ interactions (Fig. 2c).

  (ii) $C = A$ tree graphs with either one vertex with $\Delta = 2$ (Fig. 2d) or two vertices with $\Delta = 1$. In the case of neutral pion photoproduction, the counterterm graphs with one $\Delta = 2$ vertex vanish (which leaves behind finite loop graphs), while graphs with two $\Delta = 1$ vertices give rise to $1/m_N^2$ corrections proportional to the nucleon magnetic moments.

  (iii) Tree graphs with one less than the maximum number of separately connected pieces ($C = A - 1$) having only $\Delta = 0$ interactions. These graphs are of two types: two-nucleon graphs like the Feynman graph of Fig. 2e and time-ordered graphs such as those in Fig. 2f,g.

- $\nu = \mathbf{6 - 3A}$: There are many new contributions at this order, which can be constructed following the same procedure. Most of these are calculable corrections to the lower orders, but among the single-scattering diagrams for neutral pion photoproduction an undetermined counterterm appears.

In principle, all of these graphs contribute to pion photoproduction on nuclei. However, some generic simplifications arise when one sews in the nuclear wavefunctions, and when one specializes to particular channels.

First, the time-ordered graphs of the type of Fig. 2f,g get cancelled against recoil in the one-pion-exchange piece of the potential. In order to see this, consider the three diagrams of Fig. 3. These graphs all have the same spin-isospin structure: they differ only in the energy denominators. The first two graphs (Fig. 3a,b) arise when a diagram like Fig. 2a is sandwiched between wavefunctions obtained from a potential whose long-range part comes from pion exchange. They are proportional to

$$\frac{2}{w[E(\vec{p}_1 - \vec{q}) + E(\vec{p}_2 + \vec{q}) - E(\vec{p}_1) - E(\vec{p}_2)]} - \frac{1}{w^2}\left(1 + O\left(\frac{E}{w}\right)\right). \quad (5)$$

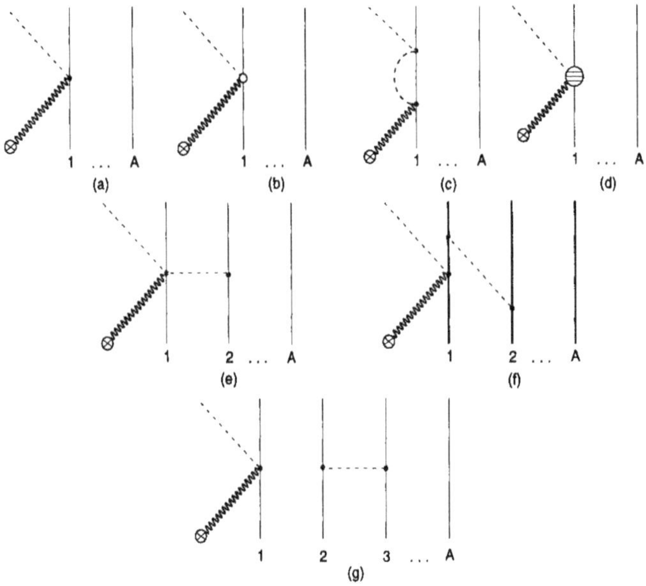

Figure 2: Characteristic graphs which contribute at each of the first three orders in pion photoproduction.

where I denoted by w $(E(\vec{p}))$ the energy of the pion (nucleon) of momentum $\vec{q}$ $(\vec{p})$. The first term corresponds to static one-pion-exchange in the potential and is big, as anticipated, because these reducible diagrams have small nucleon energy denominators. The second term is smaller because of the additional small recoil numerator. On the other hand, the remaining graph (Fig. 3c) is proportional to

$$\frac{1}{w^2}\left(1+O\left(\frac{E}{w}\right)\right), \tag{6}$$

and approximately cancels recoil in the reducible diagrams. Similar cancellations can be found among the other time-ordered diagrams. In other words, to this order in ChPT we can omit the time-ordered diagrams Fig. 2f,g by at the same time disregarding the energy dependence in the potential.

Second, in the case of neutral pion photoproduction at threshold a number of the graphs in Fig. 2 will not contribute: those where the photon line is attached to a pion, and those that go like $S \cdot q$, where q is the outgoing pion momentum and S is the covariant nucleon spin. In particular, all the leading-order graphs (Fig. 2a) vanish, which immediately suggests that the cross-section will

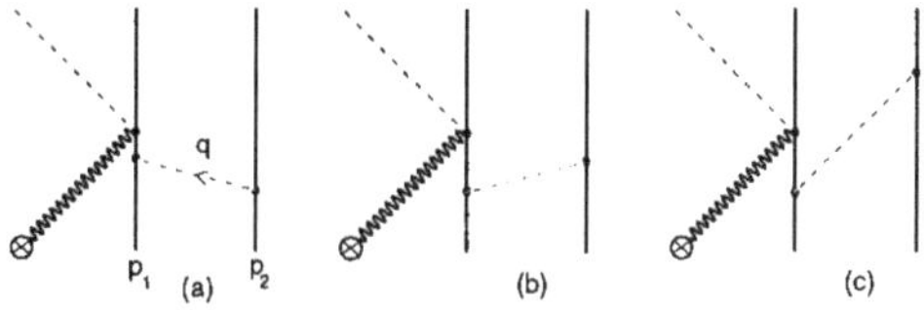

Figure 3: The energy dependence of the impulse graphs (a) and (b) approximately cancels the time-ordered two-nucleon diagram (c).

be smaller than that for charged pion production, and more sensitive to two-nucleon contributions. Moreover, the time-ordered graphs (Fig. 2f,g) also vanish, so we can expect little influence of the energy-dependent part of the potential. Finally, contributions from the delta isobar are minimized at threshold, and integrating out the delta should not damage convergence of the theory too much. As noted above, to $O(Q^4)$ there is one undetermined parameter, which is fitted to $\gamma p \to \pi^0 p$ above threshold.

For simplicity I limit myself here to threshold neutral-pion photoproduction on the deuteron [3]. As explained in the introduction, this is of special interest because it can give a handle on the elementary neutron amplitude.

Consider the reaction $\gamma(k)+d(p_1) \to \pi^0(q)+d(p_2)$ in the threshold region, $\vec{q} \simeq 0$, where the pion is in an S wave with respect to the center-of-mass (cm) frame. For real photons ($\vec{\epsilon}\cdot\vec{k}=0$, with $\vec{\epsilon}$ the photon polarization vector), the differential cross-section at threshold takes the form

$$\frac{|\vec{k}|}{|\vec{q}|}\left.\frac{d\sigma}{d\Omega}\right|_{|\vec{q}|=0} = \frac{8}{3}E_d^2. \tag{7}$$

I now present the results of the chiral expansion of the dipole amplitude E_d to $\nu=0$, that is, to relative $O(Q^4)$.

The single-scattering contribution is given by all diagrams where the photon and the pion are absorbed and emitted, respectively, from one nucleon with the second nucleon acting as a spectator (the impulse approximation). Some of these diagrams are depicted in Fig. 2. To $\nu=0$ ($O(Q^4)$) one finds

$$\begin{aligned} E_d^{ss} &= \frac{1+m_\pi/m_N}{1+m_\pi/m_d}\left\{\frac{1}{2}\left(E_{0+}^{\pi^0 p}+E_{0+}^{\pi^0 n}\right)\int d^3p\,\phi^*(\vec{p})\,\vec{\epsilon}\cdot\vec{J}\,\phi(\vec{p}-\vec{k}/2)\right.\\ &\left.-\frac{k}{m_N}\hat{k}\cdot\int d^3p\,\hat{p}\,\frac{1}{2}\left(P_1^{\pi^0 p}+P_1^{\pi^0 n}\right)\phi^*(\vec{p})\,\vec{\epsilon}\cdot\vec{J}\,\phi(\vec{p}-\vec{k}/2)\right\}, \end{aligned} \tag{8}$$

evaluated at the threshold value

$$|\vec{k}| = k_{\rm thr} = m_{\pi^0}\left(1-\frac{m_{\pi^0}^2}{2m_d}\right) = 130.1\,\text{MeV}. \tag{9}$$

Here m_d is the deuteron mass, $\vec{J} = (\vec{\sigma}_1 + \vec{\sigma}_2)/2$, and ϕ is the deuteron wave-function.

Two remarks concerning Eq. (8) are in order:

(1) The elementary S-wave pion production amplitudes to $O(Q^4)$ [10] are given in Table 1. The proton amplitude that results from a one-parameter fit above threshold compares well with experiment. The neutron amplitude is predicted to be considerably larger than the one obtained from the old "low-energy theorem" that omits the important one-loop $O(Q^3)$ single-scattering diagram (Fig. 2c).

(2) It is important to differentiate between the $\pi^0 d$ and the $\pi^0 N$ ($N = p, n$) cm systems. At threshold in the former, the pion is produced at rest; it has, however, a small three-momentum in the latter [23]. Consequently, a single-nucleon P-wave contribution appears which is proportional to the elementary amplitudes $P_1^{\pi^0 p}$ and $P_1^{\pi^0 n}$ that obey the P-wave low-energy theorems [24]

$$P_1^{\pi^0 p} = 0.480\,|\vec{q}|\,\mathrm{GeV}^{-2}, \quad P_1^{\pi^0 n} = 0.344\,|\vec{q}|\,\mathrm{GeV}^{-2}, \tag{10}$$

with

$$\vec{q} = m_\pi \left[\left(1 - \frac{m_\pi}{m_N}\right) \frac{\vec{p}}{m_N} - \frac{m_\pi}{2m_N} \left(1 - \frac{5m_\pi}{4m_N}\right) \hat{k} \right], \tag{11}$$

where $\vec{p}$ is the nucleon momentum in the $\pi^0 d$ cm system. The P-wave amplitude for neutral pion production off protons holds to within 3%. Here we neglect the energy dependence of the elementary $\pi^0 p$ and $\pi^0 n$ S- and P-wave amplitudes since the pion energy changes only by 0.4% for typical average nucleon momenta in the deuteron.

The two-nucleon contributions of $\nu = -1$ ($O(Q^3)$) that survive at threshold in the Coulomb gauge ($\epsilon \cdot v = 0$, with v the nucleons' four-velocity) are shown in Fig. 4. All vertices come from the lowest order Lagrangian $\mathcal{L}^{(0)}$. In momentum space, the corresponding structures are

$$(a): \; \frac{1}{\vec{q}^{\,2}}, \quad (b): \; \frac{(q-k)_i\, q_j}{[(\vec{q} - \vec{k}\,)^2 + m_\pi^2]\, \vec{q}^{\,2}}, \tag{12}$$

where $\vec{q} = \vec{p} - \vec{p}\,'$ with $\vec{p}$ and $\vec{p}\,'$ the cm nucleon momenta in the initial and the final state, respectively.

At $\nu = 0$ ($O(Q^4)$), we have to consider tree graphs with exactly one insertion from the Lagrangian $\mathcal{L}^{(1)}$. The relevant terms in this Lagrangian fall into two classes, the first with fixed coefficients due to Lorentz invariance, the second with four low-energy constants $c_1, \ldots, c_4$ that are related to pion-nucleon scattering and two other constants that are related to the isoscalar and isovector nucleon anomalous magnetic moments, $\kappa_v = \kappa_p - \kappa_n$, $\kappa_s = \kappa_p + \kappa_n$. The

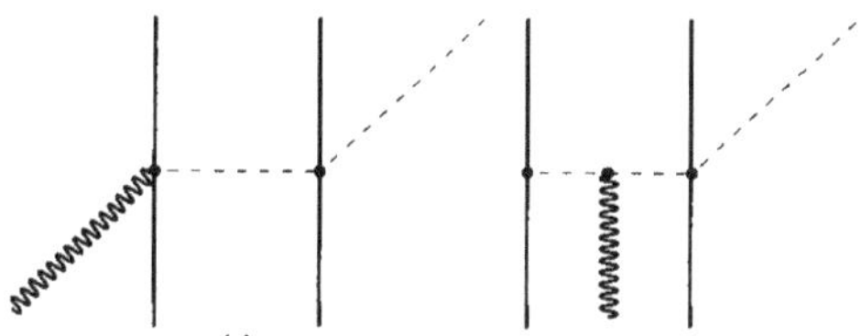

Figure 4: Two-nucleon graphs which contribute to neutral pion photoproduction at threshold to $O(Q^3)$ (in the Coulomb gauge). All vertices come from $\mathcal{L}^{(0)}$.

pertinent two-nucleon diagrams are shown in Fig. 5, the blob characterizing an insertion from $\mathcal{L}^{(1)}$. In addition, there are the recoil corrections to the graphs in Fig. 4.

One can show that all terms proportional to the low-energy constants c_i vanish at threshold either due to the selection rule $S \cdot q = 0$ or due to some isospin factor of the type $\delta^{a3}\,\epsilon^{a3b}$. We are thus left with insertions from terms $\sim 1/2m_N$, $\sim g_A/2m_N$ and $\sim \kappa_{v,s}$. All of these can be classified in momentum space according to the following structures,

$$\frac{1}{\vec{q}^{\,2}}, \frac{q_i}{\vec{q}^{\,2}}, \frac{q_i\,q_j}{\vec{q}^{\,2}}, \frac{q_i\,(p+p')_j}{\vec{q}^{\,2}}, \frac{(q-k)_i}{[(\vec{q}-\vec{k}\,)^2+m_\pi^2]}, \frac{(q-k)_i\,q_j}{[(\vec{q}-\vec{k}\,)^2+m_\pi^2]}, \frac{(q-k)_i\,(p+p')_j}{[(\vec{q}-\vec{k}\,)^2+m_\pi^2]},$$
$$\frac{(q-k)_i\,q_j\,(\vec{p}^{\,2}-\vec{p}^{\,\prime 2})}{[(\vec{q}-\vec{k}\,)^2+m_\pi^2]\,\vec{q}^{\,2}}, \frac{(q-k)_i\,q_j}{[(\vec{q}-\vec{k}\,)^2+m_\pi^2]^{3/2}}, \frac{(q-k)_i\,(q-k)_j}{[(\vec{q}-\vec{k}\,)^2+m_\pi^2]^{3/2}}. \tag{13}$$

Counting powers of momenta, one sees that the contributions from graphs 5b,l,p are divergent. This is due to the fact that in the chiral expansion one has truncated the pion-nucleon form factor and thus does not suppress the high-momentum components. This phenomenon also occurs in the calculation of the NN potential in ChPT [19,21]. As it was done there, we introduce an additional Gaussian cut-off factor of the form $F(\vec{q}^{\,2}) = \exp\{-\vec{q}^{\,2}/\Lambda_\pi^2\}$, where the cut-off Λ_π varies between the rho mass and $4\pi f_\pi = 1.2\,\mathrm{GeV}$ (since it can be related to the mass scale of the heavy particles that are integrated out of the EFT). Notice that the divergences appearing in other graphs all cancel and thus no further regularization is needed for those. For energy-independent NN potentials as used here, the contribution from graphs 5n,r should be omitted. We, however, also calculate their contribution to get a very rough check of consistency. Furthermore, we remark that the kinematic $1/m_N$ corrections to graph 5a vanish and the ones for graph 5b are taken care of by shifting the value of k_{thr} as given in Eq. (9).

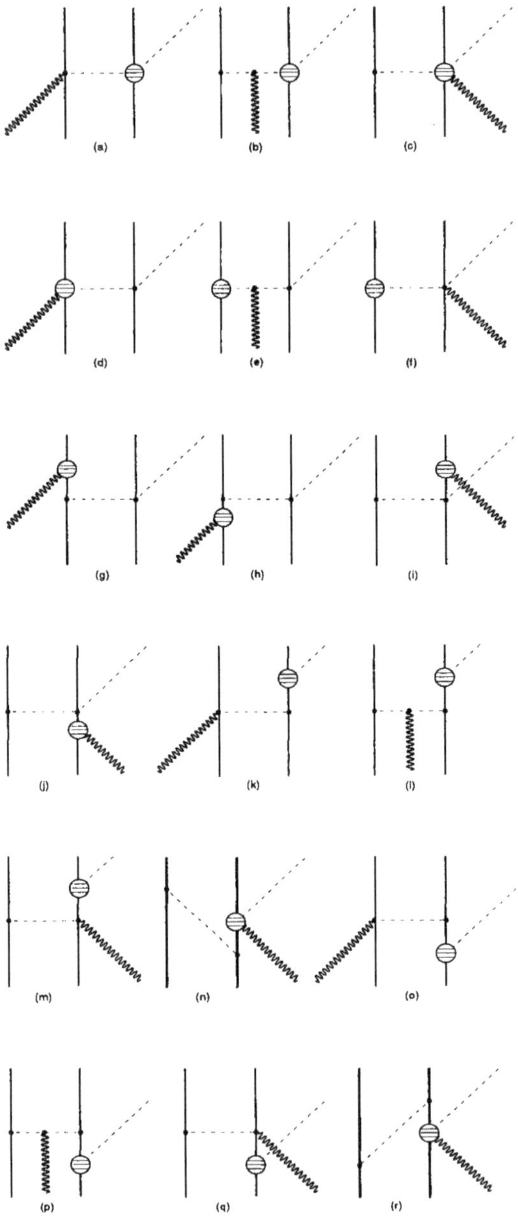

Figure 5: Two-nucleon graphs contributing at $O(Q^4)$ to neutral pion photoproduction. The hatched circles denote an insertion from $\mathcal{L}^{(1)}$. The time-ordered graphs are distinguished by bold nucleon lines.

Note that to $O(Q^4)$ there are no four-nucleon operators contributing to the deuteron dipole amplitude. The only possible graphs of this type that contribute at threshold to this order come from an $NNNN\pi$ vertex, to which a photon is attached either at the outgoing pion leg or at the vertex itself. Both vanish for neutral pions as they result from minimal substitution in a pion derivative. We therefore conclude that for threshold neutral pion photoproduction to $O(Q^4)$ there is no contribution from any four-nucleon operator and thus *no new, undetermined parameters appear.*

Using the Argonne V18 [25], the Reid [26], the Nijmegen [27] and the Paris [28] potentials, we evaluate E_d^{ss} and find

$$E_d^{ss} = 0.36 \times 10^{-3}/m_{\pi^+}, \tag{14}$$

with an uncertainty of $\delta E_d^{ss} = 0.05 \times 10^{-3}/m_{\pi^+}$ due to the various potentials used. The P-wave contribution amounts to a 3% correction to the one from the S wave, *i.e.* it amounts to a minor correction. The sensitivity of the single-scattering contribution E_d^{ss} to the elementary neutron amplitude is given by

$$E_d^{ss} = 0.36 - 0.38 \cdot (2.13 - E_{0+}^{\pi^0 n}) \ , \tag{15}$$

all in units of $10^{-3}/m_{\pi^+}$. Consequently, for $E_{0+}^{\pi^0 n} = 0$, we have $E_d^{ss} = -0.45$ which is of opposite sign to the value (14) based on the ChPT prediction for $E_{0+}^{\pi^0 n}$ (Table 1). If one were to use the empirical value for the proton amplitude (Table 1), the single-scattering contribution would be somewhat reduced.

We find for the two-nucleon (tb) contribution at $O(Q^3)$

$$E_d^{tb,3} = (-1.90\,,\,-1.88\,,\,-1.85\,,\,-1.88) \times 10^{-3}/m_{\pi^+} \ , \tag{16}$$

for the Argonne V18, Reid, super soft core [29] and Bonn [30] potentials, in order, using $g_A = 1.32$ as determined from the Goldberger-Treiman relation (for $g_{\pi NN} = 13.4$, consistent with the calculation of the elementary amplitudes [10]). We have also verified [2] that the result changes by only a few percent if the chiral potential of Ref. [21] is used. ¿From the graphs of $O(Q^4)$ we find

$$E_d^{tb,4} = (-0.25\,,\,-0.23\,,\,-0.27) \times 10^{-3}/m_{\pi^+} \ , \tag{17}$$

for the Argonne V18, Reid and super soft core potentials, in order, setting $\Lambda_\pi = 1\,\text{GeV}$. These amount to corrections of the order of 15% to the $O(Q^3)$ two-nucleon terms. If one varies the cut-off Λ_π from 0.65 to 1.5 GeV, the $O(Q^4)$ two-nucleon contribution using the Argonne V18 potential varies between -0.24 and -0.29 (in canonical units), which is a modest cut-off dependence. If one

Table 2: Values for E_d in units of $10^{-3}/m_{\pi^+}$ from one-nucleon contributions ($1N$) up to $\nu = 0$, two-nucleon kernel ($2N$) at $\nu = -1$ and at $\nu = 0$, and their sum ($1N + 2N$).

$1N$	$2N$		$1N+2N$
$\nu \leq 0$	$\nu = -1$	$\nu = 0$	$\nu \leq 0$
0.36	−1.90	−0.25	−1.79

were to include the contribution from the recoil graphs 5n,r, the numbers given in Eq. (17) would change by less than 0.5%. The result (17) is comforting because it shows that the chiral expansion of the two-nucleon contributions is well under control.

In summary, taking the Argonne V18 potential for definiteness, the chiral expansion of the electric dipole amplitude E_d is shown in Table 2. It is difficult to estimate the theoretical uncertainty. An educated guess is obtained as follows. We allow $E_{0+}^{\pi^0 p}$ to vary between −1.0 and −1.5, so that for a fixed value of $E_{0+}^{\pi^0 n} = 2.13$, E_d^{ss} varies between 0.2 and 0.4 (all numbers in canonical units). The results for the two-nucleon contributions are stable at $O(Q^3)$, whereas we assign a conservative uncertainty of ±0.1 to the corrections at next order due to the cut-off dependence. We thus obtain the $O(Q^4)$ *prediction*[3]

$$E_d = (-1.8 \pm 0.2) \times 10^{-3}/m_{\pi^+}. \tag{18}$$

Clearly, this estimate does not include contributions from higher-order effects, which might give rise to a larger uncertainty. A model-dependent estimate [31] of some $\nu = 1$ ($O(Q^5)$) terms —which are enhanced in the $m_\pi \to 0$ limit because of small energy denominators— suggests a 10% error from neglected higher orders in the kernel itself.

To see the sensitivity to the elementary neutron amplitude, we set the latter to zero and find $E_d = -2.6 \times 10^{-3}/m_{\pi^+}$ (for the Argonne V18 potential) which results in a cross-section twice as large as the one from the ChPT prediction (18). (For other values of $E_{0+}^{\pi^0 n}$, E_d can be calculated from Eq. (15).) This is consistent with previous model results[23] which by omitting the one-loop $O(Q^3)$ single-scattering diagram (Fig. 2c) get a smaller E_d^{ss} and a cross-section *about a factor of 2 larger than ours.*

This is a unique situation: the EFT makes a prediction that differs significantly from conventional models. A test of this prediction is thus an important check of our understanding of the role of QCD at low energies. Such a test was recently carried out at Saskatoon[32]. The experimental results for the pion photoproduction cross-section near threshold are shown in Fig. 6, together

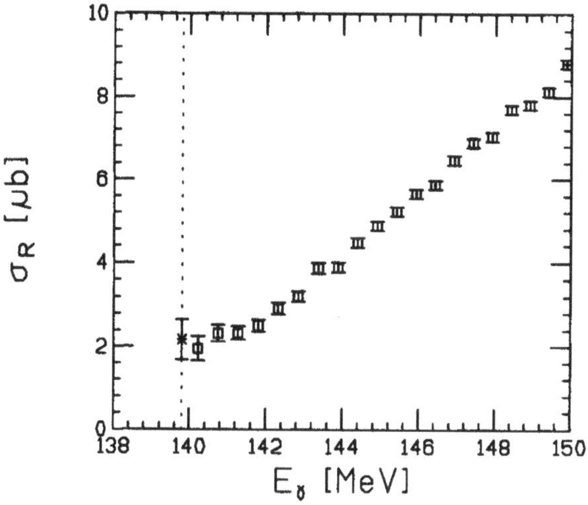

Figure 6: Reduced cross-section $\sigma_R = (k/q)\sigma$ in μb for neutral pion photoproduction as function of the photon energy in MeV. Threshold is marked by a dotted line. Squares are data points from Ref. [32] and the star is the ChPT *prediction* of Ref. [3]. Figure courtesy of Ulf Meißner.

with our ChPT prediction at threshold. Inelastic contributions have been estimated in Ref. [32] and are smaller than 10% throughout the range of energies shown. At threshold, Ref. [32] extracts

$$E_d = (-1.45 \pm 0.09) \times 10^{-3}/m_{\pi^+}. \tag{19}$$

While agreement with ChPT to order $\nu = 0$ is not better than a reasonable estimate of higher-order terms, it is clearly superior to tree-level models. This is compelling evidence of the importance of chiral loops and of the consistency of nuclear EFT. A further test will come from an electroproduction experiment currently under analysis in Mainz[33].

4 Pion Production

There is a fundamental difference between the process of pion production and the scattering scenario considered in the previous sections. In the latter, the typical momenta carried by nucleons are $O(m_\pi)$. This condition cannot be satisfied for the former reaction, even at threshold. Consider for concreteness $NN \to NN\pi$ in the cm frame. At threshold the total energy is $2m_N + m_\pi$, so that the initial kinetic energy of each nucleon is $m_\pi/2$. Since the energy is small compared to the nucleon mass, we can still use a non-relativistic framework for nucleons: the non-relativistic kinetic energy formula holds and the mass $2m_N$ plays no dynamical role. The typical momentum of real and virtual particles involved in this process is then

$$p_{\text{typ}} \sim \sqrt{m_N m_\pi}. \tag{20}$$

While in the usual power counting energy and momenta are counted as equal in irreducible diagrams, here energies are $O(m_\pi)$ but momenta are $O(\sqrt{m_N m_\pi})$. The standard power counting arguments of Sect. 2 are not valid.

Even though typical momenta of $O(\sqrt{m_N m_\pi})$ alters the power counting, it should not spoil an expansion scheme. Indeed, the scale of the typical momenta still goes to zero as we approach the chiral limit and hence a chiral expansion remains sensible. In Ref. [4] a naive extension of the usual chiral expansion was attempted. Of course, this expansion can be expected to be more slowly convergent than in the usual scattering case. The expansion parameter becomes $(m_\pi/M_{QCD})^{1/2}$ rather than m_π/M_{QCD}. The point is that since a typical momentum of the nucleons is $O(\sqrt{m_N m_\pi})$ nothing prevents a momentum transfer of $O(\sqrt{m_N m_\pi})$ in interactions. Consider for example contributions involving a single-pion exchange between nucleons. Chiral symmetry requires that (to leading order) each vertex contain a derivative and hence be proportional to the momentum transfer. In the traditional Weinberg power counting (2), a pion-nucleon vertex contributes a *relative* m_π/M_{QCD} to the total power counting of an irreducible graph. With pion production kinematics, however, those vertices would contribute $(m_\pi m_N)^{1/2}/M_{QCD} \sim (m_\pi/M_{QCD})^{1/2}$. Similarly, in the traditional power counting a meson propagator $1/(q^2 - m_\pi^2)$ contributes as $1/m_\pi^2$ while with production kinematics it goes as $1/m_\pi M_{QCD}$. In short, with these kinematics whenever a three-momentum enters into the power counting it contributes $(m_\pi M_{QCD})^{1/2}$.

This changes the usual correspondence between index Δ and order ν in irreducible diagrams. The leading nucleon propagator, for example, includes both the static term p^0 (in $\mathcal{L}^{(0)}$) and the kinetic term $p^2/2m_N$ (in $\mathcal{L}^{(1)}$) at the same order in power counting for pion production, contrary to the situation in ChPT applied to one-baryon systems. The nucleon propagator then is $\sim 1/m_\pi$. Relativistic corrections $p^4/8m_N^3 + \ldots$ are relatively smaller by $\sim m_\pi/m_N$ and can still be considered higher-order insertions. Note that this is completely consistent with our decomposition of the full amplitude into a kernel and wavefunctions obtained from a Schrödinger equation, and, contrary to what is stated in Ref. [34], does *not* imply the need of a relativistic framework. The delta propagator differs from the nucleon's by the presence of the mass difference $m_\Delta - m_N \sim 2m_\pi$. As pointed out in Ref. [4], the delta propagator is then

$\sim -1/m_\pi$ and tends to interfere destructively with the nucleon contributions to the kernel. Integrating out the delta should be a bad approximation here, even at threshold. The pion propagator, in turn, is $\sim 1/m_\pi m_N$. In interactions, each time derivative is associated with a factor of m_π, a space derivative with a factor of $\sqrt{m_N m_\pi}$. Finally, a loop brings a $(m_N m_\pi)^{3/2} m_\pi/(4\pi)^2$.

There is another subtlety in pion production and this concerns the notion of irreducibility. Nucleons can interact repeatedly by the exchange of mesons of momenta $Q \sim m_\pi$ before and after the emission of the pion. We account for this through the iteration of a potential, which produces initial and final wavefunctions, as we did before. The emission of the pion, on the other hand, involves at least one transfer of $O(m_\pi)$ energy among nucleons. Again, as before, we define a kernel where intermediate energies are of $O(m_\pi)$. The new feature here is that at least two nucleons are connected in the kernel. This is an obvious consequence of the fact that single-nucleon pion production is forbidden by four-momentum conservation. Impulse-like diagrams can be drawn, but since both intermediate states (before and after pion emission) cannot simultaneously be within $\sim m_\pi^2/m_N$ of being on-shell, they do not correspond to a decomposition of the amplitude in irreducible sub-diagrams.

¿From now on I specialize to $NN \to NN\pi$ close to threshold. We can then restrict ourselves to an S-wave pion. This immediately simplifies the theory as diagrams with the usual, all-important P-wave couplings of the external pion to nucleons (proportional to g_A) and deltas does not contribute. Of the interactions in the chiral Lagrangian $\mathcal{L}^{(0)}$, only the isovector Weinberg-Tomozawa vertex survives in lowest orders. Moreover, it does not contribute at tree level to $pp \to pp\pi^0$, since in this reaction the exchanged pion is also neutral; $pp \to pp\pi^0$ involves then at least one $\Delta \geq 1$ interaction. Results for this channel will be potentially more sensitive to short-range aspects of the QCD dynamics that contribute to parameters of higher-index operators, which are poorly constrained by chiral symmetry. Because these operators are also less well determined from other processes, the results will be under larger theoretical uncertainty.

A detailed analysis of various contributions to the kernel according to the modified power counting can be found in Refs. [4,5,6]. Contributions start at $O((m_\pi/m_N)^{1/2})$. They come from tree diagrams. First, there is a diagram of the rescattering type, where a P-wave virtual pion emission from one nucleon is followed by an S-wave rescattering on the second nucleon via the Weinberg-Tomozawa interaction. I refer to this as the "WT term". Second, there are further diagrams of the rescattering type where rescattering proceeds instead via *(i)* a nucleon pole between a P-wave interaction of the virtual pion to the nucleon and an S-wave emission of the real pion through the Galilean correction to the g_A term; or via *(ii)* a delta pole between a P-wave transition interaction of the virtual pion and an S-wave emission of the real pion through the Galilean correction to the leading $\pi N\Delta$ coupling. Third, there are diagrams just as these, but where a non-derivative four-nucleon contact interaction replaces virtual pion exchange. Diagrams with the nucleon pole are automatically calculated by sandwiching the Galilean correction to the g_A term between wavefunctions. I refer to this as the "impulse term". The corresponding delta diagrams I refer to as the "delta term". All other processes are suppressed compared to the leading terms by powers of $(m_\pi/M_{QCD})^{1/2}$. Thus, the modified power counting verifies the intuition that had we lived in a world with sufficiently light quarks the impulse term would have been among dominant contributions. The failure of the impulse term to explain the $pp \to pp\pi^0$ data by itself is interesting in that it highlights the role of chiral-symmetry breaking. At $O(m_\pi/m_N)$ two-pion exchange graphs appear. At $O((m_\pi/m_N)^{3/2})$ there are *(i)* one-pion exchange with rescattering via the Galilean correction to the Weinberg-Tomozawa term ("GC term"); *(ii)* recoil corrections to nucleon and delta pole terms ("recoil term"); *(iii)* one-pion exchange with rescattering via $\Delta = 1$ S-wave seagull operators ("seagull term") with parameters c_i; and *(iv)* short-range tree-level contributions from $NNNN\pi$ vertices in the Lagrangian $\mathcal{L}^{(2)}$. Higher orders include corrections to the $NNNN\pi$ vertices.

Casting the pion production reactions in a chiral Lagrangian framework has some obvious advantages, even if the expansion is expected to be slowly convergent.

Most important is that one can gain insight into the role of pion exchange and in particular into the confusion found in the literature regarding the on/off-shell S-wave seagull terms, which are so important for $pp \to pp\pi^0$.

(1) Although the S-wave seagull term is formally sub-leading, it is by no means obvious that it *must* be very small. Dimensionless spin and isospin factors occur in most amplitudes and can significantly enhance some partial waves; these dimensionless factors are not counted in power-counting arguments. It is also clear why this contribution is likely to be much larger than estimates based on the on-shell kinematics. The on-shell S-wave pion-nucleon scattering amplitude is anomalously small in a power-counting sense, requiring a partial cancellation among the parameters c_i [35]. However, with the kinematics of pion production, these terms do not cancel so efficiently, and one expects the on-shell estimates to be low.

(2) One can understand why the results of Ref. [17] are wrong. The key point is that the magnitude and sign of the seagull process do depend on our power-counting arguments. So far as the power counting is concerned, we have identified an interesting problem: how one proceeds when the typical momenta

are $\sim (m_\pi m_N)^{1/2}$. This is relatively straightforward with the choices of pion and nucleon fields in which the pseudo-Goldstone-boson nature of the pion is manifest; in any such basis, these fields provide non-linear realizations of chiral symmetry and pions couple derivatively in the chiral limit. As is well-known, it is possible to redefine fields without changing the physics, provided we use our field definition consistently throughout the calculation (see, *e.g.*, Ref. [36]). Different choices of interpolating fields amount, after all, to no more than different bookkeeping of where various bits of the physics reside. For example [37], one can transform to a linear basis at the expense of producing extra seagull vertices that are momentum independent but *not* proportional to the quark masses. The sole purpose of such large terms is to ensure a cancellation of large pole terms and to produce a small pion-nucleon scattering amplitude. Such a cancellation would seem magical were it not understood as a consequence of a concealed chiral symmetry; power counting in this basis becomes highly non-trivial. Moreover, failure to include such seagull vertices in a calculation would spoil chiral symmetry. Now, the pion field in Ref. [17] does not provide a non-linear realization of the symmetry, as it is based on models exploiting PCAC, while the fields we use are not. If both calculations consistently included all other processes, then the differences between Ref. [4] and Ref. [17] in the off-shell seagull contributions would be compensated for by differences in the contributions of other mechanisms. Indeed, starting from our general chiral Lagrangian and performing a simple pion-field redefinition, we can reproduce the off-shell behavior of the PCAC pion-nucleon amplitude of Ref. [17]. However, in doing so an extra, anomalously large $NNNN\pi$ contact term is generated, whose sole function is to cancel the effect of the change in off-shell behavior of the pion-nucleon amplitude brought about by the field redefinition. This is in complete analogy to the change to a linear basis mentioned above. Since the calculation of Ref. [17] does not include other mechanisms besides the impulse and an arbitrarily off-shell-extended seagull terms, it implicitly depends on the choice of pion field and violates chiral symmetry. Incidentally, the same argument applies to the two-pion-range three-nucleon force [20].

A second gain of using a ChPT framework is that one can study the relative impact of the various contributions. Exploratory studies of convergence were carried out in Refs. [4,5] for $pp \to pp\pi^0$ and in Ref. [6] for $np \to d\pi^0$, $pp \to d\pi^+$, and $pp \to pn\pi^+$. For these estimates some very drastic approximations were made.

First, phenomenological potentials were used because the chiral potential of Ref. [21] is quantitative only at energies well below the threshold for pion production. As in Sects. 2 and 3, we used two different NN potentials, the Argonne V18 and the Reid93 [38], in an attempt to estimate the uncertainty due to the mismatch between the nucleon field used in the potential and the one (based on the chiral Lagrangian) used in the calculation of the kernel. Unlike photoproduction, however, the relatively large momenta involved is not auspicious, and we indeed find that the differences among potentials are not negligible for the smallest, $pp \to pp\pi^0$ reaction.

Second, we considered only the lower NN partial waves, expected to be dominant. In $pp \to pn\pi^+$ and $pp \to d\pi^+$, the transition is mostly ${}^3P_1 \to {}^3S_1$, although for the deuteron final state the effect of the D wave is also important, and has been included. The cross-section for $np \to d\pi^0$ is directly related by isospin to that of $pp \to d\pi^+$, but it is free of Coulomb factors. In $pp \to pp\pi^0$, the transition is mostly ${}^3P_0 \to {}^1S_0$.

Third, we limited ourselves to the most natural prescription that the transferred energy is $\simeq m_\pi/2$. At sub-leading orders in the expansion, an apparent ambiguity arises concerning the correct treatment of the energy transferred in pion rescattering, as can be seen in the conflicting results of estimates of the same kernel [4,39]. This ambiguity will be resolved in a more complete EFT treatment where the potential and kernel are treated consistently.

Fourth, a meson-saturation model was used and brought substantial simplification. *(i)* In any reasonable model, the delta terms involving short-range fermion interactions turn out to be smaller than those involving pion exchange; for example, they could arise from a_1 exchange, but then the relatively high a_1 mass suppresses this contribution. For the purpose of estimating the effect of the isobar, we considered only the pion contribution. *(ii)* Ideally, of course, we should simply calculate the two-pion exchange contributions and include them in our analysis. There are two difficulties with this. The first is that we are using phenomenological potentials and we do not how to implement the traditional ChPT scheme of regulating the loop and absorbing the regulator dependence into the coefficient of a contact term in a way consistent with this potential. The second is that even if we did this in a consistent way, we would not be able to make a prediction until we fitted the contact term to some other process. Accordingly, we have not included these loop diagrams explicitly, but substituted the combined contribution of two-pion exchange and $NNNN\pi$ vertices by heavier meson-exchange diagrams. These included σ and ω exchange in Z-graphs [16] and in Roper-pole diagrams [5], and pion emission from $\rho - \omega$, $\eta - a_0$, and $\sigma - \pi'$ transitions [5].

The various contributions to the kernel obtained in this way depend on several parameters that can be estimated from other processes. Details can be found in Refs. [4,5,6]. Most crucial are the seagull parameters c_i which have been determined from pion-nucleon scattering in different ways. For illustrative purposes we used two different sets: a set from an $O(Q^3)$ fit to the σ-term, the isospin-even scattering length and the axial polarizability[35], denoted "cl" or "SeaI"; and a set extracted at $O(Q^2)$ from parameters of a sub-threshold expansion [40], denoted by "ste" or "SeaII". The spread of results from these sets gives an idea of the size of higher-order effects. The parameters used for the short-range mechanisms produce contact interactions of natural size.

The calculated cross-section for $pp \to pp\pi^0$ [5] is shown in Fig. 7, together with experimental data from Ref. [11]. The full curves represent the complete result for the Reid93 potential, the lower curve using the $O(Q^3)$ rescattering parameters, the higher the $O(Q^2)$ parameters. Also shown are the dashed curves with the corresponding results for the Argonne V18 potential. The overall theoretical uncertainty is quite large. It arises from a combination of factors. First, all contributions to the cross-section are very small. Second, there are cancellations among the various terms. In particular, we verify the expected destructive interference between impulse and delta terms which, it turns out, is sensitive to the potential employed. Rescattering contributions also tend to interfere with the impulse term, but the degree of destruction depends sensitively on the value of the seagull parameters. As a consequence the short-range contributions (particularly the σ, ω Z-graphs and $\rho - \omega$ exchange) become important. Within the large inherent theoretical uncertainty, meson exchange can explain the observed cross-section, but the situation is clearly unsatisfactory.

The calculated (without Coulomb) cross-section for $pp \to d\pi^+$ [6] is shown in Fig. 8, together with the (Coulomb corrected) experimental data from Refs. [12,13]. The theoretical curves are approximately constant, as expected from phase-space arguments. The result is dominated numerically by the WT term, as other terms are again small and partially canceling. When the small contribution from $O(Q^2)$ seagull parameters is considered, the sum of all subleading contributions is small; the sum is somewhat larger and of opposite sign to leading order when $O(Q^3)$ seagull parameters are used. The long-range nature of the dominant WT interaction ensures that dependence on the potential is small. In any case theory fails by a factor of about 2. Similar results hold for $pp \to pn\pi^+$ [6].

Although we have studied pion production within an EFT-inspired approach [c], we have incorporated most of the contributions found in the literature[15,16,42]. Reasonably-sized short-range effects can be found to fit neutral-pion production, but charged-pion production seems to require further longer-range interactions, and a better treatment of two-pion exchange contributions is probably necessary before any definite conclusions are drawn. The pandemic failure however suggests that for pion production the best organizational prin-

[c] Other EFT-inspired discussions can be found in Refs. [39,41,34].

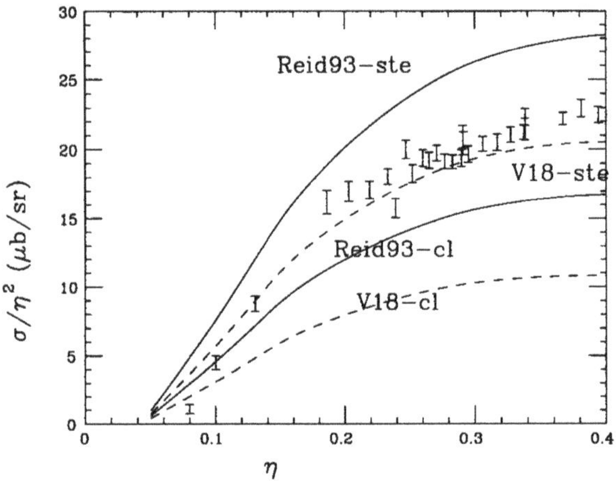

Figure 7: Reduced cross-section σ/η^2 for $pp \to pp\pi^0$ in μb/sr as function of the outgoing pion momentum η in units of the pion mass. Points are the measurements of Ref. [11] and curves are calculations of Ref. [5] for two different NN potentials and two sets of seagull parameters.

ciple has not yet been identified.

5 Conclusion

I have shown here how one can extend the EFT approach to include scattering of particles with momenta $O(m_\pi)$ on light nuclei. In the particular case of neutral pion photoproduction on the deuteron at threshold, a prediction for the cross-section was made that differed from conventional model predictions by a factor of 2; subsequent data unequivocally favored the EFT prediction. In contrast, the extension to processes of momenta $O(\sqrt{m_\pi m_N})$ is unsatisfactory, as the poor description of pion production in NN collisions in both EFT-inspired and conventional approaches shows.

Acknowledgments

I thank Silas Beane, Chang-Yong Lee, Veronique Bernard, Harry Lee, Ulf Meißner, Tom Cohen, Jim Friar, Jerry Miller, Dan Riska, and Carlos da Rocha for the enjoyable collaborations that led to the results presented here. My research is supported in part by the NSF grant PHY 94-20470.

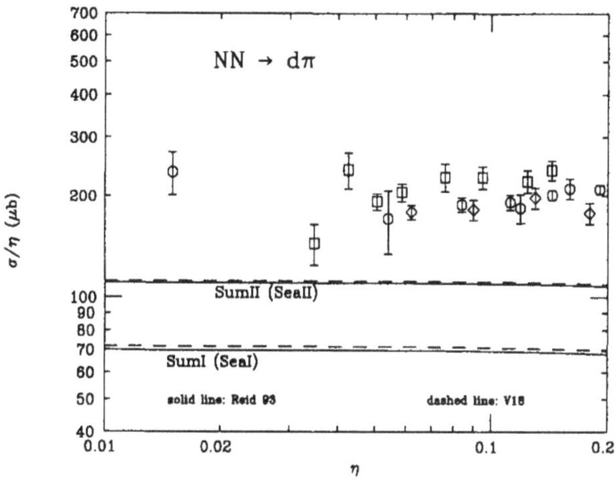

Figure 8: Reduced cross-section σ/η for $pp \rightarrow d\pi^+$ in μb as function of the outgoing pion momentum η in units of the pion mass. Points are the data from Refs. [12,13] and curves are the calculations of Ref. [6] for two NN potentials and two sets of seagull parameters.

References

1. U. van Kolck, "Overview", *these proceedings.*
2. S.R. Beane, C.Y. Lee, and U. van Kolck, *Phys. Rev.* C **52**, 2914 (1995).
3. S.R. Beane, V. Bernard, T.-S.H. Lee, U.-G. Meißner, and U. van Kolck, *Nucl. Phys.* A **618**, 381 (1997).
4. T.D. Cohen, J.L. Friar, G.A. Miller, and U. van Kolck, *Phys. Rev.* C **53**, 2661 (1996).
5. U. van Kolck, G.A. Miller, and D.O. Riska, *Phys. Lett.* B **388**, 679 (1996).
6. C.A. da Rocha, G.A. Miller, and U. van Kolck, U. of Washington preprint NT@UW-98-10, `nucl-th/9904031`.
7. S.R. Beane, *these proceedings.*
8. M. Rho, *these proceedings.*
9. M.J. Savage, *these proceedings.*
10. V. Bernard, N. Kaiser, and U.-G. Meißner, *Phys. Lett.* B **378**, 337 (1996); *Phys. Lett.* B **383**, 116 (1996).
11. H.O. Meyer *et al*, *Phys. Rev. Lett.* **65**, 2846 (1990); *Nucl. Phys.* A **539**, 633 (1992); A. Bondar *et al*, *Phys. Lett.* B **356**, 8 (1995).
12. D.A. Hutcheon *et al*, *Nucl. Phys.* A **535**, 618 (1991).

13. M. Drochner *et al*, *Phys. Rev. Lett.* **77**, 454 (1996); P. Heimberg *et al*, *Phys. Rev. Lett.* **77**, 1012 (1996).
14. W.W. Daehnick *et al*, *Phys. Rev. Lett.* **74**, 2913 (1995); J.G. Hardie *et al*, *Phys. Rev.* C **56**, 20 (1997); W.W. Daehnick *et al*, *Phys. Lett.* B **423**, 213 (1998); R.W. Flammang *et al*, *Phys. Rev.* C **58**, 916 (1998).
15. D.S. Koltun and A. Reitan, *Phys. Rev.* **141**, 1413 (1966); G.A. Miller and P.U. Sauer, *Phys. Rev.* C **44**, R1725 (1991); J.A. Niskanen, *Phys. Rev.* C **43**, 36 (1991); *Phys. Lett.* B **289**, 227 (1992); *Phys. Rev.* C **53**, 526 (1996); C. Horowitz, *Phys. Rev.* C **48**, 2920 (1993).
16. T.-S.H. Lee and D.O. Riska, *Phys. Rev. Lett.* **70**, 2237 (1992); C.J. Horowitz, D.K. Griegel, and H.O. Meyer, *Phys. Rev.* C **49**, 1337 (1994).
17. E. Hernández and E. Oset, *Phys. Lett.* B **350**, 158 (1995).
18. S. Weinberg, *Phys. Lett.* B **251**, 288 (1990); *Nucl. Phys.* B **363**, 3 (1991); *Phys. Lett.* B **295**, 114 (1992).
19. C. Ordóñez and U. van Kolck, *Phys. Lett.* B **291**, 459 (1992); U. van Kolck, U. of Texas Ph.D. dissertation (1993).
20. U. van Kolck, *Phys. Rev.* C **49**, 2932 (1994); J.L. Friar, D. Hüber, and U. van Kolck, *Phys. Rev.* C **59**, 53 (1999).
21. C. Ordóñez, L. Ray, and U. van Kolck, *Phys. Rev. Lett.* **72**, 1982 (1994); *Phys. Rev.* C **53**, 2086 (1996).
22. U. van Kolck, *Prog. Part. Nucl. Phys.* **43**, 409 (1999).
23. J.H. Koch and R.M. Woloshyn, *Phys. Rev.* C **16**, 1968 (1979); J.M. Laget, *Phys. Rep.* **69**, 1 (1981).
24. V. Bernard, N. Kaiser, and U.-G. Meißner, *Z. Phys.* C **70**, 483 (1996).
25. R.B. Wiringa, V.G. Stoks, and R. Schiavilla, *Phys. Rev.* C **51**, 38 (1995).
26. R.V. Reid, *Ann. Phys.* **50**, 411 (1968).
27. M.M. Nagels, T.A. Rijken, and J.J. de Swart, *Phys. Rev.* D **17**, 768 (1978).
28. M. Lacombe *et al*, *Phys. Rev.* D **12**, 1495 (1975); *Phys. Rev.* C **21**, 861 (1980).
29. R. de Tourrell and D.W. Sprung, *Nucl. Phys.* A **210**, 193 (1973).
30. R. Machleidt, K. Holinde, and Ch. Elster, *Phys. Rep.* **149**, 1 (1987).
31. P. Wilhelm, *Phys. Rev.* C **56**, 1215 (1997).
32. J.C. Bergstrom *et al.*, *Phys. Rev.* C **57**, 3203 (1998).
33. MAMI Proposal A1/1-96, R. Neuhausen (spokesperson).
34. V. Bernard, N. Kaiser, and U.-G. Meißner, *Eur. Phys. J.* A **4**, 259 (1999).
35. V. Bernard, N. Kaiser, and U.-G. Meißner, *Int. J. Mod. Phys.* E **4**, 193 (1995).

36. S. Coleman, J. Wess, and B. Zumino, *Phys. Rev.* **177**, 2239 (1969); C.G. Callan, S. Coleman, J. Wess, and B. Zumino, *Phys. Rev.* **177**, 2247 (1969).
37. S.A. Coon and J.L. Friar, *Phys. Rev.* C **34**, 1060 (1986).
38. J.L. Friar, G.L. Payne, V.G.J. Stoks, and J.J. de Swart, *Phys. Lett.* B **311**, 4 (1993).
39. B.Y. Park, F. Myhrer, J.R. Morones, T. Meissner, and K. Kubodera, *Phys. Rev.* C **53**, 1519 (1996); T. Sato, T.-S.H. Lee, F. Myhrer, and K. Kubodera, *Phys. Rev.* C **56**, 1246 (1997).
40. V. Bernard, N. Kaiser, and U.-G. Meißner, *Nucl. Phys.* B **457**, 147 (1995).
41. C. Hanhart, J. Haidenbauer, M. Hoffmann, U.-G. Meißner, and J. Speth, *Phys. Lett.* B **424**, 8 (1998); E. Gedalin, A. Moalem, and L. Razdolskaya, nucl-th/9803029; nucl-th/9812009; V. Dmitrašinović, K. Kubodera, F. Myhrer, and T. Sato, South Carolina preprint USC-PHYS(NT)-99-03, nucl-th/9902048.
42. T.-S.H. Lee, nucl-th/9502005; C. Hanhart, J. Haidenbauer, A. Reuber, C. Schütz, and J. Speth, *Phys. Lett.* B **358**, 21 (1995); C. Hanhart, J. Haidenbauer, O. Krehl, and J. Speth, *Phys. Lett.* B **444**, 25 (1998).

Discussion during U. Van Kolck's presentation

Chair: *F. Ravndal (Oslo University)*
Notes taken by *J. McGovern (University of Manchester)*

During the talk there was discussion of the inconsistency between the transition operator and the potential; Kaplan asked if the operator was more trustworthy, and van Kolck said that using a good wavefuntion reduced some of the errors; Bedaque asked if the operator was higher order, and Meißner asked if it was known to higher order, to which the answer was no in both cases.

Afterwards, Cohen asked if the poor agreement for $pp \to d\pi^+$ and $pp \to pn\pi^+$ was due to a problem with the calculation or to the existence of an energy scale in the process that was too large; van Kolck felt that various refinements were required before the scale was blamed. Miller asked about relativistic corrections; van Kolck pointed out that the static limit was not used, but that relativistic corrections were higher order. Lutz asked for comment on Kubedera's claim that higher loops get progressively larger; van Kolck blamed the use of the static propagator, and Meißner confirmed that higher loops are small in a relativistic calculation. Savage asked what further work would be done; van Kolck said that the wavefunction and kernel should be treated on the same footing, but that a good wavefunction would be needed first; Savage then commented that perhaps the calculation should be used to pin down counterterms rather than to make predictions.

ON MAKING PREDICTIONS WITH EFFECTIVE FIELD THEORIES IN NUCLEAR PHYSICS

MANNQUE RHO*
Service de Physique Théorique, CE Saclay
91191 Gif-sur-Yvette, France
and
School of Physics, Korea Institute for Advanced Study
Seoul 130-012, Korea
E-mail: rho@spht.saclay.cea.fr

Based on the effective field theory previously formulated by us to accurately *postdict* all low-energy two-nucleon properties as well as *predict* certain electroweak transitions in heavy nuclei, we make *parameter-free predictions* for the polarized np capture process $\vec{n} + \vec{p} \rightarrow d + \gamma$ presently being measured at the Institut Laue-Langevin in Grenoble. Other participants of this meeting are invited to make their own predictions using their preferred approaches and join the bet for the best prediction to confront the forthcoming experiment.

1 Introduction

This talk was initially meant to be given by Tae-Sun Park since he is the one doing most of the work – with a little help from three of us – but the organizers asked me (MR) to present the paper instead. In giving this talk, I would like to distinguish between the statements I am making for which the other members of the collaboration should not be held responsible and those that are endorsed by all of us. The former will be addressed (mostly in footnotes) by "I" and the latter by "We."

Since the early attempt to apply effective field theories to nuclear physics[1,2,3], there have been many papers written on the subject, the most recent development of which is being summarized in this meeting[4]. In confronting Nature with effective field theories in nuclei – one of the main themes of this meeting, one has had to be content mostly with *postdictions*, genuine *predictions* being harder to come by. The reason is simply that effective field theories involve, at each order of power counting, a certain number of parameters in the effective Lagrangian. It is believed that those parameters are in principle calculable from first principles for a given scale (*e.g.*, lattice QCD [5]) but in practice, they have to be fixed by experimental data. Once the parameters are so fixed, the Lagrangian can then be used to make predictions for other processes that

*Work presented here has been done in collaboration with T. -S. Park, K. Kubodera and D. -P. Min

involve the same parameters. Up to date, however, most of the calculations involved fixing of the parameters and only rarely could one predict and check the prediction by experiments.

In this talk, we would like to present a genuine prediction based on the formulation of an effective field theory that we have developed during the last few years. Since our formulation is available in the literature, we will not dwell on the details of the formalism but present the essence of the arguments while stressing the possible caveats involved. We illustrate how well postdict ions can be made and then how to make predictions for two specific processes, one for which data are available and hence the theory can be tested immediately and the other for which data are not yet available but will be forthcoming, offering a marvelous possibility for an honest prediction totally unbiased by available experiments.

2 The Chiral Filter

An early attempt predating the advent of QCD and effective field theories in nuclear physics was made in 1978 under a conjecture called the "chiral filter hypothesis"[6]. Based on current algebras, it was argued that corrections to the single-particle transitions in nuclei for the isovector M1 operator and the weak axial charge operator should be dominated by one-soft-pion-exchange two-body currents. As a corollary, if the soft-pion exchange two-body currents are suppressed either by symmetry or kinematics as for instance in the isoscalar EM current or the Gamow-Teller operator (*i.e.*, the space component of the axial current), the chiral filter says that corrections to the single-particle operator cannot be given by a few controlled terms, thereby making the calculations highly model-dependent. This hypothesis that appeared to be somewhat ad hoc at the time it was proposed turned out to be justified in the context of chiral perturbation theory [2]: The soft-pion exchange term is indeed the leading order correction in the chiral counting with the next-to-leading correction calculable but strongly suppressed.

There are two consequences of this chiral filter that survive in the context of modern effective field theories. One is a nontrivial postdict ion and the other is an interesting prediction that has been largely confirmed.

2.1 A strategy for effective field theory (EFT)

As it stands, there are effectively two "alternative" ways of power counting in setting up effective field theories for two-nucleon systems. One is the original Weinberg scheme [1] in which the leading four-Fermi contact interaction and a pion-exchange are treated on the same footing in calculating the "irreducible" graphs for a potential that is to be iterated to all orders in the "reducible" channel. The power counting is done only for the irreducible vertex. The other is the "power divergence subtraction" (PDS) scheme[7] in which only the leading (non-derivative) four-Fermi contact interaction is iterated to all orders with the higher-order contact interactions and the pion exchange treated perturbatively. While the PDS scheme is perhaps more systematic in the power counting, we believe that the Weinberg scheme is not only consistent with the strategy of EFT but also, in the sense developed below, more flexible and predictive with possible errors committed due to potential inconsistency in the power counting generically suppressed. In our work, this scheme is adopted.

Iterating to all orders in the reducible channel with the irreducible vertex is equivalent to solving the Schrödinger equation with a corresponding potential. This then suggests that we use, in calculating response functions to slowly varying electroweak fields that we are interested in, those wave functions computed with so-called "realistic potentials," the prime example being the Argonne v_{18} potential[8] (called in short Av18). In fact this procedure of mapping effective field theory to realistic wave functions – a hybrid approach[9] – for two-nucleon response functions employed previously by us[2] has recently been justified by means of a cutoff regularization[10]. van Kolck has presented a similar argument in support of such a hybrid procedure[11]. Now in this framework, the power counting reduces simply to a chiral counting in the irreducible vertex for the current as the current appears only once in the graphs. We are then allowed to separate the current matrix element of interest into the single-particle and exchange-current terms, with the single-particle matrix element given entirely by that with the Av18 wave function and the exchange current contribution – given by the matrix element with the same wave function – from operators computed in standard baryon chiral perturbation theory. The old soft-pion exchange term figuring in the chiral filter – if not suppressed – is just the leading order contribution in this series.

2.2 Unpolarized np capture

A good example of postdict ions that follow from the above scheme and that will be a basis for a genuine prediction described below is the process

$$n + p \rightarrow d + \gamma \qquad (1)$$

where both nucleons are unpolarized and the incoming neutron has a thermal energy[i]. This process has been computed to the next-to-next-to-leading order (NNLO) in the chiral counting in the scheme described above for the

[i] The relative momentum in the center of mass is $\sim 3.4 \times 10^{-3}$ MeV.

current [12]. The theoretical cross section $\sigma_{th} = 334 \pm 3$ mb which agrees with the experiment within the error bar consists of the leading contribution, 305.6 mb, coming from the single-particle matrix element given by the Av18 wave function and the remainder from the exchange current dominated by the soft one-pion exchange according to the chiral filter.

Two points are worth noting in this result. First, this is a bona-fide calculation and *not* a fit: Within the scheme adopted here, there are no free parameters. Secondly as shown in [10], the single-particle matrix element has a negligible uncertainty, so that the error in the theory is entirely attributed to the uncertainty in the exchange-current operator associated with the short-distance part of the interactions that cannot be accessed by chiral perturbation theory. This part introduces a scale and renormalization-scheme dependence and only those results that are not sensitive to this short-distance uncertainty can be trusted. In the framework in which the realistic wave functions figure, the short-distance scale is set by the cutoff proportional to r_C^{-1} where r_C is the "hard-core radius" that removes the part of the wave function for $r \lesssim r_c \neq 0$. The net effect of this cutoff is that in addition to cutting the radial integrals in the coordinate space, it removes *all* zero-range terms in the current operator including zero-ranged counter terms. The 1% error bar assigned to the theory for (1) represents the uncertainty in this cutoff procedure. This procedure can be justified for the process in question by using a cutoff $\sim r_C^{-1}$ and showing that the counter term "killed" by the hard core is in that uncertainty range. Given that the four-Fermi counter term is removed by the hard core which may be viewed as exploiting a scheme dependence, there are no more parameters in the theory [ii]. This procedure, familiar to nuclear physicists, will be referred to as "hard-core cutoff scheme" (or HCCS in short). Below we will see that when the chiral filter does not apply, the removal of the zero-range counter terms by the simple hard-core cutoff may be suspect.

2.3 Prediction for the axial-charge transitions in nuclei

The EFT scheme described in section 2.1 makes a rather clean prediction which has not yet been adequately appreciated in nuclear physics community. The chiral filter idea [6], by now validated in chiral perturbation theory [2], predicts that in the non-relativistic regime, the axial charge transition matrix element for the β decay process in nuclei

$$A(J^{\pm}) \to A'(J^{\mp}) + e + \nu; \quad \Delta T = 1 \tag{2}$$

[ii] See appendix B in the second reference of [12] for the zero-ranged counter term in question. In the framework of [7] which avoids scheme dependence at the expense of predictiveness, this counter term is a non-negligible parameter, rendering a bona-fide calculation infeasible [13].

should receive a huge one-soft-pion exchange current correction, amounting to more than 50 % of the single-particle matrix element. The prediction for this is quite robust and firm with very little nuclear model dependence: possible corrections from higher chiral order terms – which are calculable a priori – are estimated to be less than 10 % of the leading soft-pion terms[2,14].

How do we go about checking this prediction against experiments?

It is not possible to test this in the two-nucleon systems for which the effective field theory is most extensively developed: the matrix element for (2) cannot be measured in few-nucleon systems. The measurement can only be made in heavier nuclei. This means that to compare with experiments, we would have to account for the effect of density or multi-nucleon interactions on the transition. This introduces a subtlety which is interesting in its own right. We will show below that there is rather compelling evidence that both the chiral filter and density effect are confirmed[iii].

Experimentalists extract the axial charge matrix element $\mathcal{M}$ from their experiments and then write[17]

$$\epsilon_{MEC} = \mathcal{M}_{measured}/M_{th-1b} \tag{3}$$

where the numerator is the total "measured" value or more precisely the value extracted from experiments and the denominator is the *theoretical* single-particle matrix element of the single-particle current whose constants are *unrenormalized* by medium. The model dependence of the denominator – and to some extent the numerator – makes this quantity not entirely empirical, thus open to controversy among theorists: It would seem that the "experimental" ϵ_{MEC} would in practice depend upon the model for the wave function used for the sing-particle matrix element. On the other hand, our claim is that the theoretical expression for this quantity is well-defined and free of model dependence[iv].

[iii] The issue of effective field theory for hadrons immersed in a hadronic medium with baryon density ρ was one I (MR) would have liked to discuss in this meeting but it is out of the scope of this presentation. For a recent discussion on this subject, we simply refer to the article[15]. For the present problem, the net effect is the presence of what is known as "Brown-Rho (BR) scaling" $\Phi(\rho)$[16].

[iv] Several people in the audience voiced doubt as to whether the experimental test I discussed is truly valid. One of the reasons given is the model dependence of the so-called empirical information in ϵ_{MEC}. This objection is to some extent valid and needs to be carefully examined. The presently employed procedure is to calculate the single-particle matrix element with the "best" nuclear wave functions of the single-particle axial charge operator whose coupling constants are *unrenormalized* by medium. The question then is what does one mean by "best" wave function? Here we should stress that in the case in question, there is a reasonably satisfactory answer. Let's take Warburton's analyses[17]. Although his analyses do involve shell-model wave functions, the semi-empirical nature of the effective

Let $\mathcal{M}_{total}$ denote the total theoretical axial-charge matrix element and $\mathcal{M}_n$ with $n = 1,2$ be the matrix element of the n-body operator effective in medium. Then the prediction is that

$$\epsilon_{MEC}{}^{th} = \mathcal{M}_{total}/M_1 = \Phi^{-1}(1+\mathcal{R}), \tag{4}$$

$$\mathcal{M}_{total} = \mathcal{M}_1 + \mathcal{M}_2 \tag{5}$$

with

$$\mathcal{R} = \mathcal{M}_2/\mathcal{M}_1 = (M_2/M_1)\,(1+\mathcal{O}(10^{-1})) \tag{6}$$

where M_n is the matrix element of the n-body current with its basic coupling constants (*e.g.*, f_π, g_A etc) *unrenormalized* by medium. This relation follows within our HCCS (hard-core cutoff scheme) provided one assumes that all light-quark hadron mass M *except for the pion mass* scales in medium as

$$M^\star/M \approx f_\pi^\star/f_\pi \approx \Phi(\rho). \tag{7}$$

The expression (4) was first derived by Kubodera and Rho [18] and later corrected [14,15 v]. Now what we need is the value for Φ and $\mathcal{R}$ for a range of nuclei that have been measured. The range of nuclear density involved is $1/2 \lesssim \rho/\rho_0 \lesssim 1$ where ρ_0 is the nuclear matter density. For this range, we

transition operator method he adopted gives his results much more robustness than people normally expect from shell-model calculations. That is, after including the core polarization effects in the form of rescaling the single-particle matrix elements, there is in fact very little room for changing nuclear physics input in his analyses. As a measure of the basic soundness of his approach, we mention the following fact. As far as nuclear dynamics is concerned, the rank-zero and rank-one first-forbidden transitions share the same feature, but the exchange currents for them have very different behavior. The rank-one operators coming from the space component of the axial current has only small exchange currents whereas the time component of the axial current that contributes to the rank-zero operator is expected to have a *huge* exchange current. Now, Warburton's calculations reproduce very well the strengths of all the rank-one matrix elements with no further adjustments whereas, for the rank-zero matrix elements, he found it clearly necessary to introduce the "empirical" enhancement factor ϵ_{MEC}.

The other reason (which is theoretical) is that the BR scaling is so far implemented in the limit of infinite nuclear matter and the question arises regarding the finite size effect of the nuclei measured. My answer to this objection is that whereas experimentally the quantity ϵ_{MEC} may perhaps be somewhat model-dependent, theoretically, however, it is very well defined: It involves the BR scaling factor Φ which is smoothly varying for density up to $\rho = \rho_0$, so what matters is the average density involved and the quantity $\mathcal{R}$, the ratio of the matrix element of the exchange current over that of the single-particle operator, which is quite insensitive to nuclear models provided the same wave functions are used for both.

[v] There was an error in the original derivation due to the fact that the non-scaling of the pion mass, indicated by both theory and experiment, was not properly taken into account.

know from information gotten from giant dipole resonances and QCD sum rules in medium[15] that Φ goes as $\sim 1/(1+0.28(\rho/\rho_0))$,

$$0.78 \lesssim \Phi \lesssim 0.88 \quad \text{for} \quad 1/2 \lesssim \rho/\rho_0 \lesssim 1. \tag{8}$$

The two-body current in the ratio $\mathcal{R}$ is given within 10% by a soft-pion exchange which is completely fixed by chiral symmetry. The ratio turns out to be extremely *insensitive* to nuclear models used to compute the matrix elements and depends only slightly on density. It comes out to be[vi]

$$0.43 \lesssim \mathcal{R} \lesssim 0.61 \quad \text{for} \quad 1/2 \lesssim \rho/\rho_0 \lesssim 1. \tag{9}$$

This gives the range

$$1.63 \lesssim \epsilon^{th}_{MEC} \lesssim 2.06 \quad \text{for} \quad 1/2 \lesssim \rho/\rho_0 \lesssim 1. \tag{10}$$

A formula as simple as (4) must have a simple and clean test. It must be readily confirmed or infirmed.

We claim that there is a strong empirical support for this prediction. Indeed this prediction can be compared with the presently available "empirical" values[19,20,21,17]

$$\begin{aligned}\epsilon^{exp}_{MEC} &= 1.64 \pm 0.05 \quad (A=12)\ ; 1.62 \pm 0.05 \quad (A=50)\ ; \\ & \quad 1.95 \pm 0.05 \quad (A=205)\ ; 2.05 \pm 0.05 \quad (A=205 \sim 212).\end{aligned} \tag{11}$$

We further suggest that these constitute evidence for *both* (1) the gigantic enhancement predicted by the chiral filter, ranging from 46% to 61 % and (2) the additional enhancement predicted by BR scaling, ranging from 20 % to 45 %. There is of course the caveat that individual transitions must be subject to some finite-size effects but the point is that what one is probing here is a generic bulk property of nuclear matter – which to us is the most interesting part of the story.

3 Polarized Neutron-Proton Capture: A New Probe

We will now go outside of those processes protected by the chiral filter and make a genuine prediction even though the process, unprotected by the chiral filter, could be highly suppressed. One would think that such a prediction is out of the scope of effective field theories but surprisingly, it turns out not to be the case in our EFT scheme.

[vi] Owing to the chiral filter applicable to this process, the uncertainty in our hard-core cutoff scheme, *i.e.*, "killing" zero-range counter terms, is reflected in the 10% uncertainty in $\mathcal{R}$ mentioned above.

3.1 Selection rules

Consider the process

$$\vec{n} + \vec{p} \rightarrow d + \gamma \tag{12}$$

where now both the target proton and the projectile neutron are polarized. This process is being measured at the Institut Laue-Langevin (ILL) in Grenoble by Müller *et al.* [22] The interest in this experiment is that with the polarized target and the polarized beam, one can measure small matrix elements that are overwhelmed by the dominant isovector M1 matrix element when averaged over polarization. To see this, look at the quantum numbers involved in the process. The initial state of (12) at very low energy we are dealing with can be in either $^1S_0$ or $^3S_1$ channel. The Fermi-Dirac statistics requires that the former must be in $T = 1$ and the latter in $T = 0$ where T is the isospin. The final nuclear state in (12) is the deuteron which is in $^3S_1$ or $^3D_1$ with $T = 0$. There are then three relevant transition matrix elements with the emission of a soft photon, *i.e.*, isovector M1, isoscalar M1 (which we shall denote from now on as $\overline{\mathrm{M1}}$ to distinguish it from the isovector M1) and isoscalar E2. We shall adopt the convention of Müller et al [22] and write the transition amplitude as

$$\langle \psi_d(M_d), \gamma(\hat{k}\lambda)|T|\psi_{np}(s_p, s_n)\rangle = \chi^{\dagger}_{1M_d}\, \mathcal{M}(\hat{k}, \lambda)\, \chi_{s_p}\chi_{s_n} \tag{13}$$

with

$$\mathcal{M}(\hat{k}\lambda) = \sqrt{4\pi}\frac{\sqrt{v_n}}{2\sqrt{\omega}A_s}\left[i(\hat{k}\times\epsilon^*)\cdot(\vec{\sigma}_1 - \vec{\sigma}_2)\,\mathrm{M1}(^1S_0)\right.$$
$$\left. -i(\hat{k}\times\epsilon^*)\cdot(\vec{\sigma}_1 + \vec{\sigma}_2)\,\frac{\overline{\mathrm{M1}}(^3S_1)}{\sqrt{2}} + (\vec{\sigma}_1\cdot\hat{k}\vec{\sigma}_2\cdot\epsilon^* + \vec{\sigma}_2\cdot\hat{k}\vec{\sigma}_1\cdot\epsilon^*)\frac{\mathrm{E2}(^3S_1)}{\sqrt{2}}\right] \tag{14}$$

where M_d and λ are respectively the polarizations of the deuteron and the photon, $\hat{k}$ is the unit momentum vector of the photon, ω its energy, $\vec{\epsilon} \equiv \vec{\epsilon}(\vec{k}\lambda)$, v_n is the velocity of the neutron and A_s is the deuteron normalization factor $A_s \simeq 0.8850\,\mathrm{fm}^{-1/2}$. In the way defined, the quantities $\mathrm{M1}(^1S_0)$, $\overline{\mathrm{M1}}(^3S_1)$ and $\mathrm{E2}(^3S_1)$ all have a dimension of fm and the cross section for the unpolarized np system takes the form

$$\sigma_{unpol} = \left|\mathrm{M1}(^1S_0)\right|^2 + \left|\overline{\mathrm{M1}}(^3S_1)\right|^2 + \left|\mathrm{E2}(^3S_1)\right|^2. \tag{15}$$

The first term is the isovector M1 contribution, the second the isoscalar M1 and the last the isoscalar E2. As we shall see below, the second and third terms are strongly suppressed compared to the first, $\sim O(10^{-6})$, so the unpolarized cross section cannot "see" these terms.

3.2 Polarization observables

In order to see those small isoscalar terms, one measures polarization observables, *i.e.*, the photon circular polarization P_γ and the photon anisotropy η. For an unpolarized proton and a polarized neutron with the polarization vector $\vec{P}_n$, P_γ is given by

$$P_\gamma = \frac{|\vec{P}_n|}{2} \frac{2\sqrt{2}(\mathcal{R}_{\mathrm{M1}} - \mathcal{R}_{\mathrm{E2}}) + (\mathcal{R}_{\mathrm{M1}} + \mathcal{R}_{\mathrm{E2}})^2}{1 + \mathcal{R}_{\mathrm{M1}}^2 + \mathcal{R}_{\mathrm{E2}}^2} \tag{16}$$

where we have defined the ratios

$$\mathcal{R}_{\mathrm{M1}} \equiv \frac{\mathrm{M1}(^3S_1)}{\mathrm{M1}(^1S_0)},$$
$$\mathcal{R}_{\mathrm{E2}} \equiv \frac{\mathrm{E2}(^3S_1)}{\mathrm{M1}(^1S_0)}. \tag{17}$$

Since these ratios turn out to be $\sim 10^{-3}$, eq.(16) simplifies with high accuracy to

$$P_\gamma \simeq |\vec{P}_n| \sqrt{2}(\mathcal{R}_{\mathrm{M1}} - \mathcal{R}_{\mathrm{E2}}). \tag{18}$$

The anisotropy measures the fully polarized np system involving the polarization of the target and projectile nucleons and is given by

$$\begin{aligned}\eta &= \frac{I(90^\circ) - I(0^\circ)}{I(90^\circ) + I(0^\circ)} \\ &= pP \frac{\mathcal{R}_{\mathrm{M1}}^2 + \mathcal{R}_{\mathrm{E2}}^2 - 6\mathcal{R}_{\mathrm{M1}}\mathcal{R}_{\mathrm{E2}}}{4(1-pP) + (4+pP)(\mathcal{R}_{\mathrm{M1}}^2 + \mathcal{R}_{\mathrm{E2}}^2) + 2pP\,\mathcal{R}_{\mathrm{M1}}\mathcal{R}_{\mathrm{E2}}}\end{aligned} \tag{19}$$

where I is the photon intensity,

$$pP \equiv \vec{P}_p \cdot \vec{P}_n \tag{20}$$

and the angle in I measures the photon direction with respect the spin polarization of the neutron and the proton. Note that unless the factor $(1-pP) \sim 0$ the anisotropy η will be quadratically suppressed while P_γ is linear in the ratio. If however $(1-pP) \sim 0$, then the anisotropy could be substantial, supplying an additional formula that would allow one to extract the two ratios (17). The purpose of the experiment is to determine these two ratios. The quantity P_γ has already been measured before by Bazhenov *et al*[23], so the aim of the ILL experiment [22] is to measure the anisotropy η.

3.3 Doing EFT

As mentioned in section 2, the current involved here is not protected by the chiral filter. This means that the soft-pion exchange which is entirely given by chiral symmetry considerations cannot contribute. The corollary to the chiral filter hypothesis would then suggest that we might be opening Pandora's box. To our surprise, this does not seem to be the case for the problem at hand.

- **Power Counting**

Since the isovector M1 operator is calculated [2] very accurately to $\mathcal{O}(Q^3)$ relative to the single-particle operator (*i.e.*, $\mathrm{M1}(^1S_0) = 5.78 \pm 0.03$ fm in the notation of (15) which comes at $\mathcal{O}(Q^{-2})$) [vii], we will focus on the isoscalar operators here.

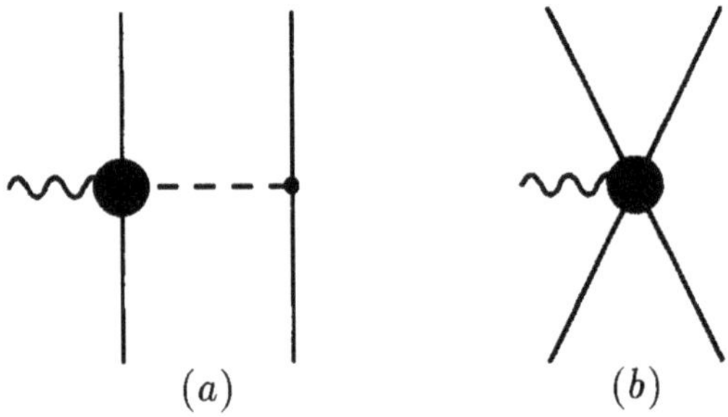

Figure 1: Generic diagrams for the two-body isoscalar current $\mathcal{B}^{\mu}_{2\mathrm{B}}$. The solid circles include counter-term insertions and (one-particle irreducible) loop-corrections. The wiggly line stands for the external field (current) and the dashed line the pion. One-loop corrections for the pion propagator and the πNN vertex are of course to be included at the same order.

We assume that as in the case studied so far [10], given the accurate wave functions, the leading single-particle matrix elements are accurately given for both $\overline{\mathrm{M1}}$ (*i.e.*, isoscalar M1) and E2 operators. The power counting that we adopt [1,2,10] then shows that while the ratio $\langle 2-\text{body}\rangle/\langle 1-\text{body}\rangle$ goes like $\mathcal{O}(Q^1)$ for the operator protected by the chiral filter (*e.g.*, the isovector M1), the ratio for the isoscalar operators goes like $\mathcal{O}(Q^n)$ with $n \geq 3$, so naively one would expect a suppression by two orders of power counting. Now the question is: How well can we pin down such two-body terms and higher-order corrections to them?

[vii] Unless otherwise specified, we will always give counting relative to the leading single-particle operator.

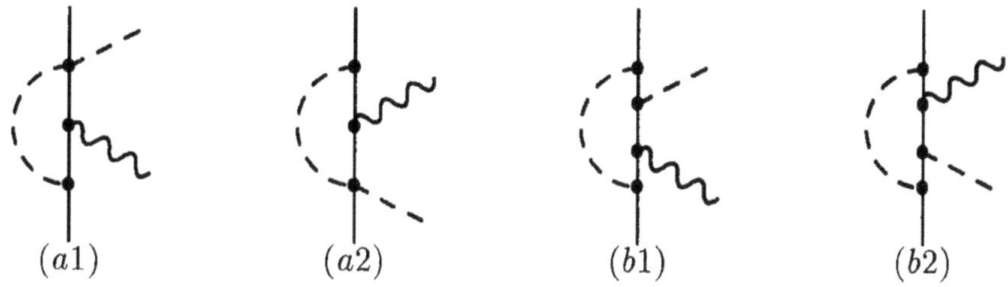

Figure 2: One-loop graphs that contribute to the $\mathcal{B}\pi NN$ vertex where $\mathcal{B}$ is the isoscalar current. They give rise to $\mathcal{O}(Q^4)$ and higher corrections to the leading order (LO) one-body term.

In our scheme, there are two classes of irreducible two-body terms in the current operator. The class A term consists of graphs with the one-pion exchange involving an $NN\pi\gamma$ vertex (see Fig.1a) and the class B term consists of graphs with two- or more-pion exchanges (see Fig.1b). Since there is no chiral-filter-protected one-pion exchange in the isoscalar vector current, the class A term receives its leading contribution from one-loop corrections (see Fig.2). The class B term is generically given by the loop graphs of Fig.3. The loop and derivatives (in Fig.3c) account for the additional power suppression.

The graphs in Figs.2 and 3 typically contribute to the isoscalar M1 operator ($\overline{\text{M1}}$) and E2 operator at $\mathcal{O}(Q^4)$. This is however not the whole story to $\mathcal{O}(Q^4)$, for there are so-called "counter terms" that can come in at $\mathcal{O}(Q^3)$. There are in fact two such terms in the case that we are concerned with. One such term is a one-pion exchange graph in Fig.1a with the $\mathcal{B}\pi NN$ vertex given by a finite counter term. This term that contributes to the $\overline{\text{M1}}$ operator but not to the E2 operator is dominated by the $\gamma\rho\pi$ coupling in the anomalous parity component of the effective chiral Lagrangian, *i.e.*, the Wess-Zumino term, which is connected with the Adler-Bell-Jackiw triangle anomaly. This term is known and so brings in no unknown parameters. The other is a four-Fermi counter term, call it g_4, in Fig.1b contributing only to $\overline{\text{M1}}$. The coefficient of this counter term, not known a priori, needs to be fixed in the usual way.

Note that there are no $\mathcal{O}(Q^3)$ counter term contributions to the E2.

- **There is No Parameter in the Theory**

We shall now argue that while there remains one parameter undetermined by the theory it can be gotten rid of in either of the two ways in which physics at short distance is treated.

We first note that there are no unknown parameters in the class A dia-

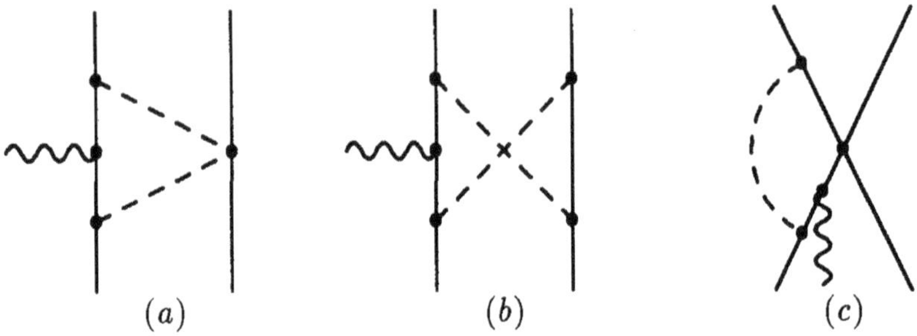

Figure 3: One-loop graphs that contribute to the two-body baryonic currents. They come at $\mathcal{O}(Q^4)$ and higher order relative to the LO one-body term. All possible insertions of the external line are understood.

grams: As mentioned, the $\mathcal{O}(Q^3)$ counter term is fixed by the Wess-Zumino term and the loop graphs are completely calculable to $\mathcal{O}(Q^4)$ without any parameters. In the class B diagrams, there are two parameters, call them V_i with $i = 1, 2$, associated with Fig.3c. One of them, V_1, contribute to both $\overline{\rm M1}$ and E2 and the other, V_2, only to $\overline{\rm M1}$. We find that the V_1 term plays no role in $\overline{\rm M1}$ or E2, as it is suppressed by some power of $Qr_c \ll 1$.

The upshot of all this is that we are finally left with the four-Fermi counter term g_4 and V_2. Furthermore, both of them are associated with zero-range terms in the coordinate space. The combination that appears here will be called, in short, the g_4"+"V_2 term. So we can combine them with the $\mathcal{O}(Q^3)$ zero-ranged operator that comes from the one-pion exchange term with the Wess-Zumino vertex (we shall call this in short the $\mathcal{O}(Q^3)$ WZ term), thereby reducing them effectively to only one parameter in $\overline{\rm M1}$. The E2 operator is completely free of parameters to $\mathcal{O}(Q^4)$.

Now in the HCCS, the single parameter of the theory does not figure since the contact operator that is multiplied by the parameter gets suppressed by the hard core.

But there is a possible caveat here: Since the leading order of this term is $\mathcal{O}(Q^3)$ whereas the finite loop corrections are of $\mathcal{O}(Q^4)$, it is not obvious that the HCCS should be reliable in the present case. One might suspect that here short-distance physics could intervene more strongly than when the chiral filter is operative. There are two possible remedies to this problem. One is to implement the operator-product-expansion (OPE) factorization in the wave function suggested by Lepage [24] and the other is to use a cutoff $\sim r_c^{-1}$ in Fourier-transforming the current operators into the coordinate-space form. We shall use the latter which is a lot simpler. We use an equivalent method which is to replace the delta function in all zero-ranged operators by the delta-shell form

$$\delta(r) \to \delta(r - r_c). \tag{21}$$

This procedure allows the $\mathcal{O}(Q^3)$ contact operator to contribute in $\overline{\mathrm{M1}}$, hence allowing to fix the unknown g_4"+"V_2 parameter by fitting the deuteron magnetic moment. It turns out however that this contact term is dominated by the known $\mathcal{O}(Q^3)$ WZ term, so in practice, the unknown parameter plays only a minor role here. We shall call this scheme the "modified hard core cutoff scheme."

3.4 Our predictions

We shall now make *our* predictions.

- **Hard Core Cutoff Scheme (HCCS)**

With the hard core in the wave function, the parameter-dependent term is killed, so we can now predict the deuteron magnetic moment μ_d, the quadrupole moment Q_d, and the ratios $\mathcal{R}_{\mathrm{M1}}$ and $\mathcal{R}_{\mathrm{E2}}$. There must of course be some dependence on the hard core size which enters as a cutoff but the consistency of EFT requires that the cutoff dependence be small.

Deuteron magnetic moment μ_d: There is a strong cancellation between the $\mathcal{O}(Q^3)$ and $\mathcal{O}(Q^4)$ two-body terms, leaving the one-body term essentially uncorrected: The net two-body correction is found to be less than 0.7% of the one-body term. The predicted values for $r_c = 0.01, 0.2, 0.4, 0.6, 0.8$ fm are (in units of nuclear magneton)

$$\mu_d = 0.8408,\ 0.8443,\ 0.8426,\ 0.8407,\ 0.8390. \tag{22}$$

The experimental value is $\mu_d^{exp} = 0.8574$.

This small discrepancy will be exploited later to fix the one parameter that figures when the zero-range operator is not killed by the hard core as in MHCCS.

Deuteron quadrupole moment Q_d: The two-body correction is equally tiny, less than 0.6% and more or less independently of r_c within the range $0.01 \lesssim r_c \lesssim 0.8$ fm. The result is (in unit of fm^2)

$$Q_d = 0.2710 \tag{23}$$

to be compared with the experiment $Q_d^{exp} = 0.2859\ \text{fm}^2$. It appears that the 5% discrepancy found here cannot be understood in low-order effective field theories.

$\mathcal{R}_{\text{M1}}$: The two-body terms of $\mathcal{O}(Q^3)$ and $\mathcal{O}(Q^4)$ in $\overline{\text{M1}}$, separately, are of the same magnitude as the one-body term, so although naively suppressed in the power counting, there is no genuine suppression according to the hierarchy of the order. However there is a considerable cancellation between the two higher-order terms, leaving the correction to be between 9% and 25% of the single-particle value. The predicted values for $r_c = 0.01, 0.2, 0.4, 0.6, 0.8$ fm are

$$-\mathcal{R}_{\text{M1}} \times 10^3 = 0.869,\ 0.788,\ 0.826,\ 0.871,\ 0.887. \qquad (24)$$

$\mathcal{R}_{\text{E2}}$: As in the quadrupole moment, the two-body correction to the E2 matrix element is small $\lesssim 0.4\%$, so the result is essentially given by the one-body term. Thus within the r_c range considered, the result is, independently of r_c,

$$\mathcal{R}_{\text{E2}} \times 10^3 = 0.242. \qquad (25)$$

Photon circular polarization P_γ: With the above values for $\mathcal{R}_{\text{M1}}$ and $\mathcal{R}_{\text{E2}}$, the predicted values for $r_c = 0.01, 0.2, 0.4, 0.6, 0.8$ fm are (for $|\vec{P}_n| = 1$)

$$-P_\gamma \times 10^3 = 1.57,\ 1.46,\ 1.51,\ 1.57,\ 1.60. \qquad (26)$$

Photon anisotropy η: Here the prediction is extremely sensitive to the value of the polarization pP. We will quote for three cases: $pP = 1$ (the ideal case), $pP = 0.96$ (the highest polarization that may be reached), and $pP = (0.5)^2$ (a case most certainly accessible to the experiment). The predicted values for $r_c = 0.01, 0.2, 0.4, 0.6, 0.8$ fm are

$$\begin{aligned} \eta[pP = 1] &= 0.57,\ 0.61,\ 0.59,\ 0.57,\ 0.56, \\ \eta[pP = 0.96] \times 10^5 &= 1.3,\ 1.1,\ 1.2,\ 1.3,\ 1.3, \\ \eta[pP = (0.5)^2] \times 10^7 &= 1.7,\ 1.5,\ 1.6,\ 1.7,\ 1.8. \end{aligned} \qquad (27)$$

- **Modified Hard Core Cutoff Scheme (MHCCS)**

We shall apply the "smoothing" (21) to the delta function and account for the term carrying information on the single parameter available in the theory. We have the possibility to fix the constant by "fine-tuning" it to the deuteron magnetic moment, that is, attributing the 5% discrepancy in μ_d from the experimental value to the counter term containing the $\mathcal{O}(Q^3)$ WZ term and the g_4"+"V_2 term. As mentioned, in practice, this term is completely dominated by the former (in fitting the deuteron magnetic moment); therefore the unknown constant plays only a minor role in the resulting isoscalar M1 operator that is to be used to compute the $\overline{\text{M1}}$ matrix element.

While μ_d is no longer predicted in this scheme (since it is used to pin down the small g_4"+"V_2 term), the deuteron quadrupole moment Q_d remains unmodified from eq.(23). The ratios $\mathcal{R}_{\text{M1}}$ and $\mathcal{R}_{\text{E2}}$ are of course predicted.

$\mathcal{R}_{\text{M1}}$: There turns out be a remarkable r_c independence for this quantity. In fact, in the range $0.01 \le r_c \le 0.8$ fm, the result is the same:

$$-\mathcal{R}_{\text{M1}} \times 10^3 = 0.500. \tag{28}$$

$\mathcal{R}_{\text{E2}}$: This quantity remains unchanged from the HCCS, (25),

$$\mathcal{R}_{\text{E2}} \times 10^3 = 0.242. \tag{29}$$

P_γ and η: Independently of r_C in the range $0.01 \le r_c \le 0.8$ fm, we find

$$-P_\gamma \times 10^3 = 1.05 \tag{30}$$

and

$$\begin{aligned} \eta[pP = 1] &= 0.80, \\ \eta[pP = 0.96] &= 0.62 \times 10^{-5}, \\ \eta[pP = (0.5)^2] &= 0.86 \times 10^{-7}. \end{aligned} \tag{31}$$

4 Conclusion: Call for a Bet

The values we have are not the final ones. First of all, they will have to be rechecked more thoroughly, and secondly, given that the isoscalar matrix elements are so suppressed relative to the isovector matrix element, it may be necessary to take into account the usually negligible isospin violation (both in the interaction and electromagnetic radiative corrections). In any event, we shall give our preliminary predictions here with the warning that they are subject to further changes. We expect to be able to publish a paper with our final numbers in the near future with the above caveats taken into account [25].

In Table 1 are summarized our predictions. Recall that there are two schemes for treating the zero-ranged counter terms. One scheme, referred to as hard core cutoff scheme (HCCS), is the usual nuclear physics practice to kill the delta function terms in the operator. The physics so purged from them is presumably shifted to the finite matrix elements affected by the "correlation hole." In this case there are no more parameters left to account for the small deviation from the experimental data in μ_d and Q_d and the results would depend on r_c; the dependence would of course be weak if the procedure were consistent with the premise of EFT. The other scheme, called modified hard core cutoff scheme (MHCCS), exploits the non-vanishing of zero-range operators to fix one parameter available in the theory by fitting the deuteron magnetic moment, which then determines completely the isoscalar M1 operator to $\mathcal{O}(Q^4)$. In this scheme, there is practically no r_c dependence and the convergence of the higher-order terms is assured by the procedure (specifically, there is no difference between the $\mathcal{O}(Q^3)$ calculation and the $\mathcal{O}(Q^4)$ calculation once μ_d is fit).

Hard-core scheme	HCCS	MHCCS
$10^3 \times P_\gamma$	-1.5 ± 0.1	-1.1
$\eta[pP = 1]$	0.58 ± 0.03	0.80
$10^5 \times \eta[pP = 0.96]$	1.2 ± 0.1	0.62
$10^7 \times \eta[pP = 0.25]$	1.6 ± 0.1	0.86
$10^3 \times \mathcal{R}_{M1}$	-0.84 ± 0.05	-0.51
$10^3 \times \mathcal{R}_{E2}$	0.24	0.24

Table 1: Predictions using two schemes for implementing the hard core: HCCS (hard-core cutoff scheme) in which zero-ranged operators are "killed" by the short-range correlation function in the wave function and MHCCS (modified hard core cutoff scheme) in which the delta function of zero-range operators is "smoothed" to the delta shell form. The "error bar" represents the variation over the range $0.01 \text{ fm} \leq r_c \leq 0.8$ fm. No error bar means that there is no r_c dependence.

There is only one polarization data available at the moment, namely, the photon circular polarization measured by the Russian group [23]

$$P_\gamma^{exp} = (-1.5 \pm 0.3) \times 10^{-3}. \tag{32}$$

As it stands, this experiment seems to favor the HCCS result. However the prediction in which we have more confidence is that of MHCCS, which puts it somewhat lower than the experimental value (32). Clearly additional measurements will be needed to confirm (32) or improve on it.

- **The Bet**

We would now like to invite other workers in the field – particularly those who are in this audience, who have worked out consistent and systematic power countings – to make similar predictions and participate in the bet for the best prediction, that is, the one that agrees best with the experimental result that is forthcoming. As for our prediction, the most interesting possibility would be that the experiment simply disagrees with our two versions of the hard-core scheme. That would sharpen our chiral filter conjecture and bring in totally new physics. [viii]

Acknowledgments

We would like to thank the organizers of this meeting for the invitation to participate in the exciting debate and Thomas Müller for discussions and correspondence on the on-going experiment at the ILL.

[viii] At the time of my talk, I made a specific proposal for the prize for the winner and I confirm that proposal here. I would propose that each participant should be willing to offer to the winner a bottle of one of the *best* wine of his/her country. As for me, I would offer a bottle of superb French wine, Chateau Mouton-Rothschild, a "premier grand cru classé."

References

1. S. Weinberg, *Phys. Lett.* B **251**, 288 (1990); *Nucl. Phys.* B **363**, 3 (1991); *Phys. Lett.* B **295**, 114 (1992).
2. M. Rho, *Phys. Rev. Lett.* **66**, 1275 (1991); T.-S. Park, D.-P. Min and M. Rho, *Physics Reports* **233**, 341 (1993); *Phys. Rev. Lett.* **74**, 4153 (1995); *Nucl. Phys.* A **596**, 515 (1996).
3. C. Ordonez and U. van Kolck, *Phys. Lett.* B **291**, 459 (1992); C. Ordonez, L. Ray and U. van Kolck, *Phys. Rev. Lett.* **72**, 1982 (1994); *Phys. Rev.* C **53**, 2086 (1996); U. van Kolck, *Phys. Rev.* C **49**, 2932 (1994).
4. S. Beane, T. Cohen, D. Kaplan, G.P. Lepage, M. Savage, J. Steele, U. van Kolck, M. Wise, *this volume.*
5. G.P. Lepage, this meeting.
6. K. Kubodera, J. Delorme and M. Rho, *Phys. Rev. Lett.* **40**, 755 (1978).
7. D.B. Kaplan, M. Savage and M. Wise, *Nucl. Phys.* B **478**, 629 (1995); *Phys. Lett.* B **424**, 390 (1998); `nucl-th/9801034`; `nucl-th/9802075`; `nucl-th/9804032`.
8. R.B. Wiringa, V.G.J. Stoks and R. Schiavilla, *Phys. Rev.* C **51**, 38 (1995).
9. K. Kubodera, "Chiral symmetry in nuclei," `nucl-th/9903057`.
10. T.-S. Park, K. Kubodera, D.-P. Min and M. Rho, *Phys. Rev.* C **58**, R637 (1998); *Nucl. Phys.* A **646**, 83 (1999).
11. U. van Kolck, "Effective field theory of nuclear forces," `nucl-th/9902015`.
12. T.-S. Park, D.-P. Min and M. Rho, *Phys. Rev. Lett.* **74**, 4153 (1995); *Nucl. Phys.* A **596**, 515 (1996).
13. M. Savage, K.A. Scaldeferri and M.B. Wise, `nucl-th/9811029`.
14. T.-S. Park, K. Kubodera and I.S. Towner, *Nucl. Phys.* A **579**, 381 (1994).
15. B. Friman, M. Rho and C. Song, `nucl-th/9809088`.
16. G.E. Brown and M. Rho, *Phys. Rev. Lett.* **66**, 2720 (1991).
17. E.K. Warburton, *Phys. Rev. Lett.* **66**, 1823 (1991); E.K. Warburton, I.S. Towner and B.A. Brown, *Phys. Rev.* C **49**, 824 (1994); E.K. Warburton and I.S. Towner, *Phys. Lett.* B **294**, 1 (1992).
18. K. Kubodera and M. Rho, *Phys. Rev. Lett.* **67**, 3479 (1991).
19. T. Minamisono *et al*, *Phys. Rev. Lett.* **82**, 1644 (1999).
20. P. Baumann *et al*, *Phys. Rev.* C **58**, 1970 (1998).
21. A. Van Geert *et al*, "Meson-exchange enhancement in the first-forbidden beta transition of ^{205}Hg," in Proceedings of Paris Conference on Nuclear

Physics, Paris, France, August, 1998.
22. T.M. Müller, private communication; T.M. Müller, D. Dubbers, P. Hautle and O. Zimmer, "Measurement of the γ anisotropy in the $\vec{p}\vec{n}, \gamma d$-process," in Proceedings of the "International Workshop on Particle Physics with Slow Neutrons," ILL, Grenoble, France, 22-24 October 1998.
23. A.N. Bazhenov *et al*, *Phys. Lett.* B **289**, 17 (1992).
24. G.P. Lepage, "How to renormalize the Schrödinger equation?", nucl-th/9706029.
25. T.-S. Park, K. Kubodera, D.-P. Min and M. Rho, in preparation.

Discussion during M. Rho's presentation

Chair: *F. Ravndal (Oslo University)*
Notes taken by *J. McGovern (University of Manchester)*

During the talk Cohen asked about the sensitivity of the "experimental" in-medium axial coupling constants to the method of extraction from the raw data; Rho and Haxton replied that it was hard to know. After the talk, Cohen asked if the absence of free parameters in the calculation was specific to the particular scheme used; Rho replied that he had not checked.

ISOSPIN VIOLATION IN THE TWO–NUCLEON SYSTEM

U.-G. MEIßNER [a]
FZ Jülich, IKP(TH), D-52425 Jülich, Germany
E-mail: Ulf-G.Meissner@fz-juelich.de

I discuss isospin violation in an effective field theory approach of the nucleon–nucleon interaction. The observed leading order charge independence breaking is explained in terms of one–pion exchange together with a four–nucleon contact term. A general classification scheme of corrections to this leading order results is presented and an explanation is offered why charge–dependent pion–nucleon coupling constants can not play a role in the observed isospin violation in the NN scattering lengths. Charge symmetry breaking, the size of $\pi\gamma$–exchange graphs and the pionless theory are also discussed.

1 Introduction

Heisenberg introduced the notion of isospin in 1930 as a label to distinguish between neutrons and protons when he was studying atomic nuclei. He noticed that the clustering of stable nuclei around $N \simeq Z$ implied a new (internal) symmetry. This was later extended to the pion triplet, which is believed to mediate the long range part of the nuclear force. Heisenberg's ideas were used by Breit and collaborators (and others) to establish the so–called charge independence, observing that after removal of the Coulomb effects between protons, the *pp*, *np* and *nn* force were equal to a good precision. A weaker condition is the so–called charge symmetry, which states that the *pp* and *nn* force should be equal with no assertion to the *np* force. Since that time, QCD has emerged as the theory of the strong interactions. In the absence of electroweak interactions, isospin symmetry is observed in the sector of the two lightest quark flavors for *equal* masses, $m_u = m_d$, i.e. the QCD Hamiltonian is invariant under a global (flavor) transformation of the type

$$q = \begin{pmatrix} u \\ d \end{pmatrix} \to q' = \mathcal{V} q = \mathcal{V} \begin{pmatrix} u \\ d \end{pmatrix} \;, \quad \mathcal{V} \in SU(2) \;. \tag{1}$$

Of course, in nature the light quark masses are not equal. In fact, they differ considerably, $m_u/m_d = 0.55$. Still, isospin is a good approximate symmetry because what counts is not the ratio of the quark masses but rather their difference compared to the typical scale of strong interactions, $(m_d - m_u)/\Lambda_{\rm QCD} \ll 1$. QCD thus provides a reason why isospin is such a good approximate symmetry. This can hardly be considered "accidental". In the presence of electro-

[a]Work presented here has been done in collaboration with E. Epelbaum.

magnetism, further isospin violation is induced since the charges of the quarks are unequal. These effects are in general small due to the explicit appearance of the fine structure constant $\alpha \simeq 1/137$. In fact, in many cases the strong and electromagnetic isospin–violating effects are of the same size. A good example is the neutron–proton mass difference. On the quark level, charge symmetry is a rotation about the 2–axis in isospin space

$$\mathcal{P}_{\rm cs}\begin{pmatrix} u \\ d \end{pmatrix} = e^{(i\pi/2)\tau_2}\begin{pmatrix} u \\ d \end{pmatrix} = \begin{pmatrix} -d \\ u \end{pmatrix} , \quad \mathcal{P}_{\rm cs}^2 = \mathbf{1} \quad . \tag{2}$$

Our aim is get some insight about isospin symmetry (charge independence) and charge symmetry in the two–nucleon system based on an effective field theory approach. On the level of the two–nucleon force, charge independence implies a potential of the form

$$V_{ij} = A + B\tau_{(i)} \cdot \tau_{(j)} \ , \tag{3}$$

where A and B are functions of the nucleon spin and momentum (or space) coordinates and '$(i), (j)$' labels the interacting nucleons. Charge symmetry allows for a more general two–body force,

$$V_{ij,\rm cs} = A + B\tau_{(i)} \cdot \tau_{(j)} + C\tau^3_{(i)} \cdot \tau^3_{(j)} \ . \tag{4}$$

If one works out the Coulomb potential in terms of the nucleon charge matrix $Q = e(1+\tau^3)/2$, it is obvious the electromagnetic effects lead to breaking of charge independence and charge symmetry,

$$V_{ij,\rm Coulomb} = \frac{e^2}{4r^2}(1+\tau^3_{(i)})(1+\tau^3_{(j)}) \ . \tag{5}$$

I now wish to show how the question of isospin violation can be addressed in an effective field theory approach to the NN interactions. There are two equivalent ways to proceed. One is to follow the suggestion of Weinberg [1] and perform the chiral expansion on the level of the potential (i.e. for the irreducible diagrams) and iterate this potential in the Lippmann–Schwinger equation to generate the bound and scattering states. The other is to directly expand the amplitude after performing a nonperturbative resummation for the S–waves.[2] The first path has been followed by van Kolck [3] and I will consider the second option here.[4] Pions are included and treated perturbatively in this approach. For the topic addressed here, the discussion whether or not pions should be treated in such a way is inessential. In fact, as shown in section 6, one can even work with a pionless theory. In that case, however, the physical interpretation of CIB and CSB becomes much less transparent.

2 Facts, explanations and mysteries

It is well established that the nucleon–nucleon interactions are charge dependent (for a review, see e.g.[5] and a recent summary is given in [6]). For example, in the $^1S_0$ channel one has for the scattering lengths a and the effective ranges r

$$\Delta a_{\rm CIB} = \frac{1}{2}\left(a_{nn} + a_{pp}\right) - a_{np} = 5.7 \pm 0.3 \text{ fm} , \tag{6}$$

$$\Delta r_{\rm CIB} = \frac{1}{2}\left(r_{nn} + r_{pp}\right) - r_{np} = 0.05 \pm 0.08 \text{ fm} . \tag{7}$$

These numbers for charge independence breaking (**CIB**) are based on the Nijmegen potential[7] and the Coulomb effects for pp (nn) scattering are subtracted based on standard methods (for a treatment of this in an EFT framework see refs.[8,9]). The charge independence breaking in the scattering lengths is large, of the order of 25%, since $a_{np} = (-23.714 \pm 0.013)\,$fm. Of course, it is magnified at threshold due to kinematic factors (as witnessed by the disappearance of the effect in the effective range). In addition, there are charge symmetry breaking (**CSB**) effects leading to different values for the pp and nn threshold parameters,

$$\Delta a_{\rm CSB} = a_{pp} - a_{nn} = 1.5 \pm 0.5 \text{ fm} , \quad \Delta r_{\rm CSB} = r_{pp} - r_{nn} = 0.10 \pm 0.12 \text{ fm} . \tag{8}$$

Combining these numbers gives as central values $a_{nn} = -18.8\,$fm and $a_{pp} = -17.3\,$fm. Both the CIB and CSB effects have been studied intensively within potential models of the nucleon–nucleon (NN) interactions. In such approaches, the dominant CIB comes from the charged to neutral pion mass difference in the one–pion exchange (OPE), $\Delta a_{\rm CIB}^{\rm OPE} \sim 3.6 \pm 0.2\,$fm. Additional contributions come from $\gamma\pi$ and 2π (TPE) exchanges. According to some calculations, their size is approximately $1/3^{\rm rd}$ of the leading OPE contribution. For the TPE, significantly smaller results can also be found in the literature, see e.g. table 3.3 in ref.[5] In case of the $\pi\gamma$ contribution, a recent calculation[10] gives a value which is more than a factor of ten smaller than the leading OPE contribution. We will come back to these issues later (see section 5). Note also that the charge dependence in the pion–nucleon coupling constants (if existing) in OPE and TPE almost entirely cancel. Naively, one would expect a possible charge dependence of the pion–nucleon coupling constants to be of a similar importance as the pion mass difference in OPE. This strong suppression of charge–dependent couplings has so far been eluded a deeper understanding. We will come back to this later on. In the meson–exchange picture, CSB originates mostly from $\rho-\omega$ mixing, $\Delta a_{\rm CSB}^{\rho-\omega} \sim 1.2 \pm 0.4\,$fm. Other contributions due to $\pi-\eta$, $\pi-\eta'$ mixing or the proton–neutron mass difference are known to be much smaller. It will be shown in what follows that EFT gives novel insight concerning the size of the TPE and $\pi\gamma$ contributions as well as the suppression of possible charge–dependent πN coupling constants.

3 Effective Lagrangian and power counting

The effective Lagrangian underlying the analysis of isospin violation in the two–nucleon system consists of a string of terms,

$$\mathcal{L}_{\rm eff} = \mathcal{L}_{\pi\pi} + \mathcal{L}_{\pi N} + \mathcal{L}_{NN} \ . \tag{9}$$

Each of these terms splits into a strong and an electromagnetic one. The strong contributions include (besides many others) the isospin–violation due to the quark mass difference $m_u - m_d$ whereas the em terms stem from the dynamics of the virtual photons (photon loops and em counterterms). To include virtual photons in the pion and the pion–nucleon system is by now a standard procedure.[11–15] We note that in the pion and pion–nucleon sector, one can effectively count the electric charge as a small momentum or meson mass. This is based on the observation that $M_\pi/\Lambda_\chi \sim e/\sqrt{4\pi} = \sqrt{\alpha} \sim 1/10$ since $\Lambda_\chi \simeq 4\pi f_\pi = 1.2\,$GeV. It is thus possible to turn the dual expansion in small momenta/meson masses on one side and in the electric coupling e on the other side into an expansion with *one* generic small parameter. We also remark that from here we use the fine structure constant $\alpha = e^2/4\pi$ as the em expansion parameter. Because of the additional $1/4\pi$ factor (from integrating out hard photons) one expects the corresponding em constants to be of order one (of natural size). A dimensional analysis of such terms in the pion–nucleon sector is given in ref.[16] and similarly for the pion sector in refs.[11,14].

We now turn to the two–nucleon sector. One can either expand the potential (irreducible diagrams) or directly the amplitudes. A complication arises due to the unnaturally large S–wave scattering lengths, $1/a \ll M_\pi$. A *systematic* power counting scheme to deal with this directly on the level of the scattering amplitude is the recently proposed power divergence subtraction method (PDS) of Kaplan, Savage and Wise (KSW) [2].[b] Essentially, one resums the lowest order local four–nucleon contact terms $\sim C_0(N^\dagger N)^2$ (in the S–waves) to generate the large scattering lengths and treats the remaining effects perturbatively, in particular also pion exchange. This means that most

[b]There exist by now modifications of this approach and it as been argued that it is equivalent to cut–off schemes. We do not want to enter this discussion here but rather stick its original version.

low–energy observables are dominated by contact interactions. The chiral expansion for NN scattering entails a new scale Λ_{NN} of the order of 300 MeV, so that one can systematically treat external momenta up to the size of the pion mass. There have been suggestions that the radius of convergence can be somewhat enlarged [17], but in any case Λ_{NN} is considerably smaller than the typical scale of about 1 GeV appearing in the pion–nucleon sector. It is straightforward to extend the underlying effective Lagrangian to include the strong isospin–violating and electromagnetic four–fermion contact interactions. Consider first the strong terms. Up to one quark mass insertion, the effective Lagrangian takes the form (note that terms involving the Pauli isospin matrices $\vec{\tau}$ can be obtained by Fierz reordering)

$$\begin{aligned}\mathcal{L}_{NN}^{\rm str} &= l_1(N^\dagger N)^2 + l_2(N^\dagger\vec{\sigma}N)^2 + l_3(N^\dagger\langle\chi_+\rangle N)(N^\dagger N) + l_4(N^\dagger\hat{\chi}_+ N)(N^\dagger N) \\ &+ l_5(N^\dagger\vec{\sigma}\langle\chi_+\rangle N)(N^\dagger\vec{\sigma}N) + l_6(N^\dagger\vec{\sigma}\hat{\chi}_+ N)(N^\dagger\vec{\sigma}N) + \ldots\ , \qquad (10)\end{aligned}$$

where $\langle\ldots\rangle$ denotes the trace in flavor space and χ_+ contains the light quark masses. $\hat{\chi}_+ = \chi_+ - \langle\chi_+\rangle/2$ is only non–vanishing for $m_u \neq m_d$ and the ellipsis denotes terms with two (or more) derivatives acting on the nucleon fields. Similarly, one can construct the em terms. The ones without derivatives on the nucleon fields read (possible terms involving the Pauli isospin matrices are of no relevance for the following)

$$\begin{aligned}\mathcal{L}_{NN}^{\rm em} &= N^\dagger\left\{r_1\langle Q_+^2 - Q_-^2\rangle + r_2\hat{Q}_+\langle Q_+\rangle\right\}N(N^\dagger N) \\ &+ N^\dagger\vec{\sigma}\left\{r_3\langle Q_+^2 - Q_-^2\rangle + r_4\hat{Q}_+\langle Q_+\rangle\right\}N(N^\dagger\vec{\sigma}N) \\ &+ N^\dagger\left\{r_5 Q_+ + r_6\langle Q_+\rangle\right\}N(N^\dagger Q_+ N) \\ &+ N^\dagger\vec{\sigma}\left\{r_7 Q_+ + r_8\langle Q_+\rangle\right\}N(N^\dagger\vec{\sigma}Q_+ N) \\ &+ r_9(N^\dagger Q_+ N)^2 + r_{10}(N^\dagger\vec{\sigma}Q_+ N)^2\ , \qquad (11)\end{aligned}$$

with $Q_\pm = uQ^\dagger u \pm u^\dagger Q u^\dagger$ and Q the nucleon charge matrix. All these terms are due to the integration of the "hard" virtual photons. The "soft" ones are contained in the pertinent covariant derivatives acting on the pion or nucleon fields, see e.g. refs.[13,15]. We remark that since Λ_{NN} is significantly smaller than Λ_χ, it does not pay to treat the expansion in the generic KSW momentum[c]

$$\tilde{Q}/\Lambda_{NN} \sim \mu/\Lambda_{NN} \sim M_\pi/\Lambda_{NN} \qquad (12)$$

[c] So that no confusion with the charge matrix can arise, we denote this parameter with an extra "tilde" contrary to common use. A pedagogical discussion of this expansion parameter can be found in ref.[18].

(with μ the scale of dimensional regularization) simultaneously with the one in the fine structure constant (as it is done e.g. in the pion–nucleon sector). Instead, one has to assign to each term a *double expansion parameter*,

$$(\tilde{Q}/\Lambda_{NN})^n\,\alpha^m\ , \quad \tilde{Q}/\Lambda_{NN} \simeq 1/3\ , \quad \alpha = 1/137\ , \tag{13}$$

with n and m integers ($m \geq 1$ and n can be negative). Clearly, even if we work at next–to–next–to leading order (NNLO) which entails a suppression by two powers in $\tilde{Q}$, we need only be concerned with the leading em terms $\sim \alpha$ since $\alpha \ll 1$. Lowest order charge independence breaking is due to a term $\sim (N^\dagger \tau^3 N)^2$ whereas charge symmetry breaking at that order is given by a structure $\sim (N^\dagger \tau^3 N)(N^\dagger N)$. In the KSW approach, it is customary to project the Lagrangian terms on the pertinent NN partial waves. Denoting by β the $^1S_0$ partial wave for a given cms energy $E_{\rm cms}$, the Born amplitudes for the lowest order CIB and CSB operators between the various two–nucleon states takes the form

$$\begin{aligned}\langle\beta, pp|\mathcal{L}^{\rm em}_{NN}|\beta, pp\rangle &= -\left(\frac{Mp}{2\pi}\right)\alpha\left(E_0^{(1)}(\mu) + E_0^{(2)}(\mu)\right)\ , \\ \langle\beta, nn|\mathcal{L}^{\rm em}_{NN}|\beta, nn\rangle &= -\left(\frac{Mp}{2\pi}\right)\alpha\left(E_0^{(1)}(\mu) - E_0^{(2)}(\mu)\right)\ ,\end{aligned} \tag{14}$$

where the coupling constant $E_0^{(1,2)}(\mu)$ will be determined later on. From the renormalization group equations (RGE) it will also be possible to derive the scaling properties of the $E_{2n}^{(1,2)}(\mu)$ for four–nucleon contact operators with $2n$ derivatives. The terms with the superscript '(1)' refer to CIB whereas the second ones relevant for CSB are denoted by the superscript '(2)'. Higher order operators are denoted accordingly. There is, of course, also a CIB contribution to the np matrix element. To be consistent with the charge symmetric calculation of ref.[2], one has to absorb its effect in the constant D_2, i.e. it amounts to a finite renormalization of D_2 and is thus not observable. In eq.(14), $p = \sqrt{ME_{\rm cms}}$ is the nucleon cms momentum and M the nucleon mass.

4 Isospin violation in nucleon–nucleon scattering

Let me now work out the pertinent $^1S_0$ phase shifts. The np case has already been calculated in ref.[2], the leading amplitude being of order $\tilde{Q}^{-1}$. We want to calculate the leading em corrections (here, "Cs" means Coulomb–subtracted and we neglect the small magnetic corrections for the nn system),

$$\Delta\mathcal{A} = \mathcal{A}_{nn} - \mathcal{A}_{np} = \mathcal{A}_{pp}^{\rm Cs} - \mathcal{A}_{np}\ . \tag{15}$$

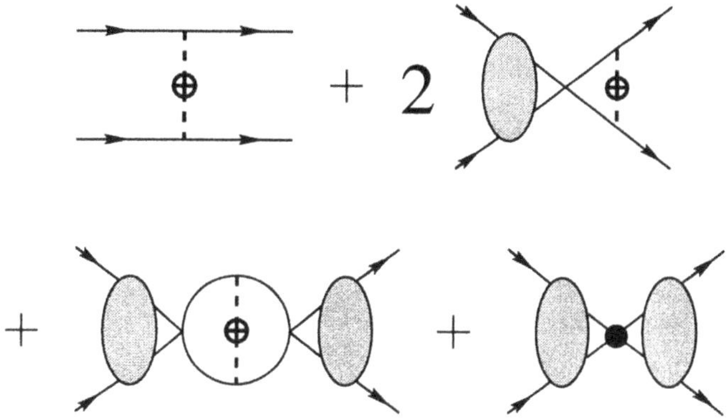

Figure 1: Relevant graphs contributing to charge independence breaking at leading order $\alpha\tilde{Q}^{-2}$. The blob stands for the resummation of the lowest order $(N^\dagger N)^2$ contact term $\sim C_0$. The open (filled) circle denotes a pion mass insertion $\sim \delta m^2$ (an insertion of the leading four–nucleon operators $\sim \alpha E_0^{(1)}$). For charge symmetry breaking, the leading contribution is given by the last diagram with the filled circle denoting an insertion $\sim \alpha E_0^{(2)}$.

The pertinent diagrams are shown in fig. 1. Besides the contact terms written down in eq.(14), we have to consider the pion mass difference in one–pion exchange. OPE diagrams with different pion masses have the isospin structure $O_{12} = \tau_{(1)a}\Delta_\pi^{ab}\tau_{(1)b} = (a+b)\,\vec{\tau}_{(1)}\,\vec{\tau}_{(2)} - b\,\tau^3_{(1)}\,\tau^3_{(2)}$ and obviously lead to CIB. From the first graph in fig. 1 one easily concludes that these effects are of order $\alpha\tilde{Q}^{-2}$ since the diagram contains two derivative πN couplings of order $\tilde{Q}$ and two pion propagators of order $\tilde{Q}^{-2}$. The notation follows closely the one of KSW in that the corresponding three amplitudes are called $\mathcal{A}_{2,-2}^{II,III,IV}$. Here, the first (second) subscript refers to the power in α ($\tilde{Q}$) and the superscripts to the first three diagrams of the figure. This gives

$$\Delta\mathcal{A}_{1,-2}^{II} = \Gamma\left[\frac{1}{4p^2}\ln\left(1+\frac{4p^2}{m^2}\right) - \frac{1}{m^2+4p^2}\right] ,$$

$$\Delta\mathcal{A}_{1,-2}^{III} = \Gamma\,\mathcal{F}\left[\frac{1}{pm}\arctan\frac{2p}{m} + \frac{i}{2pm}\ln\left(1++\frac{4p^2}{m^2}\right) - \frac{1+\frac{2ip}{m}}{m^2+4p^2}\right]$$

$$\Delta\mathcal{A}_{1,-2}^{IV} = \Gamma\,\mathcal{F}^2\left[\frac{i}{m^2}\arctan\frac{2p}{m} - \frac{1}{2m^2}\ln\left(\frac{m^2+4p^2}{\mu^2}\right) + \frac{1}{m^2} - \frac{1}{2}\frac{1+\frac{2ip}{m}}{m^2+4p^2}\right]$$

$$\Gamma = -\delta m^2 \frac{g_A^2}{2f_\pi^2} , \quad \mathcal{F} = \left(\frac{mM\mathcal{A}_{-1}}{4\pi} \right) , \quad \delta m^2 = m_{\pi^\pm}^2 - m_{\pi^0}^2 , \tag{16}$$

with $\mathcal{A}_{-1} \equiv \mathcal{A}_{0,-1}$ the leading term in the expansion of the np $^1S_0$ amplitude,

$$\mathcal{A}_1 = \frac{C_0}{1 + \frac{C_0 M}{4\pi}(\mu + ip)} . \tag{17}$$

Furthermore, $f_\pi = 92.4\,\mathrm{MeV}$ is the pion decay constant, $g_A = 1.26$ the axial-vector coupling constant and $m^2 = 2m_{\pi^\pm}^2 - m_{\pi^0}^2$ the "average" pion mass. The somewhat unusually looking expressions for m^2 and δm^2 are derived in appendix A. Note that while the diagrams II and III are convergent, IV diverges logarithmically. Therefore, the Lagrangian must contain a counterterm of the structure $E_0^{(1)}(\mu)\alpha(N^\dagger \tau^3 N)(N^\dagger \tau^3 N)$ to make the amplitude scale-independent. The leading CIB contribution thus scales as $\tilde{Q}^{-2}$ and for higher order operators with $2n$ derivatives we can establish the scaling property $E_{2n}^{(1)} \sim \tilde{Q}^{-2+n}$. This does not contradict the KSW power counting for the isospin symmetric theory since $\alpha \ll 1$. The insertion from this contact term is shown in the last diagram of fig. 1 and leads to an additional contribution to $\Delta\mathcal{A}$. In complete analogy, we can treat the leading order CSB effect which is due to an operator of the form $\alpha E_0^{(2)}(N^\dagger \tau_3 N)(N^\dagger N)$. This term is, however, finite. Putting pieces together, this gives for the nn and pp case

$$\Delta\mathcal{A}_{1,-2,\mathrm{nn}}^{VI} = -\alpha \left(E_0^{(1)} - E_0^{(2)} \right) \left[\frac{\mathcal{A}_{-1}}{C_0} \right]^2 , \tag{18}$$

$$\Delta\mathcal{A}_{1,-2,\mathrm{pp}}^{VI} = -\alpha \left(E_0^{(1)} + E_0^{(2)} \right) \left[\frac{\mathcal{A}_{-1}}{C_0} \right]^2 , \tag{19}$$

where the coupling constants $E_0^{(1,2)}(\mu)$ obey the renormalization group equations,

$$\begin{aligned} \mu\,\alpha \frac{dE_0^{(1)}(\mu)}{d\mu} &= \alpha \frac{M}{2\pi} E_0^{(1)}(\mu) C_0(\mu)\mu - \delta m^2 \frac{M g_A^2}{48\pi^2 f_\pi^2} C_0^2(\mu) , \\ \mu\,\alpha \frac{dE_0^{(2)}(\mu)}{d\mu} &= \alpha \frac{M}{2\pi} E_0^{(2)}(\mu) C_0(\mu)\mu . \end{aligned} \tag{20}$$

The relation of the pp and nn scattering lengths to the np one reads (of course, in the pp system Coulomb subtraction is assumed),

$$\frac{1}{a_{pp}} = \frac{1}{a_{np}} - \frac{4\pi\alpha(E_0^{(1)} + E_0^{(2)})}{MC_0^2} + \Delta ,$$

$$\frac{1}{a_{nn}} = \frac{1}{a_{np}} - \frac{4\pi\alpha(E_0^{(1)} - E_0^{(2)})}{MC_0^2} + \Delta \ , \qquad (21)$$

$$\Delta = \delta m^2 \frac{g_A^2(-C_0 M(m-2\mu) + 8\pi + C_0 M m \ln(m^2/\mu^2))}{12\pi m C_0 f_\pi^2} \ .$$

For the effective ranges, we have only CIB

$$r_{nn} = r_{pp} = r_{np} + \delta m^2 \frac{g_A^2(C_0 M\mu + 4\pi)(C_0 M(3\mu - 2m) + 12\pi)}{6\pi M m^4 C_0^2 f_\pi^2} \ . \qquad (22)$$

Note that the relations in eqs.(21,22) are perturbatively scale–independent and that the last one does not contain any unknown parameter. We remark that for the CIB scattering lengths difference the pion contribution alone is *not* scale–independent and can thus never be uniquely disentangled from the contact term contribution $\sim E_0^{(1)}$. While the leading OPE contribution resembles the result obtained in meson exchange models, the mandatory appearance of this contact term is a distinctively new feature of the effective field theory approach. Matters are somewhat different for CSB. Here, to leading order there is simply a four–nucleon contact term proportional to the constant $E_0^{(2)}$. Its value can be determined from a fit to the empirical value given in eq.(8). Corrections to the leading order CIB and CSB effect are classified below. We add the following remarks. Eqs.(21) and (22) are not exact but hold perturbatively. Similarly, the phase shifts are also defined perturbatively and therefore one can not have an exact effective range expansion of $\delta = \delta_0 + \delta_1$ (referring to the isospin symmetric case for the moment). If one works e.g. at next–to–leading order, the norm of the S–matrix is not exactly one, $S^\dagger S = 1 + \mathcal{O}(\tilde{Q}^2) \neq 1$. From this it follows that if one adjusts the parameters so as to reproduce the phase shifts in a certain momentum range, one can only approximately reproduce the scattering length and the effective range, $a \simeq a_{\rm exp}$ and $r \simeq r_{\rm exp}$, although the phase shifts may be very well described. This should be kept in mind.

We are now in the position to calculate the phase shifts. In total, we have five coupling constants, three of them related to the isospin–symmetric case, called C_0, C_2 and D_2. We fit the parameters $C_{0,2}, D_2$ to the np phase shift under the condition that the scattering length and effective range are exactly reproduced. The two new parameters $E_0^{(1,2)}$ are determined from the nn and pp scattering lengths. The resulting parameters at $\mu = m$ are of natural size,

$$C_0 = -3.46 \text{ fm}^2 \ , \quad C_2 = 2.75 \text{ fm}^4 \ , \quad D_2 = 0.07 \text{ fm}^4 \ ,$$
$$E_0^{(1)} = -6.47 \text{ fm}^2 \ , \quad E_0^{(2)} = 1.10 \text{ fm}^2 \ . \qquad (23)$$

The np and nn $^1S_0$ phase shifts are shown in fig. 2. To arrive at these curve, the physical masses for the proton and the neutron have been used. We stress again that the effect of the em terms $\sim E_0^{(1,2)}$ is small because of the explicit factor of α not shown in eq.(23). Having fixed these parameters, we can now predict the nn $^1S_0$ phase shift as depicted by the solid line in fig.2. It agrees with nn phase shift extracted from the Argonne V18 potential (with the scattering length and effective range exactly reproduced) up to momenta of about 100 MeV. The analogous curve for the pp system (not shown in the figure) is close to the solid line since the CSB effects are very small.

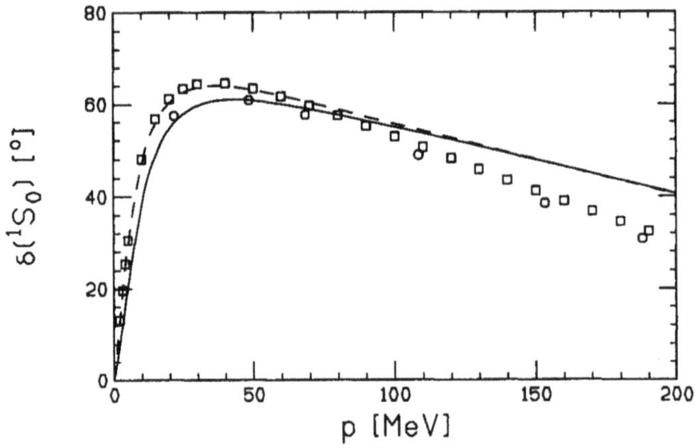

Figure 2: $^1S_0$ phase shifts for the np (dashed line) and nn (solid line) systems versus the nucleon cms momentum. The empirical values for the np case (open squares) are taken from the Nijmegen analysis. The open octagons are the nn "data" based on the Argonne V18 potential.

5 Classification scheme

In the framework presented here, it is straightforward to work out the various leading (LO), next–to–leading (NLO) and next–to–next–to–leading order (NNLO) contributions to CIB and CSB, with respect to the expansion in $\tilde{Q}$, to leading order in α and the light quark mass difference. Here, I will simply enumerate the pertinent contributions which appear at a given order. This naturally also gives an estimate about their relative numerical importance based on simple dimensional analysis. Based on such an analysis, one can shed some light on the puzzles discussed in section 2, in particular the size of the $\pi\gamma$ graphs and the role of charge–dependent πN coupling constants. Let me first consider CIB. In the classification scheme given below, TPE/3PE stands for two/three–pion–exchange.

LO $\alpha\tilde{Q}^{-2}$	Pion mass difference in OPE, four–nucleon contact interaction with no derivatives.
NLO $\alpha\tilde{Q}^{-1}$, $\varepsilon\tilde{Q}^{-1}$	Pion mass difference in TPE, four–nucleon contact interaction with two derivatives, insertions proportional to the np mass difference,
NNLO $\alpha\tilde{Q}^{0}$	Pion mass difference in 3PE, four–nucleon contact interaction with four derivatives, CIB in the pion–nucleon coupling constants, $\pi\gamma$–exchange.

Note that there are no CIB effects due to the light quark mass difference linear in $\varepsilon = m_u - m_d$ from OPE and from the four–nucleon contact terms. These type of terms can only appear to second order in ε. Effects due to the charge dependence of the pion–nucleon coupling constants, i.e. isospin breaking terms from $\mathcal{L}_{\pi N}^{\rm em}$, only start to contribute at order $\alpha\tilde{Q}^0$. Such effects are therefore suppressed by two orders of $\tilde{Q}$ compared to the leading terms, i.e. by a numerical factor of about $(1/3)^2$. This finding is in agreement with the various numerical analyses performed in potential models (if one assumes that there is indeed some charge dependence in the pion–nucleon coupling constants, see e.g. the discussion in ref.[5].) We remark that the TPE contribution is nominally suppressed by one order in the small parameter, i.e. one would expect its contribution to be one third of the leading OPE with different pion masses. This agrees with some (but not all) meson–exchange model calculations.

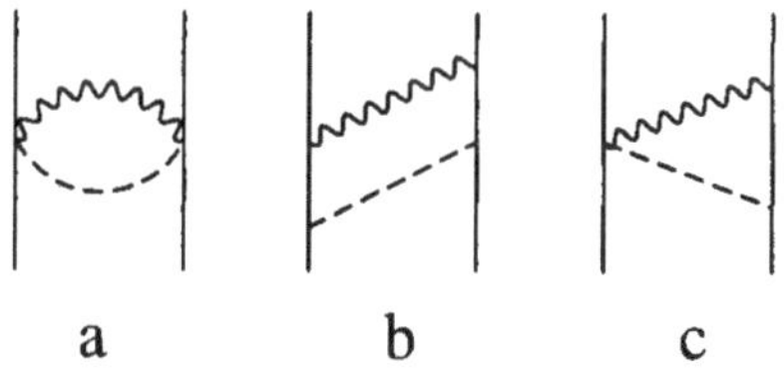

Figure 3: $\pi\gamma$ graphs. For CIB, the leading contribution comes from the three diagrams a), b) and c). Due to the Kroll–Ruderman vertex in a) and c) the exchanged pion is charged. In case of CSB, the dominant contribution comes from b) with neutral pion exchange. c) vanishes due to parity and a) is suppressed by two additional powers in $\tilde{Q}$. Crossed graphs are not shown. Also omitted are the initial and final–state interactions between the nucleon.

Some remarks on the $\gamma\pi$ graphs (some of them shown in fig. 3) are in order. First, we note that their contribution for CIB is suppressed by two powers in $\tilde{Q}$ with respect to the leading terms (the power counting for these graphs is detailed in appendix B). The larger contribution found in some meson-exchange model calculations (approximately 1/3$^{\rm rd}$ of the leading OPE contribution) is therefore at first sight at variance with dimensional analysis since one would expect a suppression of about $(1/3)^2$.[5] However, the much smaller value obtained in the more recent calculation of ref.[10] (performed in the static limit for the nucleons and Coulomb gauge for the photons), $\Delta a_{\rm CIB}(\pi\gamma)/\Delta a_{\rm CIB}({\rm OPE}) \simeq -1/14.6$, is in much better agreement with the suppression by two powers in $\tilde{Q}$. This deserves further study. Second, the $\gamma\pi$ graphs evaluated in heavy baryon chiral perturbation theory have been used in a refined potential model calculation.[19] The $^1{\rm S}_0$ np low-energy parameters were found to be very little affected. It is also instructive to see how the corrections due to the np mass difference come about. For that, let us write the nucleon mass as $M+\delta M$, where δM subsumes the em and strong contribution to the np mass splitting, i.e. these terms are $\mathcal{O}(\alpha)$ and $\mathcal{O}(\varepsilon)$. Differentiating the leading isospin-symmetric amplitude eq.(17) with respect to the nucleon mass shift gives

$$\frac{\partial \mathcal{A}_{-1}}{\partial M} = -\frac{\mathcal{A}_{-1}^2}{4\pi}\,(\mu + ip)\ , \tag{24}$$

from which it follows that

$$\delta\mathcal{A}_{-1} = -\delta M \frac{\mathcal{A}_{-1}^2}{4\pi}\,(\mu + ip) \sim \varepsilon\tilde{Q}^{-1}, \alpha\tilde{Q}^{-1}\ . \tag{25}$$

We now turn to CSB. The pattern of the various contributions looks different:

LO $\alpha\tilde{Q}^{-2}, \varepsilon Q^{-2}$	Electromagnetic four-nucleon contact interactions with no derivatives.
NLO $\alpha\tilde{Q}^{-1}, \varepsilon Q^{-1}$	Em four-nucleon contact terms with two derivatives, strong isospin-breaking contact terms w/ two derivatives, insertions proportional to the np mass difference.

We remark that the leading order CSB effects do not modify the effective range but the corresponding scattering length. Also, the leading $\pi\gamma$ graphs (with the exchanged pion now being neutral) are of order $\alpha\tilde{Q}^0$, i.e. suppressed by two orders in the KSW expansion parameter. This can be used as an argument to simply ignore such contributions in the study of CSB. Within meson-exchange models, their contribution comes out small but there seems to be no consensus about the actual size.[5] A similar hierarchy can also be derived based on the Weinberg power counting (which applies to the potential) as discussed in ref.[3].

6 The pionless theory

It is also of interest to consider a simplified theory in which even the pions are integrated out and one only deals with four–nucleon contact terms. Such an approximation is clearly only justifiable in a KSW–like approach as followed here. We do not want to enter the discussion about how to treat the pions (perturbative versus nonperturbative) but rather wish to see how accurate the pionless theory works for the question of isospin violation. In such an approach the connection to the chiral aspects of QCD is, of course, lost. Integrating out the pions modifies the coupling constants of the various contact terms like

$$\begin{array}{llll} \text{EFT with pions:} & C_0,\, C_2,\, D_2 & \text{and} & E_0^{(1)},\, E_0^{(2)}\ , \\ \text{EFT without pions:} & \tilde{C}_0,\, \tilde{C}_2 & \text{and} & \tilde{E}_0^{(1)},\, \tilde{E}_0^{(2)}\ . \end{array}$$

In the pionless theory, there is no more D_2 term. One has thus four parameters which can be fixed from the three scattering lengths and the np effective range. The values of the parameters are

$$\begin{aligned} \tilde{C}_0 &= -3.42\ \text{fm}^2\ , \quad \tilde{C}_2 = 6.05\ \text{fm}^4\ , \\ \tilde{E}_0^{(1)} &= 8.09\ \text{fm}^2\ , \quad \tilde{E}_0^{(2)} = 1.07\ \text{fm}^2\ . \end{aligned} \tag{26}$$

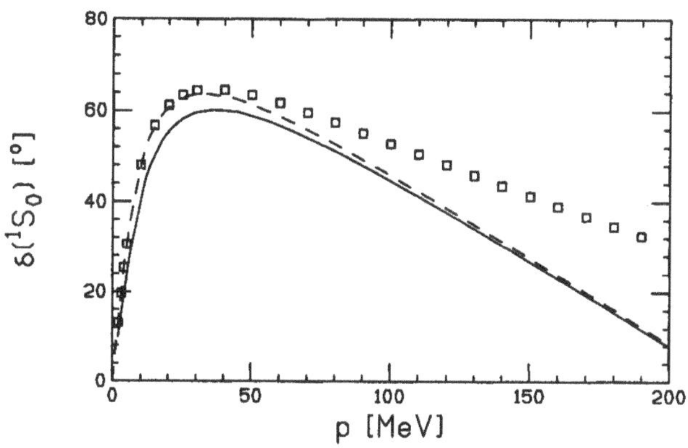

Figure 4: $^1S_0$ phase shifts for the np (dashed line) and nn (solid line) systems versus the nucleon cms momentum in the pionless theory. The empirical values for the np case (open squares) are taken from the Nijmegen analysis.

Compared to the values for the theory with pions given in eq.(23), we note that the value of $\tilde{C}_0$ almost the same as the one of C_0. This is expected since we are close to a fix point. Noticeably, there is a dramatic difference in $\tilde{E}_0^{(1)}$ compared to $E_0^{(1)}$, which is also expected since the former pion contribution is now completely absorbed in the contact term. The resulting np and nn $^1S_0$ phase shifts shown in fig. 4. The description of the phase shifts has clearly worsened, they start to deviate from the data for momenta larger than 50 MeV and are considerably off the data for higher momenta, say about 200 MeV. These results are not surprising, but it is comforting to see that the pionless theory indeed describes the small momentum region correctly. For example, to this order the nn effective range equals the one of the np system which is in agreement with the data (partial wave analysis). However, the interpretation of the CIB is now very different since it is entirely given by the isospin–violating contact interaction. This makes a comparison with the results obtained in potential models very difficult, if not impossible.

7 Summary

We have shown that isospin violation can be systematically included in the effective field theory approach to the two–nucleon system in the KSW formulation. For that, one has to construct the most general effective Lagrangian containing virtual photons and extend the power counting accordingly. We have shown that this framework allows one to systematically classify the various contributions to charge independence and charge symmetry breaking (CIB and CSB). In particular, the power counting combined with dimensional analysis allows one to understand the suppression of contributions from a possible charge–dependence in the pion–nucleon coupling constants. Including the pions, the leading CIB breaking effects are the pion mass difference in OPE together with a four–nucleon contact term. These effects scale as $\alpha\tilde{Q}^{-2}$. Power counting lets one expect that the much debated contributions from two–pion exchange and $\pi\gamma$ graphs are suppressed by factors of $1/3$ and $(1/3)^2$, respectively. This is in agreement with some, but not all, previous more model–dependent calculations. The leading charge symmetry breaking is simply given by a four–nucleon contact term. One can also use the pionless theory to describe the observed CIB and CSB, but the physical interpretation is much less clear. It would certainly be interesting to investigate such effects in the presence of external sources.

Acknowledgments

I thank the organizers for providing a stimulating atmosphere and Jerry Miller for some useful comments.

Appendix A: The average pion mass

Here, I briefly show how the average pion mass used in the np system in section 4 comes about. Consider the graphs shown in fig. 5. The strong interaction potential for OPE in the nn and pp system is given by the standard Yukawa form

$$V_{nn} = V_{pp} = \frac{g_{\pi N}^2}{4\pi} \frac{(\vec{\sigma}_{(1)} \cdot \vec{\nabla})(\vec{\sigma}_{(2)} \cdot \vec{\nabla})}{4M^2} \frac{e^{-m_0 r}}{r} , \tag{27}$$

where m_0 is a shorthand notation for the neutral pion mass.

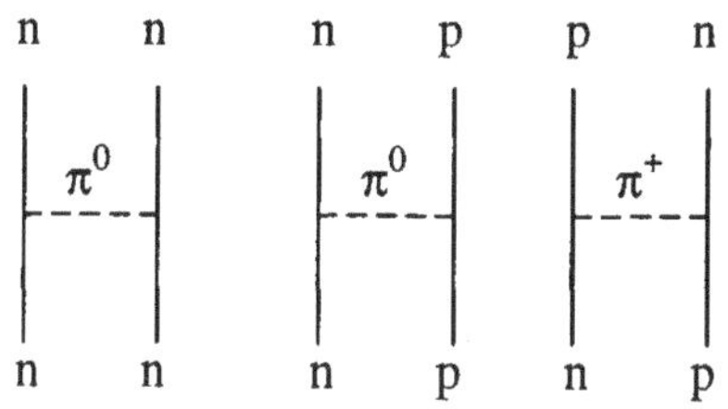

Figure 5: OPE graphs for the nn (left) and the np system (middle and right).

For the np case, we also have charged pion exchange,

$$V_{np} = V_{nn}\left[-1 + 2e^{-\delta m\, r}\right] \simeq V_{nn}\left[1 - 2\delta m\, r\right] , \tag{28}$$

with $\delta m = m_\pm - m_0$ and $m_\pm$ denotes mass of the charged pion. So effectively one has for the np OPE a mass

$$m = m_0 + 2\delta m \quad \text{or} \quad m^2 = m_0^2 + 2\delta m^2 , \tag{29}$$

since $m^2 \simeq m_0^2 + 4\delta m\, m_0 = m_0^2 + 4(m_\pm - m_0)m_0 \simeq m_0^2 + 2\delta m^2$ (note that $\delta m^2 \neq (\delta m)^2$). These are exactly the formulae used in section 4.

Appendix B: Power counting for the $\pi\gamma$ graphs

In this appendix, we will work out the explicit power counting of the various $\pi\gamma$–exchange diagrams (cf. fig. 3). First we need the counting rules for the propagators, vertices etc.: Pion propagator $\sim \tilde{Q}^{-2}$, nucleon propagator $\sim \tilde{Q}^{-1}$, pion-nucleon vertex $\sim \tilde{Q}$, photon-nucleon vertex $\sim \tilde{Q}^0$, charged pion-photon-nucleon vertex $\sim \tilde{Q}^0$, neutral pion-photon-nucleon vertex $\sim \tilde{Q}$ and loop integration $\sim \tilde{Q}^4$. This gives (displayed is the respective power in $\tilde{Q}$ for all vertices, propagators, etc for the specified diagram):

Graph	Vertices	Propagators	Loops	Total
3a ($\pi^{\pm}$)	0	-4	4	0
3a (π^0)	2	-4	4	2
3b ($\pi^{\pm}$)	2	-6	4	0
3b (π^0)	2	-6	4	0
3c ($\pi^{\pm}$)	1	-5	4	0
3c (π^0)	2	-5	4	1

For CIB, we have charged and neutral pion exchange and all $\pi\gamma$ diagrams 3a,b,c start contributing $\tilde{Q}^0$. In the case of CSB, matters are somewhat different. Graph 3b is leading at $\tilde{Q}^0$, while 3c is formally of order $\tilde{Q}$ but vanishes due to parity. The contribution of graph 3a to CSB is suppressed by two additional powers of the small momentum.

References

1. S. Weinberg, *Nucl. Phys.* B **363**, 3 (1991).
2. D.B. Kaplan, M.J. Savage and M.B. Wise, *Phys. Lett.* B **424**, 329 (1999);*Nucl. Phys.* B **534**, 329 (1998).
3. U. van Kolck, Ph.D. Thesis, University of Texas, Austin (1993); U. van Kolck, Few–Body Syst. Suppl. **9**, 444 (1995).
4. E. Epelbaum and Ulf-G. Meißner, **nucl-th/9902042**.
5. G.A. Miller, B.M.K. Nefkens and I. Slaus, *Phys. Rep.* **194**, 1 (1990).
6. S.A. Coon, **nucl-th/9903033**.
7. V.G.J. Stoks et al., *Phys. Rev.* C **49**, 2950 (1994).
8. X. Kong and F. Ravndal, **nucl-th/9811076**.
9. B.R. Holstein, **nucl-th/9901041**.
10. J.L. Friar and S.A. Coon, *Phys. Rev.* C **53**, 588 (1996).
11. R. Urech, *Nucl. Phys.* B **433**, 234 (1995).
12. H. Neufeld and H. Rupertsberger, *Z. Phys.* C **68**, 91 (1995).
13. Ulf-G. Meißner, G. Müller and S. Steininger, *Phys. Lett.* B **406**, (1997); (E) *ibid* B **407**, 434 (1997).
14. M. Knecht and R. Urech, *Nucl. Phys.* B **519**, 329 (1998).
15. Ulf-G. Meißner and S. Steininger, *Phys. Lett.* B **419**, 403 (1998).
16. G. Müller and Ulf-G. Meißner, **hep-ph/9903375**.
17. T. Mehen and I.W. Stewart, *Phys. Lett.* B **445**, 378 (1999); **nucl-th/9809095**.
18. D.B. Kaplan, **nucl-th/9901003**.
19. U. van Kolck et al., *Phys. Rev. Lett.* **80**, 4386 (1998).

Discussion during U-G. Meißner's presentation

Chair: *F. Ravndal (Oslo University)*
Notes taken by *J. McGovern (University of Manchester)*

The comments on this talk focused on the corrections made to the data with which comparison of the theoretical results was being made. Birse asked for details of the Coulomb subtraction, to which Ravndal briefly responded that standard techniques were used; Birse said that there was a logarithmic ambiguity in the scattering length, but Ravndal assured him that the procedure was model independent. Birse reiterated that comparison with the unmassaged data should be possible within the framework of ChPT, but Meißner said there were problems with divergences. Cohen suggested that comparison of cross-sections rather than phase shifts should be done.

SESSION 5

Session Chair: *M. Birse (University of Manchester)*
Notes taken by: *M. N. Butler (St. Mary's University)*

A SURPRISING RENORMALIZATION ON THE WAY TO N-BODY POWER COUNTING

P. F. BEDAQUE
Institute for Nuclear Theory
University of Washington, Seattle, WA 98195, USA
E-mail: bedaque@mocha.phys.washington.edu

We address the renormalization of the three-body system with short-range forces in the presence of fine tuning in the two-body sector. There is a cutoff dependence that does not appear in any order in perturbation theory. We argue that this cutoff dependence can be absorbed in a single three-body force counterterm and compute the running of the three-body force with the cutoff. This result extends the applicability of effective field theory to three nucleons systems using either the KSW scheme or an effective theory where pions are integrated out, as well as to other systems as He4 atom clusters.

There are nowadays (at least) three different effective field theories used in nuclear physics, each with its own power counting scheme. The first one contains no explicit pions and is limited to momenta $p \ll m_\pi$[1]. We will refer to this theory as the "pionless" theory. The second one is the "Kaplan-Savage-Wise" (KSW) effective theory and is supposed to be valid up to higher scales [2]. It contain pions explicitly and so can exploit the restrictions coming from chiral symmetry. One of its features is that pion exchanges are perturbative and, in particular, do not contribute at leading order. The third one is the initial scheme of Weinberg[3]. It also uses explicit pions but the power counting is different from the KSW one in the sense that it doesn't take into account the fine tunning existent in the two body s-wave scattering. In the Weinberg scheme the whole ladder of pion exchanges contribute already at leading order. The first two of these schemes, the pionless theory and the KSW scheme, are identical at leading order that includes only contact, non-derivative interactions between nucleons. Furthermore this two body interaction is fine tuned in such a way as to produce unnaturally large scattering lengths and shallow two nucleon bound states. The question I will address here is how to extend the pionless and the KSW theories to the many body sector at leading order.

In systems with three or more particles operators describing multinucleon forces play a role. A very naive estimate of the size of this multi-nucleon forces that does not take into account the fine tunning existent in the two body sector predicts them to be small and contributing only at high orders. In the presence of this fine tunning though this cannot be the whole story. A very old variational argument implies that the binding energy of *three* bosons goes to minus infinity as $1/R^2$ when the range R of the interactions goes to zero, if the binding of the *two* body system is kept constant as this limit is taken [4]. From the point of view of effective field theory that means that if we include only the two body interactions and choose a cutoff $\Lambda \gg 1/a$ (a is the two body scattering length) the three particle binding energy will be proportional to $1/\Lambda^2$. This suggests that the renormalization is not being properly done since the binding is strongly cutoff dependent. We will see below that a similar cutoff dependence appears even at low energy scattering of a particle by the bound state of other two. This cutoff dependence is coming from modes with momenta much larger than $1/a$ where we do not trust the effective theory. Our ignorance about the scale $p \sim \Lambda$ is encoded in the coefficients of operators in the lagrangian that are chosen in such a way as to cancel the contribution coming from high momenta inside the loops and to reproduce the experimental results. It is shown below that the cutoff dependence appearing in the three body system can be absorbed in one contact, non-derivative, three body force. This leaves *one* extra free parameter, already at leading order, that has to be fixed by experiment. This means that different models with the same two body scattering length will not predict similar three body bound states, but their predictions for the three particle bindings and scattering lengths will lie on a line, as observed some time ago.

Let us now make all statements above more precise. To simplify matters let us consider scalar particles with mass M and two body scattering length a whose interaction with each other have the range $R \ll a$. We are interested in momenta $p \ll R, M$ so the most general lagrangian one can write is

$$\mathcal{L} = \psi^\dagger(i\partial_0 + \frac{\vec{\nabla}^2}{2M})\psi - \frac{C_0}{2}(\psi^\dagger\psi)^2 - \frac{D_0}{6}(\psi^\dagger\psi)^3 + \ldots, \tag{1}$$

where the ellipsis denote terms with more derivatives and/or fields; those with more fields will not contribute to the three-body amplitude, while those with more derivatives are suppressed at low momentum. Whether the term proportional to D_0 should be included in the leading order lagrangian above is, of course, the point we want to settle. It is convenient in three body calculations to work with a lagrangian equivalent to Eq. (1):

$$\begin{aligned}\mathcal{L} &= \psi^\dagger(i\partial_0 + \frac{\vec{\nabla}^2}{2M})\psi + \Delta d^\dagger d - \frac{g}{\sqrt{2}}(d^\dagger\psi\psi + \text{h.c.})\\ &+ h d^\dagger d \psi^\dagger\psi + \ldots\end{aligned} \tag{2}$$

with the parameters in Eq. (reflagNd) satisfying

$$C_0 = g^2/\Delta \tag{3}$$

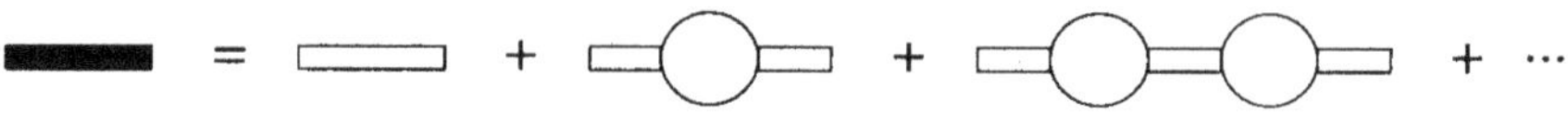

Figure 1: Dressed d field propagator.

$$D_0 = -3hg^2/\Delta^2 \tag{4}$$

and Δ is an arbitrary scale included only to give the field d the usual mass dimensions. The equivalence between the lagrangians in Eqs. (1) and (2) is shown by performing a gaussian path integral over the field d and a subsequent field redefinition. At tree level the field d is not propagating but, already at leading order in the momentum expansion, its propagator is corrected by the bubble chain depicted in Fig. 1 that produces a pole in the d field propagator. This is a familiar story for the EFT'ers in this workshop: each bubble produces a large (divergent) momentum independent piece whose value depends on the regulator used and a non analitic finite part determined by the unitarity cut.

If the theory is natural the only poles in this propagator are of the order $1/MR^2$ and for momenta $p \ll 1/R$ only the momentum independent piece has to be resummed to all orders. However in a theory with a fine tuned pole close to threshold the whole bubble chain must be summed resulting in

$$iD(p) = -i\frac{4\pi}{Mg^2}\frac{1}{-\frac{4\pi\Delta}{Mg^2} + c\Lambda - i\sqrt{Mp^0 - \frac{\vec{p}^2}{4}}}. \tag{5}$$

where c is a number dependent on the regulator and subtraction scheme used and Λ is the arbitrary scale introduced in the renormalization procedure. From the expression above we see that in order for this propagator to have a pole at $p_0 = -1/Ma^2$ $g^2/\Delta = C_0$ should be chosed such that

$$-\frac{4\pi\Delta}{Mg^2} + c\Lambda = -\frac{4\pi}{C_0} + c\Lambda = -\frac{1}{a} \tag{6}$$

Notice that $C_0(\Lambda \sim 1/R)$ has to be carefully arranged in order to produce a value of $a \gg R$. Also it is clear from Eq. (5) that for the range of momenta we are interested in ($p \sim 1/a$) the non analytic piece from the bubble graphs has indeed to be summed to all orders.

Let us now consider the scattering of one particle out of the s-wave bound state of other two. The field d has the quantum numbers of two ψ particles in a s-wave so it can be used as an interpolating field for the bound state. The

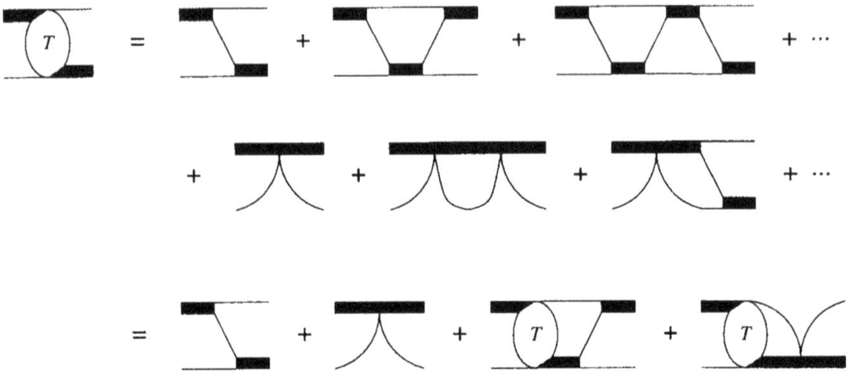

Figure 2: The amplitude T for particle/bound-state scattering as a sum of dressed one nucleon exchange and three-body-force diagrams (first and second lines) and as an integral equation (third line).

scattering amplitude is given by

$$T(k) = \sqrt{Z}t(k,k)\sqrt{Z}, \tag{7}$$

where $t(k,k)$ is the sum of diagrams with two d and two ψ external legs and Z is the wave function renormalization constant defined by

$$\frac{1}{Z} = i\,\frac{\partial}{\partial p_0}(iD(p))^{-1}\bigg|_{p_0=-1/ma^2} = \frac{M^2g^2a}{8\pi}. \tag{8}$$

The function $t(k)$ depends on the parameters g and Δ separatelly but the scattering matrix $T(k)$ depends only on the ratio of parameters $g^2/\Delta = C_0$ since it is a physical quantity independent of the particular normalization of the field d chosen here.

To determine which diagram must be included in leading order we follow very simple rules. Wc ount the external momenta k and $1/a$ as quantities of the same order denoted by $Q \sim k \sim 1/a$. Each dNN vertes brings a power of g, each ψ propagator a power of M/Q^2 and each d propagator a power of $1/(Mg^2Q)$. The loop integrals are estimated by scaling all momenta by Q and all energies by Q^2/M, resulting on dimensionless integrals independent of Q or a. The result of this rescaling is that each loop brings a power of Q^5/M. Each diagram is then multiplied by $Z \sim 1/(M^2g^2a)$. Using these rules we find that the diagrams of Fig. 2 not involving the $d^\dagger d\psi^\dagger\psi$ are all of the same order $\sim 1/MQ$. There is an infinite number of diagrams given by a double sum. This involves a double resummation. First the bubble chain of Fig. 1 is summed, what can be done analytically since it is a geometric series, resulting in Eq. (5). After this first resummation all the diagrams are ultraviolet finite by power counting. The resummed d propagator is then inserted inside the loops shown in Fig. 2 the infinite describing the multiple nucleon exchanges is summed. This last step cannot be done analytically so we derive an integral equation whose solution gives the amplitude we are after. The equation is shown diagramatically in Fig. 2 and it is the Fadeev equation for zero range forces derived a long time ago. To be precise, let us fix the kinematics in these graphs in such a way that the incoming d line carries momentum $(3k^2/4M - B, \vec{k})$, the incoming ψ line momentum $(k^2/2M, -\vec{k})$ the outgoing d line momentum $(3k^2/4M - B + \epsilon, \vec{p})$ and the outgoing ψ line momentum $(k^2/2m - \epsilon, -\vec{p})$, with $B = 1/(Ma^2)$ being minus the binding energy of two particles. Denoting by $t(\vec{k}, \vec{p}, \epsilon)$ the blob of Fig. 2 with this kinematics we have

$$\begin{aligned} it(\vec{k},\vec{p},\epsilon) &= -2g^2 iS(-k^2/4M - B + \epsilon, \vec{p}+\vec{k}) + ih \\ &= \int \frac{d^4q}{(2\pi)^4} it(\vec{k},\vec{q},\epsilon+q_0) iD(k^2/2M - \epsilon - q_0, -\vec{q}) \\ &\quad (-2g^2 iS(-k^2/4M - B + 2\epsilon + q_0, \vec{p}+\vec{q}) + ih) \end{aligned} \tag{9}$$

We also included the three body force diagrams for future reference. Let us now rewrite this equation somewhat. First the q_0 integral is performed, ϵ is set to $\epsilon = k^2/2M - p^2/2M$, we integrate over the angle between k and p to project the s-wave and the s-wave componente of $t(\vec{k}, \vec{p}, k^2/2M - p^2/2M)$ is denoted simply by $t(k,p)$. The function $t(k,p)$ is a off shell amplitude since the d outgoing line is off shell. It goes on-shell for $p = k$ and the scattering amplitude is obtained by $T(k) = Zt(k,k)$. Finally, instead of using $t(k,p)$ we use the function $a(k,p)$ defined by

$$t(k,p) = -\frac{1}{4\pi a}(k^2 - p^2) D(3k^2/4 - p^2/2 - B, p) \frac{a(k,p)}{1 - ika(k,k)}. \tag{10}$$

The point of this definition is that $a(k,p)$ (K-matrix) is real below the breakup threshold even for off shell momenta and that will make our discussion easier. It is normalized is such a way that, on shell, $a(k,k) = k\text{cotg}\delta$, where δ is the s-wave phase shift. After all these manipulation the Eq. (9) translates in the following equation for $a(k,p)$ [5,7,8]:

$$a(k,p) = M(k,p;k) + \frac{2\lambda}{\pi}\int_0^\Lambda dq\; M(q,p;k)\frac{q^2}{q^2-k^2} a(k,q), \tag{11}$$

with the kernel

$$M(q,p,k) = \frac{4}{3}\left(\frac{1}{a_2} + \sqrt{\frac{3p^2}{4} - mE}\right)\left[\frac{1}{pq}\ln\left(\frac{q^2+qp+p^2-mE}{q^2-qp+p^2-mE}\right) + \frac{h}{mg^2}\right]. \tag{12}$$

$E = 3k^2/4M - B$ is the total energy, Λ is a cutoff and $\lambda = 1$. In the case of nucleons in the $L = 0$, $J = 3/2$ channel an identical equation is derived with $\lambda = -1/2$. Eq. (11) is the zero range version of the Fadeev equation [6]. The presence of an ultraviolet cutoff seems unnecessary since we mentioned that all the diagrams are finite but it is here that the surprise lies. For the sake of argument, let us assume that the three body forces are as small as naive dimenisonal analysis indicate and can be dropped. Eq. (11) can be solved numerically in the $\lambda = -1/2$ case relevant for neutron-deuteron scattering in the quartet channel. The solution is independent of the cutoff as long as $\Lambda \gg 1/a$, as one would expect from the observation that this equation sums diagrams that are ultraviolet finite. However Eq. (11) is solved with $\lambda = 1$, the case relevant for the scalar case, a strong dependence on the cutoff is observed. This is shown on Fig. 3. Notice that it is not only the high momentum, off shell part that is cutoff dependent, but also the on shell amplitude. As a matter of fact any given amplitude can be obtained by moving the cutoff around. In the limit $\Lambda \to \infty$ the solution does not converge to any value, but oscillates indefinitelly.

A strong cutoff dependence indicates that much of the contribution to the amplitude comes from high momentum modes. In perturbative calculations this is a familiar situation and it happens everytime there is an ultraviolet divergent graph. However the contribution from the high momentum modes comes not only from the loops, but also from counterterms. The separation between loops contributions and counterterms is dependent on the details of the regulator and renormalization scheme used but the sum of these two contributions is cutoff independent. Since the precise value of the counterterm is not known *a priori* the total contribution coming from high energies is incalculable and has to be determined by experiment. A similar situation occurs here and this cutoff dependence should be accompanied by contributions coming from graphs involving a three body force counterterm. The difference is in the unusual origin of this cutoff dependence. We will need to understand this better to be able to determine the counterterm needed.

For this we borrow some insight on the Eq. (11) from [9,10]. Let us look at the asymptotic behavior of $a(k,p)$ for large $p \gg 1/a$ and fixed $k \sim 1/a$ in the case $h = 0$ and $\lambda \to \infty$. Most of the contribution to the integral comes from the high momentum region $q \gg 1/a$ and, dropping terms suppressed by $k/p, pa$

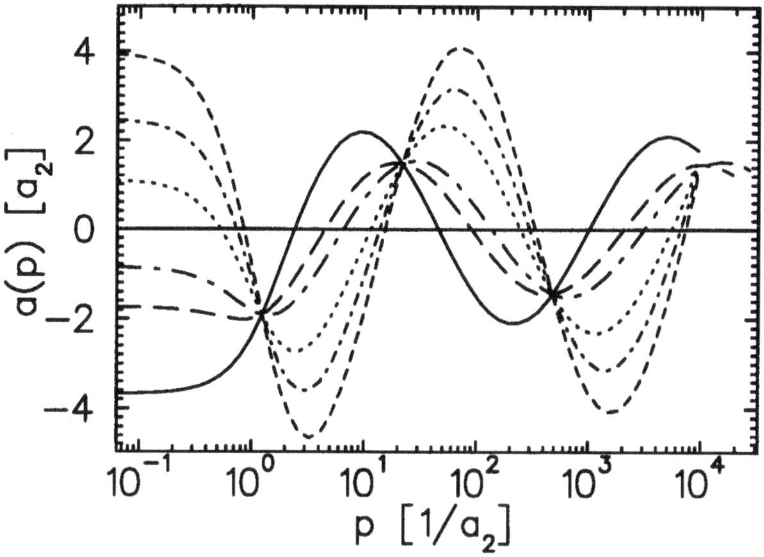

Figure 3: Amplitude $a(0,p)$ as function of the momentum p. Full, dashed, and dash-dotted curves are for $H = 0$ and $\Lambda a = 1.0,\ 2.0,\ 3.0 \times 10^4$, respectively. Dotted, short-dash-dotted, and short-dashed curves are for $\Lambda a = 1.0 \times 10^4$ and $H = -6.0,\ -2.5,\ -1.8$, respectively.

we have the equation that determines the asymptotic behavior of $a(k,p)$:

$$a(k,p) = \frac{4\lambda}{\sqrt{3}\pi} \int_0^\infty \frac{dq}{q} \ln\left(\frac{q^2+pq+p^2}{q^2-pq+p^2}\right) a(k,q). \tag{13}$$

This equation has two symmetries. It is obviouslly scale invariant so if $a(k,p)$ is a solution so is $a(k,xp)$ for any x. It is also invariant by inversions so $a(k,\mu^2/p)$ is also a solution for any μ. Scale invariance suggests the ansatz

$$a(k,p) \sim p^s. \tag{14}$$

Inserting this ansatz into Eq.(13) we find a condition on s

$$1 - \frac{8\lambda}{\sqrt{3}s} \frac{\sin\frac{\pi s}{6}}{\cos\frac{\pi s}{2}} = 0. \tag{15}$$

We should consider only solutions of Eq.(15) with $\Re s < 1$ to ensure convergence of the integral in Eq. (11). There is also the condition that $\Re s < 1$ coming from the convergence of the intergal. For $\lambda < \lambda_c = \frac{3\sqrt{3}}{4\pi} \simeq 0.4135$, Eq. (15) has only real roots. The largest one (with $\Re s < 1$) determines the ultraviolet behavior of $a(k,p)$. That is the case when $\lambda = -1/2$, as is relevant for nucleons in the $L = 0, J = 3/2$ channel, and $a(k, p \ll 1/a) \sim p^{-2.17\ldots}$ at high $p \gg 1/a$. The proportionality constant is not determined by this analysis but by matching this asymptotic behavior to the solution in the infrared, that is, by solving the full Eq.(11). In the case of three scalar particles $\lambda = 1$ and there are two imaginary solutions $s = \pm i s_0 = \pm i\ 1.0064\ldots$. They come in pairs due to the inverison symmetry. The amplitude in the ultraviolet is a linear combination of p^{is_0} and p^{-is_0}

$$a(k,p) \sim \cos(s_0 \ln p + \delta) \tag{16}$$

and there is, besides of an overall constant determined by matching with the solution in the infrared, an arbitrary phase not determined by the integral equation. This suggests that the original integral equation Eq. (11) has a one parameter family of solutions when $h = 0, \Lambda = \infty$. A more formal proof of this nonuniqueness can be given by showing that a solution to the *homogeneous* version of Eq. (11) admits a solution for any value of k [10]. By adding this homogeneous solution (multiplied by an arbitrary constant) to a particular solution of Eq. (11) we caan construct another solution. Since homogeneous solution has the same asymptotic behavior shown in Eq.(16) the new solution obtained will have the ultraviolet behavior shown in Eq.(16), but with a different phase.

If we keep the ultraviolet cutoff finite the inversion symmetry is broken, a particular phase is chosen and there is a unique solution to the integral equation Eq. (11). The particular phase chosen is however dependent on the cutoff. This is what the numerical solutions in Fig. 2 show. There is a complicated form for $a(k,p)$ for small $p \sim 1/a$ and for large $p \sim \Lambda$ but, in the intermediate region $1/a \ll p \ll \Lambda$, $a(k,p)$ has the simpler form (16) with a particular phase dependent on the cutoff used.

In one loop perturbative calculations cutoff dependent terms are insensitive to the infrared scales since they are due to high momentum modes and can be expanded in power of the external momenta. Since the contribution coming from counterterms included at tree level is also analytic in the external momenta they can be arranged to cancel the cutoff dependent loop contribution. In the case discussed here the cutoff dependence coming from loops (that is, the solution of Eq. (11) with $h = 0$) is *not* analytic in p. The only way the inclusion of a non zero h can cancel this cutoff dependence is if the three body terms are also included at all orders in Eq.(11) and generate the same p dependence as shown in Eq.(16). The three body force term will be leading if

$h \sim Mg^2/\Lambda^2$

$$h = Mg^2 H(\Lambda)/\Lambda^2 \tag{17}$$

with $H(\Lambda) \sim 1$, what corresponds to

$$D_0(\Lambda) = -\frac{3MC_0^2}{\Lambda^2} H(\Lambda). \tag{18}$$

The essential question now is whether the inclusion of the three body force term can generate a contribution to cancel the cutoff dependence of the phase in Eq. (11). To answer this question let us look at Eq. (11) with a finite cutoff and h given by Eq.(17). Its solution in the infrared region $p \sim 1/a$ and on the ultraviolet region $p \sim \Lambda$ is complicated but in the intermediate region the arguments leading to Eq. (16) are still valid. The difference is that now the arbitrary phase $\delta = \delta(\Lambda, H(\Lambda))$ is determined by the value of the cutoff Λ and the coefficient $H(\Lambda)$. Adjusting $H(\Lambda)$ in order to cancel the explicit dependence of δ on Λ we achieve cutoff independence of the asymptotic phase and, consequently, of the function $a(k,p)$ for $p \ll \Lambda$ (what includes the on-shell point $p = k$). In other words, we want to choose $H(\Lambda)$ such that

$$a(k, 1/a \ll p \ll \Lambda) \sim \cos(s_0 \ln p + \delta(\Lambda, H(\Lambda))) = \cos(s_0 \ln(p/\bar{\Lambda})) \tag{19}$$

where $\bar{\Lambda}$ is a constant fixed by experiment. Notice that since δ is not a function of the momentum k, fixing the cutoff dependence for one value of k fixes it for all the other values. To determine the Λ dependence of $H(\Lambda)$ we will write Eq.(11) for two different values of the cutoff Λ and Λ' [8].

$$\begin{aligned}
a_\Lambda(k,p) = M(k,p,kk) &+ \frac{2}{\pi}\int_0^\mu dq\, a_\Lambda(k,q)\frac{q^2}{q^2-k^2}M(q,p,k) \\
&+ \frac{2}{\pi}\int_\mu^\Lambda dq\, a_\Lambda(k,q)(\frac{1}{q^2}+\frac{H(\Lambda)}{\Lambda^2}) \\
a_{\Lambda'}(k,p) = M(k,p,k) &+ \frac{2}{\pi}\int_0^\mu dq\, a_{\Lambda'}(k,q)\frac{q^2}{q^2-k^2}M(q,p,k) \\
&+ \frac{2}{\pi}\int_\mu^\Lambda dq\, a_{\Lambda'}(k,q)(\frac{1}{q^2}+\frac{H(\Lambda)}{\Lambda^2}) \\
&+ \mathcal{X}
\end{aligned} \tag{20}$$

with

$$\mathcal{X} = \frac{2}{\pi}\int_\mu^\Lambda dq\, a_{\Lambda'}(k,q)(\frac{H(\Lambda')}{\Lambda'^2} - \frac{H(\Lambda)}{\Lambda^2}) + \frac{2}{\pi}\int_\Lambda^{\Lambda'} dq\, a_{\Lambda'}(k,q)(\frac{1}{q^2}+\frac{H(\Lambda')}{\Lambda'^2}), \tag{21}$$

where μ is a scale satisfying $1/a \ll \mu \ll \Lambda$. We want to choose $H(\Lambda)$ as a function of Λ in order to keep the phase of $a(k,p)$ in the intermediate regime fixed and set by $\bar{\Lambda}$ as the cutoff is varied

$$a_{\Lambda'}(k,p) = a_{\Lambda}(k,p) \sim \cos(s_0 \ln(p/\bar{\Lambda})). \tag{22}$$

If $H(\Lambda)$ is chosen is such a way that $\mathcal{X} = \mathcal{O}(\mu/\Lambda)$ the equations satisfied by $a_{\Lambda}(k,p)$ and $a'_{\Lambda}(k,p)$ are, up to higher order terms, identical and thus have the same solution (the solution is unique with a finite cutoff !). This can be accomplished if $H(\Lambda)$ is chosen as

$$H(\Lambda) = -\frac{\sin(s_0 \ln(\Lambda/\bar{\Lambda}) - \mathrm{arctg}(\frac{1}{s_0}))}{\sin(s_0 \ln(\Lambda/\bar{\Lambda}) + \mathrm{arctg}(\frac{1}{s_0}))}. \tag{23}$$

$H(\Lambda)$ is plotted in Fig. 3. An independent determination of the running of H was also done by numerically finding a value of $H(\Lambda)$ that produces some "experimental" value of the scattering at zero energy. This can always be done since there is one free parameter (H) to fit one quantity (the zero energy scattering). With these values of H the scattering at finite values of the energy is computed. By our previous arguments they should be cutoff independent and this is observed numerically as Fig. 4 shows.

Our arguments up to now followed a convoluted path: we tried not to use the three body force, found a problem, went back and fixed it with the inclusion of the counterterm. Let us now rephrase the result in a straighforward manner. Solving the integral equation (11) with a particular value of the cutoff and imposing that its solution say, at zero energy, agrees with the experiment we obtain a value for h. Is this a large value or it is small and can be neglected without much change in the solution? Looking at Fig. 3 we can read the answer. It depends on the particular values of the cutoff and of the zero energy scattering amplitude. By chance we may find that the cutoff is such that h is tiny. That is unlikely though and a small change in Λ renormalizes h away from zero. The *natural* value for the three body force is larger and given by Eq. (18). In this sense we can say that the three body force is large, enhanced by powers of a compared to the estimate of a dimensional analysis argument that does not take into account the fine tuned two-body sector. There is one important details though. In perturbative calculations coupling constants are given by their naive dimensional analysis size times a dimensioless number that runs slowly (logarithmically) with the cutoff. Here this dimensionless number (H) is a wild function containing zeros and even diverging at some points. Thus one should be aware that numerically h can be much larger or much smaller than estimated just because H can be close to a zero or a pole.

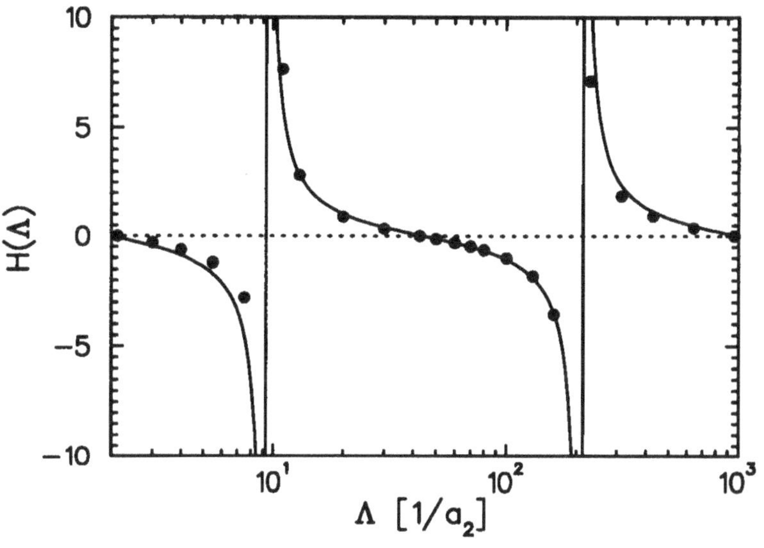

Figure 4: Three-body force H as a function of the cutoff Λ for $a_3 = 1.56a$: numerical solution (dots) and Eq.(17) with $\bar{\Lambda}a = 19.5$ (solid line).

Different potential models that predict the same two body scattering length (or even the same scattering at finite energies) can have different behavior off shell. Of course this is a statement dependent on the particular interpolating field one uses but it made people worry in the past that two body scattering data would not be enough to make predictions on multi-particle systems. This "off shell" effects can be traded by multi-body forces. Probably the easiest way to see this is to observe that off shell Greens' functions are changed by non-linear field redefinitions and that non-linear field redefinitions generate multi-body terms from the quadratic ones in the lagrangian. By computing two different properties of three body systems, say the binding energy of the three particle ground state B_3 and the scattering length of particle-bound state a_3, with different phase equivalent potentials, one can determine the importance of "off shell " effects. The result is that the predictions of different models neither accumulate around one point in the $a_3 \times B_3$ plane (as they would if off shell effects were irrelevant) nor scatter all over the plane, but lie on a well defined line (known as Phillips' line). In other words, the off shell ambiguity is, at leading order, determined by one parameter. This numerical result is nicely explained by the effective theory approach (using a different language

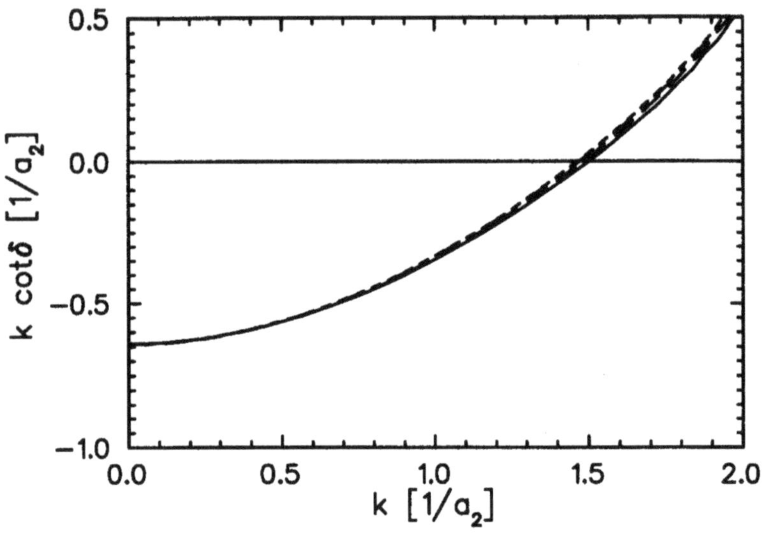

Figure 5: Energy dependence for $\bar{\Lambda}a = 19.5$: $k\cot\delta$ as function of k for different cutoffs ($\Lambda a = 42.6, 100.0, 230.0, 959.0$).

Efimov has previously given an explanation for the Phillips line that I believe is physically equivalent [11]).

Let me know try to answer a few questions asked during the workshop. The first one is: Eq. (11) was derived as a way to sum a series of diagrams. Since they are finite, why not compute and sum them ? Since no renormalization is needed, no new free parameter will need to be fixed by experiment. The answer to that is simply to observe that, eventhough we motivated Eq. (11) through diagrams, it is only the Schroedinger equation in disguise and its validity goes beyond perturbation theory. I could have derived it without reference to diagrams the same way that the Schwinger-Dyson equation is derived in textbooks, manipulating the path integral. Thus all its solutions should be taken seriously, not only the particular solution obtained by the sum of diagrams with infinite cutoff. This is reminiscent of the instanton contributions in QCD. We believe them because they solve the equations of motion, eventhough they do not show up in any diagram. It is also unclear whether the infinite series of diagrams can be summed, or even Borel summed. Maybe the non uniqueness of the solution of the integral equation with infinite cutoff shows up in its perturbative series as some disease of the series, but that is not known for sure.

The second issue raised calls into question the fact that we have used two different regulators, one to compute the d propagator and another one for the three particle integrals. This is strange, but a legitimate way to proceed. To see that just consider the changes that would be introduced by keeping a finite cutoff in the bubble chain of Fig. 1. The only change in Eq. (11) would have been through terms suppressed by q/Λ that are irrelevant except for $q \sim \Lambda$. This would not change our analysis leading to the cutoff dependence that was based in the behavior of the solution in the intermediate range $1/a \ll q \ll \Lambda$. Of course the precise form of the three body force (Eq. (23)) would change but the physical results would be the same. In other words, those two different procedures correspond to two different regulators, so they split the high energy contributions into loop plus counterterm in different ways,but the final result is the same.

The third question is whether four, five, ..., -body forces will also be leading. I do not know the answer to that yet but, at least in the nuclear case, I guess there is at most a four-body force. The reason is simply Fermi statistics: five or more body force terms will have to involve derivatives and should be suppressed. In the three body sector we noticed that only in the unique channel where all three particles can be in a s-wave in relation to the center of mass (the $L = 0$, $J = 1/2$ channel) is that we find the thre body force to be large. There one channel like that in the four body sector and none for more than four nucleons.

To finalize, eventhough we have discussed only the spinless case here, the study of three nucleon systems in the $L = 0, J = 3/2$ channel shows remarkable simillarity to the scalar case, even at the quantitative level. This should be expected since the renormalization of the three body force is related to the ultraviolet behavior of the theory and, at high momenta, the difference between the two two-body scattering lengths a_t (triplet) and a_s (singlet) are irrelevant $(p \gg 1/a_s, 1/a_t)$. Thus, in the ultraviolet, nucleon interactions have a $SU(4)$ symmetry. We can make a three particle state that belongs to the $\bar{\mathbf{4}}$ of $SU(4)$ (totally antisymmetric) and is totally symmetric in space. Thus the equation for this channel is identical, in the ultraviolet, to the scalar case discussed above.

The power counting of next-to-leading order corrections and beyond is tricky. In the cutoff scheme used here the leading operators (two and three body forces) are renormalized at every order in the momentum expansion and it is not obvious that those corrections do not necessitate higher order counterterms. For the leading correction (due to the range of two-body forces), in the pionless theory, one can show that this is the case and the divergences appearing are absorbed in a sub-leading piece of the non-derivative three-body force included above[12,13]. Probably the easiest way of establishing the power counting to all orders is through the use of dimensional regularization, but I have not been able to implement it in the non-perturbative kind of calculation discussed here.

Acknowledgments

The work presented in this talk was done in collaboration with U. van Kolck and H.-W. Hammer. Discussions with the EFT'ers of Seattle and numerous participants of this workshop are very much appreciated.

This research was supported in part by the U.S. Department of Energy grants DOE-ER-40561.

References

1. U. van Kolck, *Nucl. Phys.* A **645**, 273 (1999) and references therein; nucl-th/9902015 (to appear in *Prog. Part. Nucl. Phys.*).
2. D.B. Kaplan, M.J. Savage and M.B. Wise, *Phys. Lett.* B **424**, 390 (1998); *Nucl. Phys.* B **534**, 329 (1998).
3. S. Weinberg, *Phys. Lett.* B **251**, 288 (1990); *Nucl. Phys.* B **363**, 3 (1991); *Phys. Lett.* B **295**, 114 (1992); C. Ordóñez and U. van Kolck, *Phys. Lett.* B **291**, 459 (1992); C. Ordóñez, L. Ray, and U. van Kolck, *Phys. Rev. Lett.* **72**, 1982 (1994); *Phys. Rev.* C **53**, 2086 (1996). U. van Kolck, *Phys. Rev.* C **49**, 2932 (1994).
4. L.H. Thomas, *Phys. Rev.* **47**, 903 (1935).
5. G.V. Skorniakov and K.A. Ter-Martirosian, *Sov. Phys. JETP* **4**, 648 (1957).
6. An excellent place to learn about the Fadeev equation and the three body problem in general is R.D. Amado, in *The Three Body Problem in Elementary Particle Physics and Scattering Theory, Vol. 2*, ed. M. Chretien and S.S. Schweber (Brandeis, New York).
7. P.F. Bedaque and U. van Kolck, *Phys. Lett.* B **428**, 221 (1998); P.F. Bedaque, H.-W. Hammer, and U. van Kolck, *Phys. Rev.* C **58**, R641 (1998).
8. P.F. Bedaque, H.-W. Hammer, and U. van Kolck, *Phys. Rev. Lett.* **82**, 463 (1999).
9. R.A. Minlos and L. Fadeev, *Sov. Phys. JETP* **14**, 1315 (1962).
10. S. Danilov, *Sov. Phys. JETP* **13**, 349 (1961).
11. V.N. Efimov, *Sov. J. Nucl. Phys.* **12**, 589 (1971); *Phys. Rev.* C **47**, 1876 (1993).

12. H.-W. Hammer, *this volume.*
13. P.F. Bedaque, H.-W. Hammer, and U. van Kolck, in progress.

Discussion during P. Bedaque's presentation

Chair: *M. Birse (University of Manchester)*
Notes taken by: *M. N. Butler (Saint Mary's University)*

During the talk, there was some discussion on the nature of the integral equation solutions for the three-body system. Jerry Miller pointed out that the $J = 1/2$ state can't be treated exactly as bosonic, because of spin asymmetry. This was acknowledged to be an analogy, only. There was significant discussion regarding the convergence and uniqueness of the integral equation solutions, most noticeably with respect to the cutoff Λ. Mark Wise and Jambul Gegelia were concerned that solutions would be cutoff independent, and unique for any energy. It was stated that unique solutions exist for different values of the cutoff, and there is one solution for each energy. Gegelia, further, raised the question of summing finite diagrams first, vs. taking the approach of using the integral equation to solve the three-body system. Bedaque pointed out that the integral equations are more than the sum of diagrams, and represent a non-perturbative approach to the three-body system.

In the questions after the talk, there was some discussion of the nature of the three-body force - the counterterm required to remove the cutoff dependence. Tom Cohen queried the connection to Weinberg's dimensional analysis, which stated that three-body forces are small. However, Bedaque found results proportional to a^2, which could make three-body forces large if the binding is small. Bedaque argued that Weinberg was thinking about pions, which was a very different situation, but Mark Wise noted that Weinberg, in fact, used naive dimensional analysis. Gegelia continued his argument that finite diagrams should be summed, first, as they would not require a cutoff. Bedaque emphasised again that the integral equations gave more than the summing of diagrams, and are a better approach.

At the end, Miller asked about whether the 'serious problems' mentioned in Bedaque's summary contradicted his statement that many processes can be calculated. Bedaque responded that much of their work so far was on processes not requiring extra renormalization. If they do that, then many more processes can be studied. Thomas Mehen pointed out that you could use a cutoff even with the KSW method.

EFFECTIVE THEORY FOR NEUTRON-DEUTERON SCATTERING AND THE TRITON

H.-W. HAMMER
TRIUMF, 4004 Wesbrook Mall, Vancouver, B.C.
Canada V6T 2A3
E-mail: hammer@triumf.ca

We apply the effective field theory approach to the three-nucleon system. In particular, we consider neutron-deuteron scattering and the triton. Precise predictions for $S = 3/2$ scattering are obtained in a straightforward way. In the $S = 1/2$ channel, however, a unique nonperturbative renormalization takes place which requires the introduction of a three-body force at leading order. We also show that invariance under the renormalization group explains some universal features of the three-nucleon system.

1 Introduction

Effective field theories (EFT) are a powerful concept designed to explore a separation of scales in physical systems.[1] For example if the momenta k of two particles are much smaller than the inverse range of their interaction $1/R$, observables can be expanded in powers of kR. Following the early work of Weinberg,[2] EFT's have become quite popular in nuclear physics.[3,4] Their application, however, is complicated by the presence of shallow (quasi) bound states, which create a large scattering length $a \gg R$. In this finely tuned case, the perturbative expansion in kR already breaks down at rather small momenta $k \sim 1/a$. In order to describe bound states with typical momenta $k \sim 1/a$, the range of the EFT has to be extended. A certain class of diagrams has to be resummed which generates a new expansion in kR where powers of ka are kept to all orders. For the two-nucleon system a power counting that incorporates this resummation has been found recently.[5,6,7] Pions are included and treated perturbatively in this scheme. It has successfully been applied to NN scattering and deuteron physics.[8]

The three-nucleon system is a natural testground for the understanding of the nuclear forces that has been reached in the two-nucleon system. However, the extension of these ideas is not straightforward, as the three-nucleon system shows some remarkable universal features. It has been found that different models of the two-nucleon interaction that are fitted to the same low-energy two-nucleon (NN) data predict different but correlated values of the triton binding energy B_3 and the S-wave nucleon-deuteron (Nd) scattering length $a_3^{(1/2)}$ in the spin $S = 1/2$ channel; all models fall on a line in the $B_3 \times a_3^{(1/2)}$

plane, the Phillips line.[9] Other universal features of three-body systems are the existence of a logarithmic spectrum of bound states that accumulates at zero energy as the two-particle scattering length a_2 increases (the Efimov effect [10]), and the collapse of the deepest bound state when the range of the two-body interaction r_2 goes to zero (the Thomas effect [11]).

We will show that the renormalization of the three-nucleon system is non-perturbative and requires a three-body force at leading order. In addition to the description of experimental data from a few low-energy parameters, EFT allows to understand the above mentioned universal features in a unified way.[12,13,14]

2 Neutron-Deuteron System

In order to avoid the difficulties due to the long range Coulomb force, we concentrate here on the neutron-deuteron (nd) system. We also restrict ourselves to scattering below the deuteron breakup threshold where S-waves are dominant. There is only one scale at low energy, $\sqrt{MB_2} \approx 40$ MeV, where B_2 is the binding energy of the deuteron and M the nucleon mass. Since $\sqrt{MB_2}$ is small compared to the pion mass, the pions can be integrated out and only the nucleon degrees of freedom remain. The leading order of this pionless theory is equivalent to the leading order in the KSW counting scheme [6] where pions are included perturbatively. As a consequence, the extension of our results to higher energies is well defined. Recently, a number of two nucleon observables have been studied in the pionless theory as well.[15]

There are two S-wave channels for neutron-deuteron scattering, corresponding to total spin $S = 3/2$ and $S = 1/2$. For scattering in the $S = 3/2$ channel all spins are aligned and the two-nucleon interactions are only in the $^3S_1$ partial wave. The interaction is repulsive and the Pauli principle forbids the three nucleons to be at the same point in space. As a consequence, this channel is insensitive to short distance physics and very precise predictions are obtained in a straightforward way. There is also no three-body bound state in this channel. The $S = 1/2$ channel is more complicated. The two-nucleon interaction can take place either in the $^3S_1$ or in the $^1S_0$ partial waves. This leads to an attractive interaction which sustains a three-body bound state, the triton. The $S = 1/2$ channel also shows a strong sensitivity to short distance physics as the Pauli principle does not apply. As a consequence, it displays the Thomas [11] and Efimov [10] effects. The generic features of this channel are very similar to the system of three spinless bosons. There is a strong cutoff dependence even though all Feynman diagrams are finite. As we will show, the renormalization requires a leading order three-body force counterterm.[13,14]

Since the details of the renormalization in the three-body system are discussed in Bedaque's talk,[12] we will mainly focus on the application of these ideas to the nd system.

Let us start from the assumption that the three-body force is of natural size. The lowest order effective Lagrangian is then given by

$$\mathcal{L} = N^\dagger \left(i\partial_0 + \frac{\vec{\nabla}^2}{2M} \right) N - C_0^t \left(N^T \tau_2 \vec{\sigma} \sigma_2 N\right)^\dagger \cdot \left(N^T \tau_2 \vec{\sigma} \sigma_2 N\right) \qquad (1)$$
$$-C_0^s \left(N^T \sigma_2 \boldsymbol{\tau} \tau_2 N\right)^\dagger \cdot \left(N^T \sigma_2 \boldsymbol{\tau} \tau_2 N\right) + \ldots ,$$

where the dots represent higher order terms supressed by derivatives and more nucleon fields. $\vec{\sigma}$ ($\boldsymbol{\tau}$) are Pauli matrices operating in spin (isospin) space, respectively. The contact terms proportional to C_0^t (C_0^s) correspond to two-nucleon interactions in the $^3S_1$ ($^1S_0$) NN channels. Their renormalized values are related to the corresponding two-body scattering lengths a_2^t and a_2^s by $C_0^{s,t} = \pi a_2^{s,t}/(2M)$. Since no derivative interactions are included, this Lagrangian generates only two-nucleon interactions of zero range. For practical purposes, it is convenient to rewrite this theory by introducing "dibaryon" fields with the quantum numbers of two nucleons.[16] In our case, we need two dibaryon fields: (i) a field $\vec{T}$ with spin (isospin) 1 (0) representing two nucleons interacting in the $^3S_1$ channel (the deuteron) and (ii) a field $\boldsymbol{S}$ with spin (isospin) 0 (1) representing two nucleons interacting in the $^1S_0$ channel. Using a Gaussian path integration, it is straightforward to show that the Lagrangian from Eq. (1) is equivalent to

$$\mathcal{L} = N^\dagger \left(i\partial_0 + \frac{\vec{\nabla}^2}{2M} \right) N + \Delta_T \vec{T}^\dagger \cdot \vec{T} + \Delta_S \boldsymbol{S}^\dagger \cdot \boldsymbol{S} \qquad (2)$$
$$-\frac{g_T}{2} \left(\vec{T}^\dagger \cdot N^T \tau_2 \vec{\sigma} \sigma_2 N + h.c. \right) - \frac{g_S}{2} \left(\boldsymbol{S}^\dagger \cdot N^T \sigma_2 \boldsymbol{\tau} \tau_2 N + h.c. \right) + \ldots .$$

At first it may look like the Lagrangian, Eq. (2), contains more parameters than the original one, Eq. (1). However, the scales Δ_T and Δ_S are arbitrary and included in Eq. (2) only to give the dibaryon fields the usual mass dimension of a heavy field. They can easily be removed by rescaling the dibaryon fields. All observables depend only on the ratios $g_{T,S}^2/\Delta_{T,S}$. Since the theory is nonrelativistic, all particles propagate forward in time, the nucleon tadpoles vanish, and the propagator for the nucleon fields is

$$iS(p) = \frac{i}{p_0 - p^2/2M + i\epsilon} . \qquad (3)$$

The dibaryon propagators are more complicated because of the coupling to two-nucleon states. The bare dibaryon propagator is simply a constant, $i/\Delta_{S,T}$, but the full propagator gets dressed by nucleon loops to all orders as illustrated in Fig. 1. The nucleon loop integral has a linear UV divergence which can

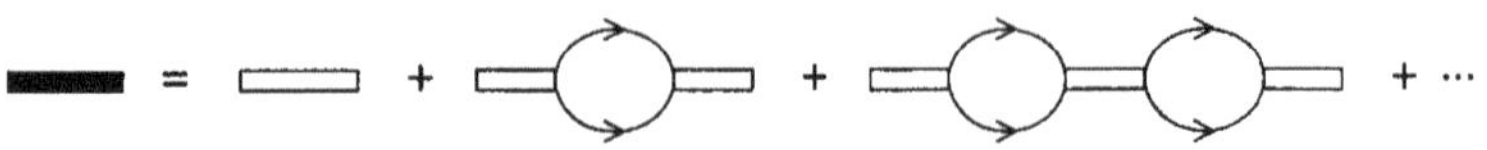

Figure 1: Dressing of the bare dibaryon propagator.

be absorbed in $g^2_{T,S}/\Delta_{T,S}$, a finite piece determined by the unitarity cut, and subleading terms that have already been omitted in Eq. (2). Summing the resulting geometric series leads to

$$iD_{S,T}(p) = \frac{-i}{-\Delta_{S,T} + \frac{Mg^2_{S,T}}{2\pi}\sqrt{-Mp^0 + \frac{\vec{p}^{\,2}}{4} - i\epsilon} + i\epsilon}\,, \tag{4}$$

where $g_{T,S}$ and $\Delta_{T,S}$ now denote the renormalized parameters. The NN scattering amplitude in the respective channel is obtained by attaching external nucleon lines to the dressed propagator. In the center of mass frame, the S-wave amplitude in the ${}^3S_1$, ${}^1S_0$ channels for energy $E = k^2/M$ is

$${}^{3,1}T(k) = \frac{4\pi}{M}\left(-\frac{2\pi\Delta_{S,T}}{Mg^2_{S,T}} - ik\right)^{-1}, \tag{5}$$

and the renormalized parameters $g_{T,S}$ and $\Delta_{T,S}$ can be determined from

$$a_2^{s,t} = \frac{Mg^2_{S,T}}{2\pi\Delta_{S,T}}\,. \tag{6}$$

3 $S = 3/2$ nd-Scattering

For $S = 3/2$ scattering only the dibaryon field $\vec{T}$ contributes. The first few diagrams that contribute are shown in the first line of Fig. 2. Since the leading piece of all diagrams is of the order $\sim Mg_T^2/Q^2$, they have to be resummed for typical momenta $Q \sim 1/a_2^t$. This is conveniently achieved by solving the integral equation given by the second line in Fig. 2. Performing the integration over the time component of the four-momentum flowing in the loop

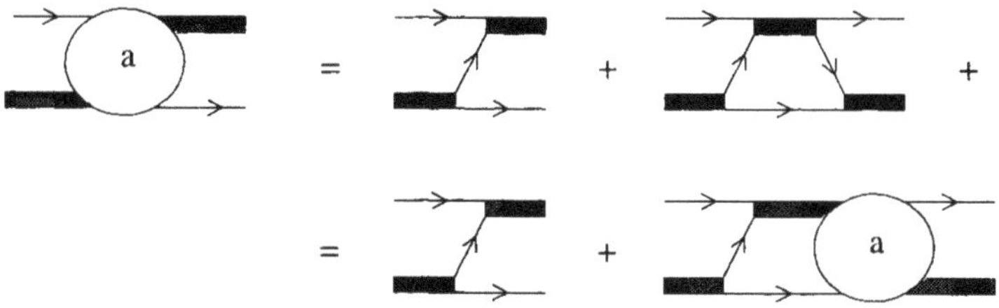

Figure 2: The first few diagrams contributing to $S = 3/2$ nd-scattering (first line) and the integral equation which sums the diagrams to all orders (second line).

and projecting onto the S-waves, we obtain the scattering amplitude $a(k,p)$,[17]

$$\frac{3}{4}\left(1/a_2^t + \sqrt{3p^2/4 - ME}\right)^{-1} a(p,k) \tag{7}$$
$$= -K(p,k) - \frac{2}{\pi}\int_0^\infty \frac{q^2\, dq}{q^2 - k^2 - i\epsilon} K(p,q)\, a(q,k)\,,$$

where

$$K(p,q) = \frac{1}{2pq}\ln\left(\frac{q^2 + pq - p^2 - ME}{q^2 - pq - p^2 - ME}\right) \tag{8}$$

and $ME = 3k^2/4 - 1/(a_2^t)^2$ is the total energy. $k\,(p)$ denote the incoming (outgoing) momenta in the center of mass frame. The amplitude $a(p,k)$ is normalized such that the on-shell value $a(k,k) = (k\cot\delta - ik)^{-1}$ where δ is the elastic scattering phase shift. The solution of Eq. (7) can be obtained numerically and is very insensitive to the high momentum modes in the integral equation. If, for example, the integral equation is cut off at some finite momentum Λ, the low energy behavior of the solution remains unchanged. The EFT in this channel is very predictive. The solution of Eq. (7) is shown by the dashed line in Fig. 3. When the corrections of the effective range in the two-nucleon interaction are taken into account up to $\mathcal{O}((r_2/a_2)^2)$, one obtains the solid curve [20] which agrees nicely with the phase shift analysis of van Oers and Seagrave (dots).[18] For the scattering length the agreement is even better. The calculation to $\mathcal{O}((r_2/a_2)^2)$ gives [21] $a_3^{(3/2)} = (6.33 \pm 0.10)$ fm to be compared with the experimental value [19] $a_3^{(3/2)} = (6.35 \pm 0.02)$ fm. Furthermore, the EFT expansion in powers of r_2/a_2 converges well, as the contributions from $\mathcal{O}((r_2/a_2)^0)$ to $\mathcal{O}((r_2/a_2)^2)$ to the scattering length are $a_3^{(3/2)} = (5.09 + 0.91 + 0.33)$ fm, in order. The extension beyond the deuteron

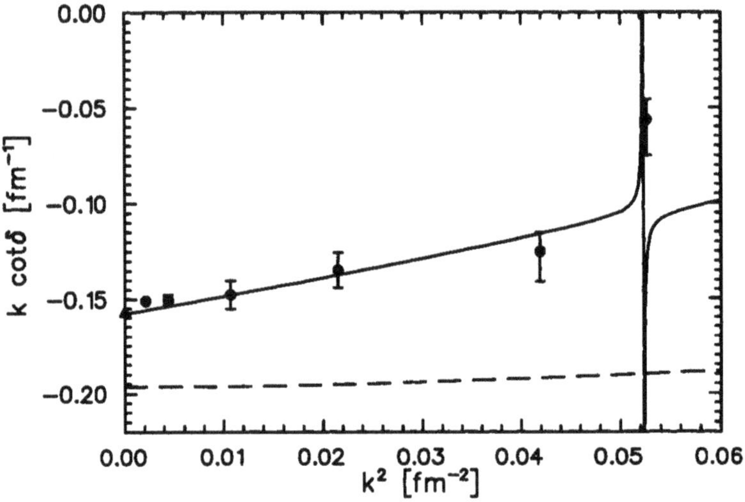

Figure 3: $k \cot \delta$ for $S = 3/2$ nd-scattering to order $(r_2/a_2)^0$ (dashed line) and $(r_2/a_2)^2$ (solid line). The dots [18] and the triangle [19] indicate the experimental values.

breakup threshold and the inclusion of pions in KSW counting has recently been carried out by Bedaque and Grießhammer.[22]

4 $S = 1/2$ nd-Scattering and the Triton

In the $S = 1/2$ channel, the situation is more complicated. The two-nucleon interactions can now take place both in the $^3S_1$ and $^1S_0$ partial waves. As shown in Fig. 4, there are two coupled amplitudes, a and b. The amplitude

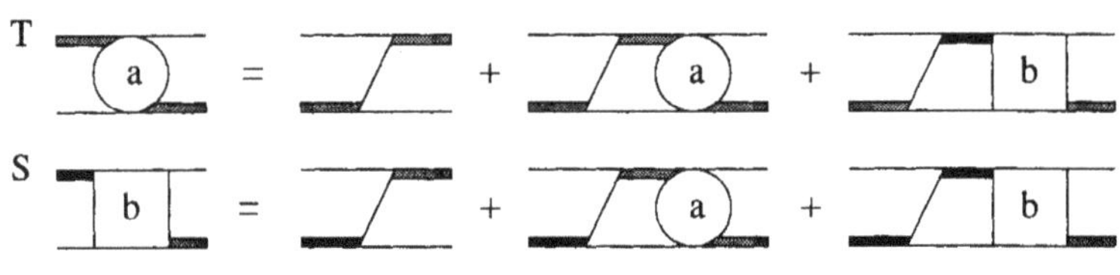

Figure 4: Coupled integral equations for $S = 1/2$ nd-scattering. $\vec{T}$ (S) dibaryon is indicated by grey shaded (black) thick line.

a which has both an incoming and outgoing dibaryon field $\vec{T}$ gives the phase

shifts for $S = 1/2$ nd-scattering. However, a is coupled to the amplitude b which has an incoming dibaryon $\boldsymbol{S}$ and an outgoing dibaryon $\vec{T}$. Although only a corresponds to $S = 1/2$ nd-scattering, both amplitudes have the quantum numbers of the triton. From the Lagrangian, Eq. (2), one obtains the coupled integral equations for the two amplitudes. After the integration over the time component of the loop-momentum and the projection onto the S-waves has been carried out, we have

$$\frac{3}{2}\left(1/a_2^t + \sqrt{3p^2/4 - ME}\right)^{-1} a(p,k) \tag{9}$$
$$= K(p,k) + \frac{2}{\pi}\int_0^\Lambda \frac{q^2\,dq}{q^2 - k^2 - i\epsilon} K(p,q)[a(q,k) + 3b(q,k)]$$
$$2\,\frac{\sqrt{3p^2/4 - ME} - 1/a_2^s}{p^2 - k^2}\, b(p,k) \tag{10}$$
$$= 3K(p,k) + \frac{2}{\pi}\int_0^\Lambda \frac{q^2\,dq}{q^2 - k^2 - i\epsilon} K(p,q)[3a(q,k) + b(q,k)]\,.$$

As before, k (p) denote the incoming (outgoing) momenta in the center of mass frame and $ME = 3k^2/4 - (1/a_2^t)^2$ is the total energy. The kernel $K(p,q)$ is given in Eq. (8). a_2^t (a_2^s) are the scattering lengths in the $^3S_1$ ($^1S_0$) NN channels, respectively. The amplitude $a(p,k)$ is normalized such that $a(k,k) = (k\cot\delta - ik)^{-1}$ with δ the elastic scattering phase shift in the $S = 1/2$ channel. Furthermore, we have introduced a momentum cutoff Λ in the integral equations. Eqs. (9, 10) have previously been derived using different methods.[17] In the limit $\Lambda \to \infty$ these equations do not have a unique solution because the phase of the asymptotic solution is undetermined.[23] For a finite Λ this phase is fixed and the solution is unique. However, the equations with a cutoff have the same disease as in the boson case: a strong cutoff dependence that does not appear in any order perturbation theory.[12,13,14] The amplitude $a(p, k = \text{const.})$ shows a strongly oscillating behavior. Varying the cutoff Λ slightly changes the asymptotic phase by a number of $\mathcal{O}(1)$ and results in large changes of the amplitude at the on-shell point $a(p = k)$. This cutoff dependence is not created by divergent Feynman diagrams. It is a nonperturbative effect and appears although all individual diagrams are UV finite.

In order to control this strong Λ dependence, it is useful to note that the integral equations (9, 10) are $SU(4)$ symmetric in the ultraviolet (UV). Therefore it is sufficient to consider the $SU(4)$ limit ($a_2^t = a_2^s = a_2$), since the cutoff dependence is a problem rooted in the UV behavior of the amplitudes. Furthermore, we note that the equations for $a_+ = [a + b]$ and $a_- = [a - b]$

decouple in the $SU(4)$ limit. The equations for a_+ and a_- are given by

$$\frac{3}{4}\left(1/a_2+\sqrt{3p^2/4-ME}\right)^{-1} a_+(p,k) \tag{11}$$
$$= 2K(p,k) + \frac{2H(\Lambda)}{\Lambda^2} + \frac{2}{\pi}\int_0^\Lambda \frac{q^2\,dq}{q^2-k^2-i\epsilon}\left\{2K(p,q)+\frac{2H(\Lambda)}{\Lambda^2}\right\} a_+(q,k)$$

$$\frac{3}{4}\left(1/a_2+\sqrt{3p^2/4-ME}\right)^{-1} a_-(p,k) \tag{12}$$
$$= -K(p,k) - \frac{2}{\pi}\int_0^\Lambda \frac{q^2\,dq}{q^2-k^2-i\epsilon} K(p,q)\, a_-(q,k)\,,$$

where we have introduced a contact three-body force $H(\Lambda)$ which runs with the cutoff Λ into the equation for a_+. Let us disregard the three-body force for a moment. The equation for a_+ is exactly the same equation as in the case of spinless bosons while the equation for a_- is the same equation as in the $S=3/2$ channel. The $S=3/2$ equation is well behaved and its solution is very insensitive to the cutoff as discussed in the previous section. Consequently, the observed cutoff dependence stems solely from the equation for a_+. As an example, the cutoff dependence of $a_+(p,k=0)$ is shown by the solid, dashed, and dash-dotted curves in Fig. 5 for three different cutoffs, $\Lambda = 1.0, 2.0, 3.0\times 10^4 a_2^{-1}$. But we already know the solution to this problem: a three-body force counterterm that runs with the cutoff Λ.[13] This is exactly the three-body force we have introduced into Eq. (11). The dotted, short-dash-dotted, and short-dashed curves in Fig. 5 show the effect of the three-body force $H(\Lambda)$ on $a_+(p,k=0)$ for $\Lambda = 10^4 a_2^{-1}$ and $H = -6.0, -2.5, -1.8$. It is clearly seen that the variation of H for a constant Λ has the same effect on the amplitude as varying the cutoff. Consequently, we can compensate the changes in the asymptotic phase when Λ is varied by adjusting the three-body force term appropriately. (A more detailed discussion of the renormalization procedure can be found in Bedaque's talk.[12])

We can obtain an approximate expression for the running of $H(\Lambda)$ from invariance under the renormalization group. Requiring that the equation for a_+ does not change its form when the high momentum modes are integrated out, we find

$$H(\Lambda) = -\frac{\sin(s_0\ln(\Lambda/\Lambda_\star) - \mathrm{arctg}(1/s_0))}{\sin(s_0\ln(\Lambda/\Lambda_\star) + \mathrm{arctg}(1/s_0))}\,, \tag{13}$$

where $s_0 \approx 1.0064$.[12,13,14] $H(\Lambda)$ contains one new dimensionful parameter, Λ_*, which must be determined from experiment. The running of the three-body force $H(\Lambda)$ according to Eq. (13) is shown by the solid line in Fig. 6. The dots are obtained by adjusting $H(\Lambda)$ such that the low energy solution of Eq.

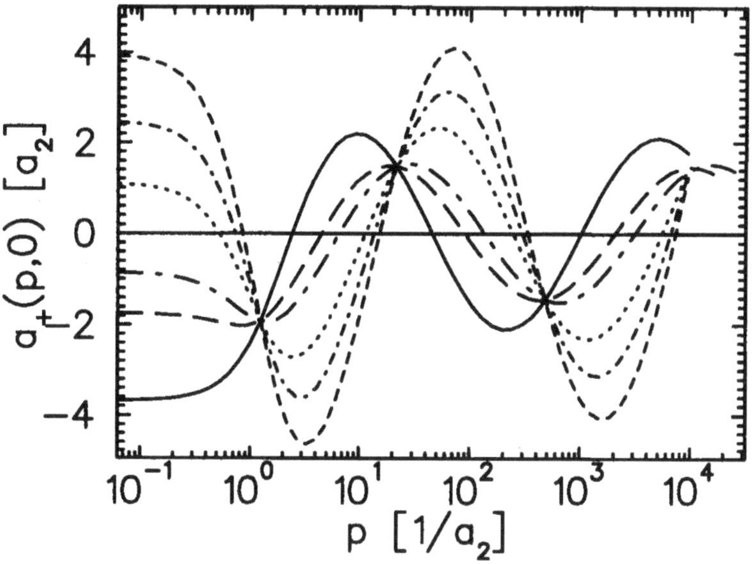

Figure 5: Cutoff dependence of $a_+(p,k=0)$. Solid, dashed, and dash-dotted curves are for $H=0$ and $\Lambda = 1.0, 2.0, 3.0 \times 10^4 a_2^{-1}$, respectively. Dotted, short-dash-dotted, and short-dashed curves show the effect of the three-body force for $\Lambda = 10^4 a_2^{-1}$ and $H = -6.0, -2.5, -1.8$, respectively.

(11) remains unchanged when Λ is varied. The observed agreement provides a numerical justification for our proceeding. The three-body force is periodic with $H(\Lambda_n) = H(\Lambda)$ for $\Lambda_n = \Lambda \exp(n\pi/s_0) \approx \Lambda(22.7)^n$. Since it enters only in the equation for $a_+ = a + b$, the three-body force is also $SU(4)$ symmetric.

Formally, the three-body force term in Eq. (11) is obtained by adding

$$\mathcal{L}_3 = -\frac{2MH(\Lambda)}{\Lambda^2}\Big(g_T^2 N^\dagger (\vec{T}\cdot\vec{\sigma})^\dagger (\vec{T}\cdot\vec{\sigma}) N + \frac{1}{3} g_T g_S \left[N^\dagger (\vec{T}\cdot\vec{\sigma})^\dagger (\boldsymbol{S}\cdot\boldsymbol{\tau}) N + h.c. \right] + g_S^2 N^\dagger (\boldsymbol{S}\cdot\boldsymbol{\tau})^\dagger (\boldsymbol{S}\cdot\boldsymbol{\tau}) N \Big), \quad (14)$$

to the Lagrangian, Eq. (2). Eq. (14) represents a contact three-body force written in terms of dibaryon and nucleon fields. Via a Gaussian path integra-

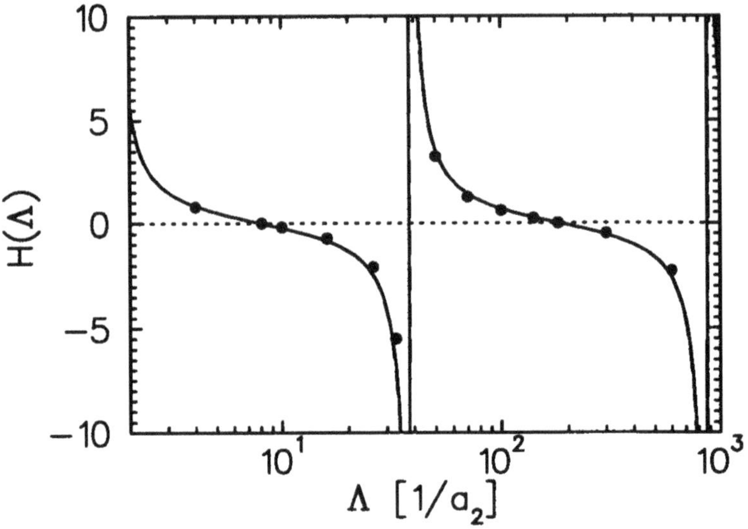

Figure 6: Running of $H(\Lambda)$ for $\Lambda_* = 0.9\ \mathrm{fm}^{-1}$: (a) from Eq. (13) (solid line), (b) from numerical solution of Eq. (11) (dots).

tion it is equivalent to a true three-nucleon force,

$$\begin{aligned}\mathcal{L}_3 = &-\frac{2MH(\Lambda)}{\Lambda^2}\left(\frac{g_T^4}{4\Delta_T^2}(N^T\tau_2\sigma_k\sigma_2N)^\dagger(N^\dagger\sigma_k\sigma_lN)(N^T\tau_2\sigma_l\sigma_2N)\right.\\&+\frac{1}{3}\frac{g_T^2}{2\Delta_T}\frac{g_S^2}{2\Delta_S}\left[(N^T\tau_2\sigma_k\sigma_2N)^\dagger(N^\dagger\sigma_k\tau_lN)(N^T\sigma_2\tau_l\tau_2N)+h.c.\right]\\&\left.+\frac{g_S^4}{4\Delta_S^2}(N^T\sigma_2\tau_k\tau_2N)^\dagger(N^\dagger\tau_k\tau_lN)(N^T\sigma_2\tau_l\tau_2N)\right).\end{aligned} \tag{15}$$

By performing a Fierz rearrangement, it can then be shown that the three terms in Eq. (15) are equivalent. As a consequence, there is only one three-body force which is also $SU(4)$ symmetric. Naive power counting would suggest that the three-nucleon force scales with $1/(Mm_\pi^4)$. The three-body force from Eqs. (14, 15), however, is enhanced by the renormalization group flow by two powers of a_2. Using Eq. (6), it is found to scale as $a_2^2/(Mm_\pi^2)$ which makes it leading order.

Recently Mehen, Stewart, and Wise [24] found an approximate $SU(4)$ symmetry in the two-nucleon sector for $1/a_2 \ll p \ll m_\pi$ and also noticed that the only S-wave four-nucleon force that can be written down is $SU(4)$ symmetric. Furthermore, there are no contact interactions with more than four nucleons without derivatives because of the Pauli principle. It is also reasonable to assume that the low-energy dynamics of nuclei is dominated by S-wave interactions. Therefore our $SU(4)$ symmetric three-body force together with the findings of Mehen et al.[24] gives an explanation for the approximate Wigner $SU(4)$ symmetry in nuclei.[25]

Now we are in the position to solve the full equations for the broken $SU(4)$ case. Introducing the three-body force from above into Eqs. (9, 10), we obtain

$$\frac{3}{2}\left(1/a_2^t + \sqrt{3p^2/4 - ME}\right)^{-1} a(p,k) = K(p,k) + \frac{2H(\Lambda)}{\Lambda^2} \tag{16}$$
$$+\frac{2}{\pi}\int_0^\Lambda \frac{q^2\,dq}{q^2-k^2-i\epsilon}\left[K(p,q)[a(q,k)+3b(q,k)] + \frac{2H(\Lambda)}{\Lambda^2}[a(q,k)+b(q,k)]\right]$$

$$2\,\frac{\sqrt{3p^2/4 - ME} - 1/a_2^s}{p^2-k^2}\, b(p,k) = 3K(p,k) + \frac{2H(\Lambda)}{\Lambda^2} \tag{17}$$
$$+\frac{2}{\pi}\int_0^\Lambda \frac{q^2\,dq}{q^2-k^2-i\epsilon}\left[K(p,q)[3a(q,k)+b(q,k)] + \frac{2H(\Lambda)}{\Lambda^2}[a(q,k)+b(q,k)]\right] .$$

We need one three-body datum to fix the three-nucleon force parameter Λ_*. We choose the experimental value for the $S = 1/2$ nd-scattering length,[19] $a_3^{(1/2)} = (0.65 \pm 0.04)$ fm and find $\Lambda_* = 0.9\ \text{fm}^{-1}$. (For the special cutoffs with vanishing $H(\Lambda)$, we recover the results of Ref.[26].) Although one three-body datum is needed as input, the EFT has not lost its predictive power. We can still predict (i) the energy dependence of $S = 1/2$ nd-scattering and (ii) the binding energy of the triton.

The resulting energy dependence of $S = 1/2$ nd-scattering for three different cutoffs is shown in Fig. 7. It is clearly seen that the introduction of the three-body force renders the low-energy amplitude cutoff independent. The scattering length is reproduced exactly because it was used to fix Λ_*. The agreement for finite momentum is at least encouraging. Our experience from the $S = 3/2$ channel is that the range corrections improve the agreement considerably (cf. Fig. 3). The dashed curve in Fig. 7 gives a crude estimate of these corrections. In the zero range approximation, we have the relation

$$\sqrt{MB_2} = 1/a_2^t, \tag{18}$$

which holds only approximately in nature. In our calculations we take the deuteron binding energy B_2 from experiment and determine a_2^t from Eq. (18). The dashed curve in Fig. 7 is obtained by taking the experimental value of a_2^t

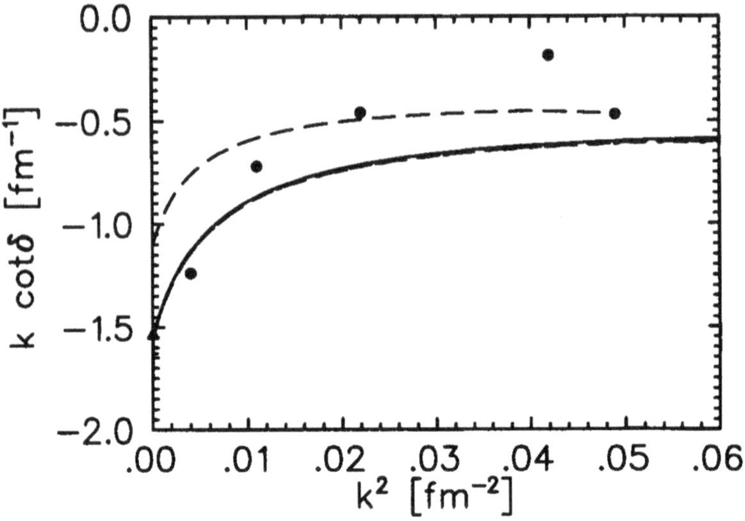

Figure 7: Energy dependence for $S = 1/2$ nd-scattering with three different cutoffs $\Lambda = 1.9, 6.0, 11.6\ \text{fm}^{-1}$ and $\Lambda_* = 0.9\ \text{fm}^{-1}$ (hidden under the solid curve). Dashed curve gives estimate of range corrections. Experimental points are from van Oers and Seagrave[18] (dots) and Dilg et al. [19] (triangle).

as input and leaving all other parameters unchanged. From the estimated size of the range corrections, we anticipate an improved agreement once the range corrections are included. (One should also keep in mind that the experimental phase shift analysis[18] does not give any error estimate).

The triton binding energy is obtained from the solution of the homogeneous equations corresponding to Eqs. (16, 17) for $E = -B_3$. We find $B_3 = 8.0$ MeV for the triton binding energy, to be compared with the experimental result $B_3^{exp} = 8.48$ MeV which is known to very high precision. For a leading order calculation, the agreement is very good. The theory without pions seems to work for triton physics. However, to draw definite conclusions one has to calculate the range corrections for both the energy dependence and the triton binding energy.

In Fig. 8 we show the bound state spectrum as a function of the cutoff Λ. The shallowest bound state is the triton. Its binding energy is cutoff independent. However, as Λ is increased new deeper bound states appear whenever $H(\Lambda)$ goes through a pole. These new bound states appear with infinite binding energy directly at the pole. When the cutoff is increased further, their

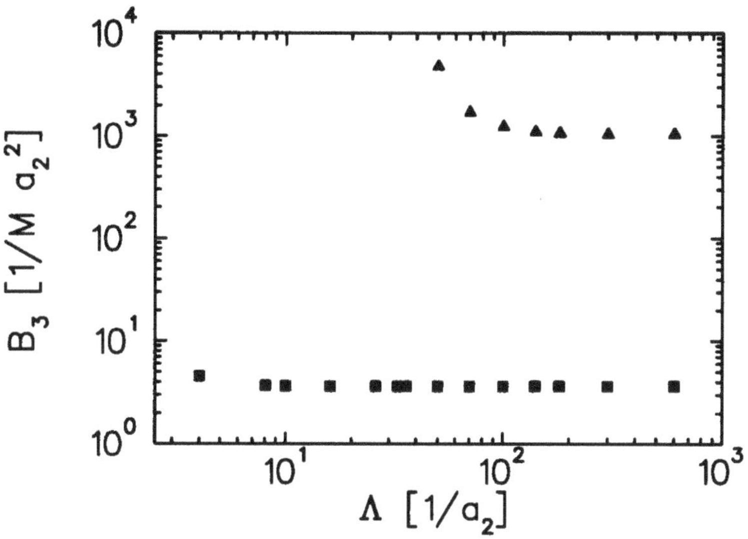

Figure 8: Three-nucleon bound state spectrum for $\Lambda_* = 0.9\ \mathrm{fm}^{-1}$. The shallowest bound state corresponds to the triton.

binding energy reduces and becomes cutoff independent as well. The poles of $H(\Lambda)$ can be parametrized as

$$\Lambda_n = \underbrace{f(\Lambda_* a_2)}_{\mathcal{O}(1)} \exp(n\pi/s_0) a_2^{-1} \,. \tag{19}$$

One counts n bound states for $\Lambda_{n-1} < \Lambda < \Lambda_n$. However, only the states between threshold and $\Lambda \sim m_\pi \sim 1/r_2$ are within the range of the EFT. By solving Eq. (19) for the number of bound states n, we then obtain

$$\#_{BS} = \frac{s_0}{\pi} \ln\left(\frac{a_2}{r_2}\right) + \mathcal{O}(1) \,, \tag{20}$$

and recover the well-known Efimov effect.[10] In the limit $a_2 \to \infty$, an infinite number of shallow three-body bound states accumulates at threshold. That these bound states are shallow follows from the fact that the binding energy is naturally given in units of $1/(Ma_2^2)$ which vanishes as $a_2 \to \infty$. Furthermore, we also recover the Thomas effect.[11] In a hypothetical world where $\Lambda \sim m_\pi \sim 1/r_2 \to \infty$, the range of the EFT increases and deeper and deeper physical

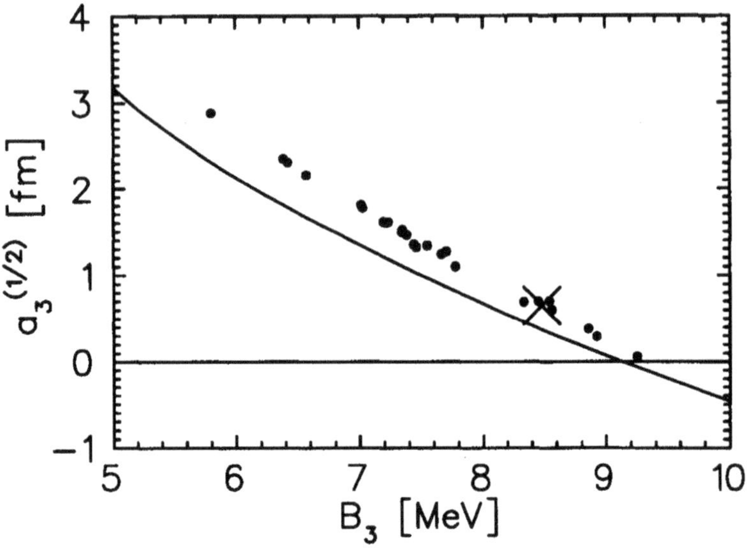

Figure 9: Phillips line: $S = 1/2$ nd scattering length $a_3^{(1/2)}$ as function of the triton binding energy B_3: EFT to leading order (solid line); potential models (dots); experiment (cross).

bound states appear. Consequently, there is an infinitely deep bound state for $\Lambda \to \infty$. However, for $m_\pi \sim 1/r_2$ as in the real world the deep bound states are outside the range of the EFT and their presence does not influence the physics of the shallow ones.

Moreover, the variation of the parameter Λ_* gives a natural explanation for the Phillips line.[9] No matter how good the agreement with the actual experimental number is, values of B_3 and $a_3^{(1/2)}$ in different potential models are correlated and fall on the Phillips line. Obviously, there is a correlation between B_3 and $a_3^{(1/2)}$. In the EFT framework, different models correspond in leading order to different values of Λ_*. Varying Λ_* then generates the observed Phillips line. The dynamics of QCD chooses a particular value which up to higher order corrections is $\Lambda_* = 0.9\ \text{fm}^{-1}$. In Fig. 9 we show the Phillips line obtained in the EFT compared with results from various potential model calculations[27] and the experimental values for B_3 and $a_3^{(1/2)}$. Our Phillips line is slightly below the one from the potential models. We expect this discrepancy to be reduced once range corrections are taken into account.

5 Conclusions

We have studied the three-nucleon system using EFT methods. While for nd-scattering in the $S = 3/2$ channel precise predictions are obtained in a straightforward way, the $S = 1/2$ channel is more complicated. It displays a strong cutoff dependence even though all individual diagrams are UV finite. In this channel a nonperturbative renormalization takes place similar to the case of spinless bosons. This renormalization requires an $SU(4)$ symmetric three-body force which is enhanced to leading order by the renormalization group flow. Together with the recent results of Mehen et al.[24] this gives an explanation for the approximate $SU(4)$ symmetry in nuclei. Furthermore, we find that the Phillips line is a consequence of variations in the new dimensionful parameter Λ_* which is introduced by the three-body force. Λ_* is not given from two-nucleon data alone and has to be determined from a three-body datum. In the appropriate limits for the two-body parameters a_2 and r_2, we also recover the well known Thomas and Efimov effects.

The leading order of the EFT gives a quantitative description of the triton binding energy and the energy dependence for $S = 1/2$ nd-scattering. The theory shows the potential for a realistic description of the triton once the range corrections are included. Immediate applications of the EFT include polarization observables in nd-scattering and triton properties such as its charge form factor. The incorporation of the long range Coulomb force would widen the possible applications considerably as pd-scattering and the physics of ^{3}He becomes accessible.

Finally, since the pionless theory is equivalent to the leading order in KSW counting,[6] the success in the three-nucleon system opens the possibility of applying the EFT method to a large class of systems with three or more nucleons.

Acknowledgements

This work was done in collaboration with P.F. Bedaque and U. van Kolck, whom I thank for many valuable discussions. I would also like to thank Jim Friar for the potential model data. This research was supported by the Natural Sciences and Engineering Research Council of Canada.

References

1. G.P. Lepage, in *From Actions to Answers, TASI'89*, ed. T. DeGrand and D. Toussaint (World Scientific, Singapore, 1990); D.B. Kaplan, `nucl-th/9506035`; H. Georgi, *Ann. Rev. Part. Sci.* **43**, 209 (1994).

2. S. Weinberg, *Phys. Lett.* B **251**, 288 (1990); *Nucl. Phys.* B **363**, 3 (1991).
3. *Nuclear Physics with Effective Field Theory*, ed. R. Seki, U. van Kolck, and M.J. Savage (World Scientific, Singapore, 1998)
4. U. van Kolck, nucl-th/9902015.
5. U. van Kolck, in *Proceedings of the Workshop on Chiral Dynamics 1997, Theory and Experiment*, ed. A. Bernstein, D. Drechsel, and T. Walcher (Springer-Verlag, Berlin, Heidelberg, 1998); *Nucl. Phys.* A **645**, 273 (1999).
6. D.B. Kaplan, M.J. Savage, and M.B. Wise, *Phys. Lett.* B **424**, 390 (1998); *Nucl. Phys.* B **534**, 329 (1998).
7. J. Gegelia, *Phys. Lett.* B **429**, 227 (1998), nucl-th/9802038, nucl-th/9805008.
8. M.J. Savage, *this volume* and references therein.
9. A.C. Phillips, *Nucl. Phys.* A **107**, 209 (1968).
10. V.N. Efimov, *Sov. J. Nucl. Phys.* **12**, 589 (1971); *Phys. Rev.* C **47**, 1876 (1993).
11. L.H. Thomas, *Phys. Rev.* **47**, 903 (1935).
12. P.F. Bedaque, *this volume.*
13. P.F. Bedaque, H.-W. Hammer, and U. van Kolck, *Phys. Rev. Lett.* **82**, 463 (1999); *Nucl. Phys.* A **646**, 444 (1999).
14. H.-W. Hammer, in *Proceedings of BARYONS 98* (World Scientific, Singapore, to appear), nucl-th/9811047.
15. J.-W. Chen, G. Rupak, and M.J. Savage, nucl-th/9902056.
16. D.B. Kaplan, *Nucl. Phys.* B **494**, 471 (1997).
17. G.V. Skorniakov and K.A. Ter-Martirosian, *Sov. Phys. JETP* **4**, 648 (1957).
18. W.T.H. van Oers and J.D. Seagrave, *Phys. Lett.* B **24**, 562 (1967).
19. W. Dilg, L. Koester, and W. Nistler, *Phys. Lett.* B **36**, 208 (1971).
20. P.F. Bedaque, H.-W. Hammer, and U. van Kolck, *Phys. Rev.* C **58**, R641 (1998).
21. P.F. Bedaque and U: van Kolck, *Phys. Lett.* B **428**, 221 (1998).
22. P.F. Bedaque and H.W. Grießhammer, in preparation.
23. G.S. Danilov and V.I. Lebedev, *Sov. Phys. JETP* **17**, 1015 (1963); G.S. Danilov, *Sov. Phys. JETP* **13**, 349 (1961).
24. T. Mehen, I.W. Stewart, and M. Wise, hep-ph/9902370
25. E.P. Wigner, *Phys. Rev.* **51**, 106 (1937).
26. V.F. Kharchenko, *Sov. J. Nucl. Phys.* **16**, 173 (1973).
27. J.L. Friar, private communication; C.R. Chen, G.L. Payne, J.L. Friar, and B.F. Gibson, *Phys. Rev.* C **44**, 50 (1991).

Discussion during H. Hammer's presentation

Chair: *M. Birse (University of Manchester)*
Notes taken by: *M. N. Butler (Saint Mary's University)*

There was continued discussion on the three-body counterterms, carried over from the talk by Paulo Bedaque. Mark Wise asked about derivative forces present at next-to-next-to-leading and, specifically, whether one could write down a two-derivative three-body term. Bira van Kolck agreed that one could, but dimensional analysis shows that such a term is of higher order. One sees two-derivative operators only in two-body terms at this order. Dan Phillips asked about the presence of three-body terms in potential model calculations. van Kolck noted here that the potential model calculations that come close to experiment have three-body terms adjusted to reproduce experimental results. Phillips noted that this was not true in all cases. van Kolck agreed that by cooking up some arbitrary off-shell behavior one can get the right binding, but that this is not any different than adjusting a three-body term.

At the end of the talk, Jerry Miller asked about the connection to potential model three-body forces, noting that they came, traditionally, from compositeness which made the three-body terms small (large mass scale and large energy denominator). He noted that a leading three-body term in EFT was a new feature. Hammer emphazised that the three-body term in EFT is *not* the three-body force for potentials. The cutoffs don't match and it cannot be treated separately to the two-body force. Tom Cohen noted too that, from counting, the size of the three-body term was suprprising. He wondered whether an a priori estimate of, for example, form factors could be made given this enhancement effect? Hammer responded that, unless there is evidence otherwise, one can just use naive counting to deal with the size of the three-body term. Paulo Bedaque noted that one doesn't know yet what will happen.

EFFECTIVE FIELD THEORY FOR NUCLEAR MATTER

M. LUTZ
GSI, 64291 Darmstadt, Germany
E-mail: m.lutz@gsi.de

We apply the relativistic chiral Lagrangian to the nuclear equation of state. An effective chiral power expansion scheme, which is constructed to work around nuclear saturation density, is presented. The leading and subleading terms are evaluated and are shown to provide an equation of state with excellent saturation properties. Our saturation mechanism is found to probe details of the nuclear pion dynamics.

1 Introduction

The nuclear equation of state awaits a fundamental theoretical description. It is therefore important to develop the appropriate form of effective chiral perturbation theory (EχPT), a promising tool of modern nuclear physics, and apply it to the nuclear many body problem systematically.

The key element of any microscopic theory for the nuclear equation of state is the elementary nucleon-nucleon scattering process. In the context of chiral perturbation theory this problem was first addressed by Weinberg who proposed to derive a chiral nucleon-nucleon potential in time ordered perturbation theory [1,2]. In [3,4] we proposed an alternative scheme with chiral power counting rules applied directly to the nucleon-nucleon scattering amplitude. This approach applies the manifest covariant version of the chiral Lagrangian and relies on the crucial observation that the chiral power counting rules can be generalized for 2-nucleon reducible diagrams. Non-perturbative effects like the pseudo bound state pole in the $^1S_0$ channel are generated by properly renormalized local two-nucleon vertices. A similar approach was suggested by Kaplan, Savage and Wise [5] in the context of the chiral potential approach. Unfortunately the straightforward application of the extended counting rules to the s-wave nucleon-nucleon scattering phase shifts is hampered by a poor convergence and inconsistency with empirical data[6]. In particular, unitarity is violated strongly even at small energy where one would expect this scheme to work well[4]. This is, at first glance, a surprising result since it is long known that for the one pion exchange Yukawa potential perturbation theory works fine if the coupling strength is put to its empirical value[7,8,9]. In fact in the $^1S_0$ channel, the failure of a naive expansion is linked to the presence of a second strong subthreshold singularity reflecting a subtle interplay of intermediate attraction and short range repulsion[4]. A properly modified unitary implementation of the extended chiral counting rules describes empirical s-wave elastic and inelastic nucleon-nucleon scattering in the spin singlet channel accurately up to $E_{\rm lab} \simeq 600$ MeV [4].

We point out that counting rules for the vacuum nucleon-nucleon scattering amplitude greatly facilitate the construction of chiral power counting rules for the nuclear matter problem [10]. Basically the density expansion or equivalently the multiple scattering expansion can be combined smoothly with the chiral expansion where one identifies $k_F \sim Q$ as a further small scale. Such an identification leads to a systematic resummation of the density expansion avoiding the expansion in large ratios like k_F/m_π but exploiting the chiral mass gap and expand in small ratios like k_F/m_χ or m_π/m_χ with $m_\chi \simeq m_N$.

2 Primer on relativistic chiral nucleon-nucleon scattering

In this section we recall basic elements required for a successful application of the relativistic chiral Lagrangian to the baryon sector. The systematic construction of the Lagrangian density in accordance with chiral symmetry was presented in great detail by Weinberg using stereographic coordinates [11]. There are two problems when evaluating the relativistic chiral Lagrangian in the baryon sector. First, any covariant derivative acting on the nucleon field produces the large nucleon mass and therefore must be assigned the minimal chiral power zero. Thus an infinite tower of interaction terms needs to be evaluated at given finite chiral order [12,13]. Second, the straightforward evaluation of relativistic diagrams involving nucleon propagators generates positive powers of the large nucleon mass from loop momenta larger than the nucleon mass. The chiral power counting rules are spoiled. The relativistic scheme appeared impractical and the heavy mass formulation of χPT, which overcomes both problems by performing a non relativistic $1/m$ expansion at the level of the Lagrangian density, was developed [14] and applied extensively in the one-nucleon sector [15].

We point out that in the two-nucleon channel it will be advantageous to work with the relativistic form of the chiral Lagrangian [3,4]. One reason is that in the heavy baryon formulation two-particle reducible diagrams are ill defined. A second reason is that a proper treatment of the pion production process $pp \to pp\,\pi^0$ requires relativistic kinematics [4]. Note that an alternative scheme to tame potentially ill defined two-particle reducible diagrams is the use of time-ordered perturbation theory as suggested by Weinberg [1]. Since covariance is not manifest in time-ordered perturbation theory we prefer the relativistic scheme. Here we outline how the two aforementioned problems can be overcome within the framework of the relativistic chiral Lagrangian. Of course we expect the relativistic scheme to be equivalent to the more familiar heavy fermion formulation of chiral perturbation theory in the one-nucleon sector. Rather than performing the $1/m$ expansion at the level of the Lagrangian density we suggest to work out this expansion explicitly at the level of individual relativistic Feynman diagrams. Therewith we avoid the heavy baryon chiral Lagrangian with its known artifacts in the two-nucleon sector. Here it is convenient to consider the chiral Lagrangian (at an intermediate stage) to represent a finite cutoff theory. The finite cutoff, $\Lambda \ll m$, is required to restrict the nucleon virtuality inside a given loop diagram such as to justify the $1/m$ expansion. Obviously a finite cutoff has to be introduced with great care not to break any chiral Ward identity. Note that one may alternatively apply dimensional regularization. We emphasize that obviously if dimensional regularization is applied one has to *first* expand in $1/m$ and then perform the loop integration. The two steps do *not* commute here.

We need to outline how to arrange higher order terms of the relativistic chiral Lagrangian in the nucleon sector. To have a practical scheme one would like to have only a finite number of interaction terms with a given number of 'small' derivatives involved. The known problem of the 'large' nucleon mass generated by time derivatives acting on a nucleon field has to be taken care of. This problem is cured easily as follows. Consider a covariant derivative $\mathcal{D}_\mu(x)$ acting on a nucleon field $N(x)$. We need to construct a 'counter interaction' which leads to a small renormalized vertex. This rearrangement is illustrated at hand of the simple case when the Lorentz index μ is saturated by another covariant derivative acting on the same nucleon field. In this case the counter interaction is readily found

$$\begin{aligned} \mathcal{D}^\mu(x)\,\mathcal{D}_\mu(x)\,N(x) \;&\to \left(\mathcal{D}^\mu(x)\,\mathcal{D}_\mu(x) + m^2\right)\,N(x) \\ &\to -\left(p^2 - m^2\right) \sim Q \end{aligned} \tag{1}$$

and the 'renormalized' vertex is proportional to the virtuality of the nucleon 4-momentum which is of 'minimal' order Q. It is important to observe that this counter interaction must have occurred already in the chiral Lagrangian to some lower order since it involves fewer derivatives. Note that a similar construction works if two covariant derivatives act on different nucleon fields.

We are left to consider the case in which the Lorentz index μ of the covariant derivative $\mathcal{D}_\mu$ is contracted either with the Lorentz index of a Dirac matrix or the Lorentz index of a derivative acting on a pion field. Interaction vertices in which a covariant nucleon derivative $\mathcal{D}_\mu$ is saturated with a pionic four vector D^μ or with the Lorentz index of its covariant derivative behave well since they are already suppressed by the small derivative acting on the pion field. Here one has to assign the nucleon derivative the chiral power Q^0. The large nucleon mass present is not hazardous provided that one may identify the natural scale $\bar{\Lambda}$ of chiral interactions with the nucleon mass. With $\bar{\Lambda} \simeq m_\rho$ this seems indeed reasonable. The remaining case in which the nucleon derivative couples to the Lorentz index of a Dirac matrix does not generate an infinite tower of interaction terms since at a given number of nucleon fields, n_i, the number of available Lorentz indices provided by Dirac matrices is at most n_i.

The above arguments show how to organize the infinite tower of chiral interaction terms such that there is always only a finite number of terms with a given number of 'small' derivatives. Thus at any given chiral order only a finite number of diagrams need to be evaluated. The result of this reorganization of the relativistic chiral Lagrangian leads to the applicability of Weinberg's counting rules provided that individual diagrams are expanded properly.

In the two-nucleon sector the chiral Lagrangian leads to a well defined expansion scheme for the Bethe-Salpeter kernel K. For the unitary iteration of K, as induced by the Bethe-Salpeter equation, Weinberg's counting rules are not applicable due to the s-channel unitarity cut. Nevertheless a generalized counting for two-particle reducible diagrams can be established. For a given diagram each pair of intermediate nucleons causes a reduction of one chiral power as compared to the 'naive' chiral power [3]. The argument goes as follows: consider the once iterated Bethe-Salpeter kernel

$$K\,G\,K \sim \int \frac{d^4 l}{(2\pi)^4} \frac{2\,m\,K(l)}{\left(\frac{1}{2}\,W - l\right)^2 - m^2 + i\,\epsilon} \frac{2\,m\,K(l)}{\left(\frac{1}{2}\,W + l\right)^2 - m^2 + i\,\epsilon} \tag{2}$$

with $W^2 = s$. The object G represents the two-nucleon propagator and K the Bethe-Salpeter kernel evaluated according to Weinberg's counting rules to a given chiral order. It is instructive first to study the s-channel spectral density, $\rho(s)$, of this contribution. Since by assumption the Bethe-Salpeter kernel K is 2-nucleon irreducible the s-channel spectral density picks up strength only from the pinch singularity generated by the two nucleon propagators in (2). In the center of mass frame with $\vec{W} = 0$ one finds:

$$\rho(s) = \Im\, K\,G\,K \sim \frac{p}{\sqrt{m^2 + p^2}} \frac{m^2}{4\,\pi} \int \frac{d\Omega_{\vec{l}}}{4\,\pi} K(\vec{l}\,)\,K(\vec{l}\,)\Big|_{l_0=0,\,|\vec{l}|=p} \tag{3}$$

with[a] $s = 4\left(m^2 + p^2\right)$. Expression (3) can now be used to introduce the chiral

[a]Of course the iterated Bethe-Salpeter kernel does not necessarily satisfy a s-channel dispersion relation. However, we point out that only the part which provides strength for the s-channel spectral density requires special attention and modified power counting rules. Obviously a part which does not generate strength for the s-channel spectral density has *no* pinch singularity leading to a s-channel cut so that standard counting rules apply.

power, ν, of the s-channel spectral density, $\rho(s)$,

$$\nu = 1 + 2\,\nu_K \tag{4}$$

with the chiral power ν_K of the Bethe-Salpeter kernel K as given by Weinberg's power counting rules. In (4) we exploit the important observation that the small scale momentum $p \sim Q$ occurs in the spectral density.

To have a useful scheme we also need to derive counting rules for the real part of the scattering amplitude. Here we evoke causality which relates the real part of a given Feynman diagram to its imaginary part by means of a dispersion relation

$$I(s) = \Re K\,G\,K = \frac{1}{\pi}\mathcal{P}\int_{4\,m^2}^{\infty} d\,\bar{s}\,\frac{\rho(\bar{s})}{\bar{s}-s-i\,\epsilon}\,. \tag{5}$$

Of course we must not assume the dispersion relation (5) to be finite and well defined. It may be ultraviolet divergent[b]. However, we recall that we are free to consider our chiral Lagrangian as a finite cutoff theory. So let us introduce a covariant cutoff Λ into the dispersion relation (5). If the cutoff Λ is to be interpreted as a bound for the maximal virtuality of nucleons one should first remove the nucleon rest mass. Therefore we take the cutoff Λ to restrict the momentum $p \leq \Lambda$ rather than the Mandelstam variable s:

$$I(p,\Lambda) = \frac{1}{\pi}\mathcal{P}\int_{0}^{\Lambda^2} d\,\bar{p}^2\,\frac{\rho(\bar{p})}{\bar{p}^2-p^2}\,. \tag{6}$$

The natural guess for the chiral power of the integral $I(p)$ would be the chiral power of its imaginary part. However, the integral (6) as it stands does not permit this chiral power yet. Though the spectral density $\rho(p)$ exhibits only small scales as shown above, the cutoff scale Λ prevents the conclusion that the integral behaves like a small scale to the power $1+2\,\nu_K$. We point out that the counting rule (13) can nevertheless be applied to the real part of the loop function, however, only after an appropriate number of subtractions at $p^2 = -\mu^2 < 0$ with $\mu \sim m_\pi$. The subtraction is required to render the dispersion integral finite in the large cutoff limit:

$$I_S^{(n)}(p,\Lambda;\mu) = \frac{1}{\pi}\mathcal{P}\int_{0}^{\Lambda^2} d\,\bar{p}^2\,\frac{\rho(\bar{p})}{\bar{p}^2-p^2}\left(\frac{p^2+\mu^2}{\bar{p}^2+\mu^2}\right)^n\,. \tag{7}$$

[b]Note that for a given contribution one has to specify whether the dispersion relation holds at fixed Mandelstam variable t or u. This technical detail, however, does not affect our argument.

The subtraction polynomial can always be absorbed into the local two-nucleon vertices. Note that the performed subtractions do not necessarily simplify our attempt to apply dimensional counting rules since this procedure introduces a new scale, μ, the subtraction point. Nevertheless we now can exploit the freedom to choose the subtraction point. If we insist on a 'small' subtraction scale $\mu \sim m_\pi$ of the order of the pion mass one can apply dimensional power counting since the integral is finite in the large cutoff limit by construction. Thus we arrive at our desired chiral counting rule (13) for 2-particle reducible diagrams: each pair of nucleon propagators which exhibits a s-channel unitarity cut gets a chiral power -3 in contrast to the 'naive' power -2. We stress, first, that this counting rule holds *only if* one introduces a 'small' subtraction scale $\mu \sim m_\pi$ at the level of the Bethe-Salpeter equation and second, that the subtraction scale is *independent* of the intrinsic cutoff. We will refer to this counting rule as the 'L'-counting rule since for the one pion exchange interaction the chiral power of the n-th iterated interaction is simply given by the number of loops. In [4] the reader may find explicit examples which confirm our L-counting rule (4).

The non-perturbative structures like the pseudo bound state in the spin singlet channel are generated naturally since the bare 2-nucleon vertex is renormalized strongly so that it effectively carries chiral power minus one. This can be seen as follows. Consider the leading order interaction

$$\mathcal{L}_{4N}(x) = \frac{1}{4}\, g\, \left(\bar{N}(x)\, \gamma_5\, C\, \tau \tau_2\, \bar{N}^t(x)\right) \left(N^t(x)\, \tau_2\, \tau\, C^{-1} \gamma_5\, N(x)\right) . \qquad (8)$$

with the charge conjugation matrix $C = i\,\gamma_0\,\gamma_2$ and the isospin Pauli matrices τ_i. At tree level the coupling g can be interpreted in terms of the s-wave scattering lengths $a(^1S_0)$ with $2\,m\,g = -4\,\pi\,a(^1S_0)$. We assume here that the energy dependent 4-nucleon bare coupling strengths, $g(p)$, is natural. According to our subtraction scheme, which is required for the manifestation of the L-counting rule, the bare coupling g is renormalized by the one loop bubble J at leading order:

$$g^{-1}(p^2) - J(p^2, \Lambda) = g_R^{-1}(p^2, \mu, \Lambda) - J_R(p, \mu) \, . \qquad (9)$$

where the loop function J is given by

$$\begin{aligned} J(p^2, \Lambda) &= -i\, \mathrm{tr} \int \frac{d^4 l}{(2\pi)^4} C^{-1} \gamma_5\, S_F\left(l + \tfrac{1}{2} W\right) \gamma_5\, C\, S_F\left(-l + \tfrac{1}{2} W\right) \\ &= \frac{1}{\pi} \int_0^{\Lambda^2} \frac{d\bar{p}^2\, \rho_J(\bar{p})}{\bar{p}^2 - p^2 - i\,\epsilon} \end{aligned} \qquad (10)$$

in terms of the relativistic nucleon propagator $S_F(p)$ and $W^2 = 4\,(m^2+p^2) = s$. The renormalized loop $J_R(p,\mu) = m\,(\mu + i\,p)/(2\,\pi)$ follows after the proper non-relativistic expansion of the spectral density:

$$\rho_J(p) = \frac{m\,p}{2\pi}\sqrt{1+\frac{p^2}{m^2}} = \frac{m\,p}{2\pi}\left(1+\frac{p^2}{2\,m^2}+\mathcal{O}\left(\frac{p^4}{m^4}\right)\right) . \tag{11}$$

It is important to observe that for $p \leq \Lambda$ the renormalized coupling function can be expanded in powers of the momentum p with

$$\begin{aligned} g_R^{-1}(p^2,\Lambda;\mu) &= g^{-1}(p^2) - \frac{m}{2\pi^2}\int_0^{\Lambda^2} \frac{d\,\bar{p}^2\,\bar{p}}{\bar{p}^2+\mu^2} + \frac{m}{2\pi^2}\int_{\Lambda^2}^{\infty} \frac{d\,\bar{p}^2\,\bar{p}}{\bar{p}^2-p^2}\,\frac{p^2+\mu^2}{\bar{p}^2+\mu^2} \\ &= g^{-1}(p^2) - \frac{m}{\pi^2}\left(\Lambda - \frac{\pi}{2}\,\mu\right) + \frac{m\,p^2}{\pi^2\,\Lambda}\left(1+\mathcal{O}\left(\frac{p^2}{\Lambda^2}\right)\right) \end{aligned} \tag{12}$$

where we dropped $1/m$ correction terms. Our result (12) demonstrates that all cutoff dependence can safely be absorbed in the bare coupling $g(p)$ provided that it is legitimate to identify the typical cutoff Λ with the natural scale of the bare coupling function.

Equation (12) exhibits an interesting effect. Starting with a perfectly natural coupling function $g(p^2)$ it potentially generates an anomalously large renormalized coupling function $g_R(p^2,\Lambda;\mu)$, however, only if the subtraction scale $\mu \sim m_\pi$ is taken to be small. Then for an attractive natural coupling $g < 0$ the g^{-1} term is potentially canceled by the $m\,\Lambda$ term such that the renormalized coupling g_R will be anomalously large. We stress that this is a familiar phenomenon since it is long understood that an attractive system may dynamically generate 'new' scales which are anomalously small (like for example the binding energy of a shallow bound state). What is intriguing here is the transparent and simple mechanism provided by (12) to generate such a 'small' scale naturally [3,4,17].

We close this section with a few simple examples demonstrating how to expand diagrams induced by the relativistic chiral Lagrangian and correct an erroneous statement made during my talk. It is sufficient to discuss scalar diagrams since any diagram with Dirac structure can always be reduced to scalar loop function by simple algebra. Consider for example the scalar master loop functions which one naturally encounters in the evaluation of the relativistic pion nucleon box diagram:

$$\begin{aligned} I_{3,N}(s) &= i\int \frac{d^4l}{(2\pi)^4}\,S_\pi(l+\tfrac{1}{2}\,Q)\,S_N(P-l)\,S_N(K+l)\,, \\ I_{4,N}(s,t) &= -i\int \frac{d^4l}{(2\pi)^4}\,S_\pi(l+\tfrac{1}{2}\,Q)\,S_\pi(l-\tfrac{1}{2}\,Q)\,S_N(P-l)\,S_N(K+l) \end{aligned} \tag{13}$$

with $S_i(p) = 1/(p^2 - m_i^2 + i\,\epsilon)$. For technical convenience we introduce the four vectors P, K and Q as follows:

$$p_1 = K + \tfrac{1}{2}Q\,, \qquad p_1' = K - \tfrac{1}{2}Q\,,$$
$$p_2 = P - \tfrac{1}{2}Q\,, \qquad p_2' = P + \tfrac{1}{2}Q\,. \tag{14}$$

with $(P+K)^2 = s$ and $Q^2 = t$ but $P \cdot Q = K \cdot Q = 0$. The master functions are easily calculated by means of dispersion techniques:

$$I_{3,N}(s) = \int_{4\,m^2}^{4(m^2+\Lambda^2)} \frac{d\,s'}{\pi}\,\frac{\rho_{3,N}(s')}{s'-s}\,,$$
$$I_{4,N}(s,t) = \int_{4\,m^2}^{4(m^2+\Lambda^2)} \frac{d\,s'}{\pi}\,\frac{\rho_{4,N}(s',t)}{s'-s}\,. \tag{15}$$

The finite cutoff $\Lambda < m$ is kept at this intermediate stage (it will be removed later) in order to mathematically justify the $1/m$ expansion. We derive the s-channel spectral densities:

$$\rho_{3,N}(s) = \frac{1}{16\,\pi}\,\frac{1}{\sqrt{4\,m^2-u-t}}\,\frac{1}{\sqrt{-u-t}}\,\ln\left(\frac{m_\pi^2-u-t}{m_\pi^2}\right)\,,$$
$$\rho_{4,N}(s,t) = \frac{1}{4\,\pi}\,\frac{1}{\sqrt{-t}}\,\frac{1}{\sqrt{4\,m^2-u-t}}\,\frac{1}{\sqrt{b(s,t)}}\,\mathrm{artanh}\sqrt{\frac{t(u+t)}{b(s,t)}} \tag{16}$$

with $u+t = 4\,m^2 - s$ and

$$b(s,t) = 4\,m_\pi^4 - 4\,m_\pi^2\,(u+t) + t\,(u+t)\,. \tag{17}$$

The s-channel spectral densities are expanded in powers of the inverse nucleon mass. At leading order we then derive:

$$I_{3,N}(s) = \frac{1}{16\pi m}\,\frac{1}{\sqrt{-u-t}}\left(\arctan\left(\frac{\sqrt{-u-t}}{m_\pi}\right) + \frac{i}{2}\,\ln\left(1-\frac{u+t}{m_\pi^2}\right)\right)\,,$$

$$I_{4,N}(s,t) = \frac{1}{8\,\pi m}\,\frac{1}{\sqrt{-t\,b(s,t)}}\left(i\,\mathrm{artanh}\left(\sqrt{\frac{t(u+t)}{b(s,t)}}\right)\right.$$
$$+\,\frac{1}{2}\,\arctan\left(\frac{\sqrt{-t\,b(s,t)}}{4\,m_\pi^3} - \frac{\sqrt{-u-t}}{m_\pi}\left(1-\frac{t}{4\,m_\pi^2}\right)\right)$$
$$\left.+\,\frac{1}{2}\,\arctan\left(\frac{\sqrt{-t\,b(s,t)}}{4\,m_\pi^3} + \frac{\sqrt{-u-t}}{m_\pi}\left(1-\frac{t}{4\,m_\pi^2}\right)\right)\right) \tag{18}$$

where we expand in Λ/m, u/m and t/m. Note that the $(1/m)^n$ correction terms of $I_{3,N}(s)$ and $I_{4,N}(s,t)$ can be evaluated in closed form (up to a scheme dependent subtraction polynomial) by simply replacing the $1/m$ factor in (18) by the factor $2/\sqrt{s}$. This is an immediate consequence of the representation (15). We find this example highly instructive since it clearly demonstrates how the difference of the full relativistic loop function and the $1/m$ expanded loop is lumped efficiently into the subtraction polynomial.

It is instructive to confront our relativistic scheme with the potential approach. Inherent of the potential approach is the use of the static pion exchange. The pion propagator is expanded

$$S_\pi(l_0,\vec{l}\,) = \frac{-1}{l^2+m_\pi^2} - \frac{l_0^2}{(l^2+m_\pi^2)^2} + \mathcal{O}\left(l_0^4\right) \tag{19}$$

where it is commonly argued that terms involving the energy transfer l_0^2 are suppressed by $1/m$. Mathematically it is not quite obvious that such an expansion is justified and in accordance with covariance since the expansion is highly divergent due to multiple powers of the energy transfer l_0. We evaluate the master pion nucleon box function $I_{4,N}(u,t)$ in the limit of static pions

$$I_{4,N}^{(static)}(\vec{p},\vec{p}\,') = \frac{1}{4\,m}\int \frac{d^3l}{(2\pi)^3}\,\frac{1}{(\vec{l}+\vec{p}\,)^2+m_\pi^2}\,\frac{1}{(\vec{l}+\vec{p}\,'\,)^2+m_\pi^2}\,\frac{1}{l^2-p^2-i\,\epsilon} \tag{20}$$

where we chose the center of mass frame with $p^2 = p'^2$ and $\vec{p}\cdot\vec{p}\,' = p^2\,\cos\theta$ and $s = 4\,(m^2+p^2)$. Note that we dropped terms suppressed by $1/m^3$. Explicit evaluation of the integral (20) leads in fact to the same expression derived before in (18). It remains to be shown that also the $(1/m)^n$ correction terms evaluated in such a scheme agree with those of the relativistic approach. Note that in any case it becomes more and more tedious to evaluate the $(1/m)^n$ correction terms using (19).

In the algebraic reduction of the pion nucleon box diagram one encounters a further master loop function

$$I_{3,\pi}(t) = i\int \frac{d^4l}{(2\pi)^4}\,S_\pi(l+\tfrac{1}{2}\,Q)\,S_\pi(l-\tfrac{1}{2}\,Q)\,S_N(l+K) \tag{21}$$

also relevant for the pion-nucleon scattering process. The triangle loop, $I_{3,\pi}(t)$, is finite and can be evaluated by means of a dispersion integral:

$$I_{3,\pi}(t) = \int_{4\,m_\pi^2}^{4\left(m_\pi^2+\Lambda^2\right)} \frac{d\,t'}{\pi}\,\frac{\rho_{3,\pi}(t')}{t'-t} \tag{22}$$

with the t-channel spectral density:

$$\rho_{3,\pi}(t) = \frac{1}{8\,\pi}\,\frac{\Theta\Big(t-4\,m_\pi^2\Big)}{\sqrt{t(4\,m^2-t)}}\,\arctan\left(\sqrt{4\,m^2-t}\,\frac{\sqrt{t-4\,m_\pi^2}}{t-2\,m_\pi^2}\right) \tag{23}$$

for $t > 4\,m_\pi^2$. Naively one may consider the spectral density in the heavy nucleon mass limit with

$$\rho_{3,\pi}(t) = \frac{1}{32\,m\,\sqrt{t}}\,\Theta\Big(t-4\,m_\pi^2\Big) \tag{24}$$

and derive

$$I_{3,\pi}(t) = \frac{1}{16\,\pi\,m}\,\frac{1}{\sqrt{-t}}\,\arctan\left(\frac{\sqrt{-t}}{2\,m_\pi}\right) \tag{25}$$

for $t < 0$. We point out, however, that the $1/m$ expansion of the loop integral $I_{3,\pi}(t)$ is subtle. This can be anticipated from the fact that the $1/m$ expansion of the t-channel spectral density leads, at subleading orders, to singularities in the dispersion integral at $t' = 4\,m_\pi^2$ [c]. Note that the 'naive' leading order result (25) is in agreement with the corresponding expression derived in the heavy baryon framework [16,18]. An alternative way to perform the non-relativistic expansion is through the exact identity

$$I_{3,\pi}(t) = I_{3,N}(4\,m^2-t) - \left(m_\pi^2 - \tfrac{1}{2}\,t\right) I_{4,N}(4\,m^2-t,t) \tag{26}$$

which holds for the full relativistic loop functions. It expresses the desired loop integral $I_{3,\pi}(t)$, in terms of the master loop functions $I_{3,N}(s)$ and $I_{4,N}(s,t)$ which posses a well defined $1/m$ expansion. The non-relativistic expansion of $I_{3,N}(s)$ and $I_{4,N}(s,t)$ then confirms (25). We suggest that the proper $(1/m)^n$ correction terms of $I_{3,\pi}(t)$ are in fact accessed via (26). In terms of the t-channel dispersion relation (22) this amounts to only including the $1/m$ correction terms from the factor $1/\sqrt{4\,m^2-t}$ but use the large nucleon mass limit of the arctan-term in (23).

The few examples presented here clearly demonstrate the usefulness of the relativistic chiral Lagrangian. The required $1/m$ expansion of Feynman diagrams is straightforward and in fact performed most economically applying dispersion relation techniques introduced by Cutkosky [19]. These techniques keep covariance manifestly and are applied easily also to higher loop diagrams involving pion production cuts [10]. Note that dealing with relativistic Feynman diagrams avoids the cumbersome splitting of pions into static, soft and radiation pions as suggested by Mehen and Stewart [20].

[c] Note that such singularities disappear in the chiral limit with $m_\pi = 0$.

3 Chiral expansion scheme for nuclear matter

An attempt to apply the generalized chiral power counting rules of[3,4] to the nuclear many body problem quickly reveals that even though the pion dynamics remains perturbative the local two-nucleon interaction requires extensive summations. This is an immediate consequence of the chiral power given to the renormalized coupling $g_R \sim Q^{-1}$. In nuclear matter at small density with $k_F \simeq 100$ MeV and $\rho = 2\,k_F^3/(3\,\pi^2)$ it is not sufficient to sum the particle-particle ladder diagrams of the local two-nucleon interaction. We find that one must sum simultaneously also the particle-particle and hole-hole ladder diagrams including all interference terms where a self consistently dressed nucleon propagator must be used. According to the generalized counting rules the pions can then be evaluated perturbatively on top of the above described parquet resummation for the local two-nucleon interaction[d]. In terms of vacuum scales this leads to an expansion of the energy per particle, $\bar{E}(k_F)$, of the form

$$\bar{E}(k_F) = \sum_n \bar{E}_n\left(\frac{k_F}{m_\pi}, \frac{k_F}{\Lambda_S}\right)\left(\frac{k_F}{\Lambda_L}\right)^n \tag{27}$$

with typical small scales, Λ_S, like $\sqrt{m_N\,\epsilon_D}$ where $\epsilon_D \simeq 2$ MeV is the deuteron binding energy, and typical large scales $\Lambda_L \simeq 4\,\pi\,f_\pi \simeq 1$ GeV. The expansion coefficients $\bar{E}_n$ are complicated and hitherto unknown functions of the Fermi momentum k_F. They can be computed in terms of the free space chiral Lagrangian furnished with a systematic summation technique[21].

In this work we pursue a somewhat less microscopic approach in spirit close to the Brueckner scheme but more systematic in the sense of effective field theory. Since the typical small scale Λ_S is much smaller than the Fermi momentum $k_{F,0} \simeq 265$ MeV at nuclear saturation density, one may expand the coefficient functions $\bar{E}_n$ around $k_{F,0}$ in the following manner

$$\bar{E}_n\left(\frac{k_F}{m_\pi}, \frac{k_F}{\Lambda_S}\right) = \bar{E}_n\left(\frac{k_F}{m_\pi}, \frac{k_{F,0}}{\Lambda_S}\right) + \sum_{k=1}^{\infty} \bar{E}_n^{(k)}\left(\frac{k_F}{m_\pi}\right)\left(\frac{\Lambda_S}{k_F} - \frac{\Lambda_S}{k_{F,0}}\right)^k . \tag{28}$$

Note that we do not expand in the ratio m_π/k_F. If one expanded also in this ratio m_π/k_F one would arrive at the Skyrme phenomenology[22] applied

---

[d]Our expectation that the nuclear equation of state can be calculated microscopically solving the parquet theory for the local interaction and then include pions perturbatively may be somewhat too naive. A refined formulation which incorporates relevant subthreshold singularities of the vacuum nucleon-nucleon scattering amplitude (see[4]) may lead to a more involved treatment of pionic effects.

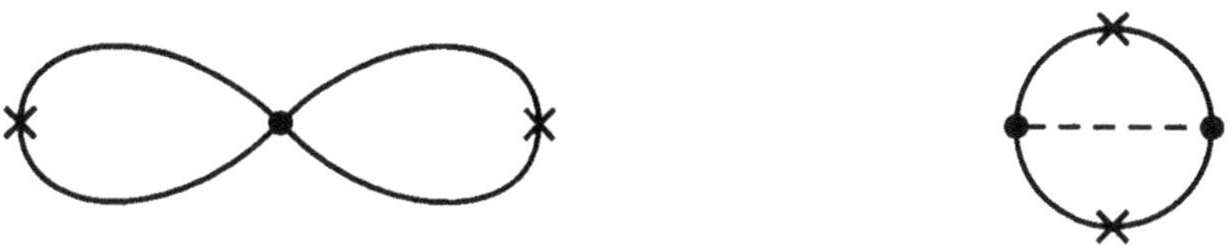

Figure 1: Leading contribution of chiral order Q^6.

successfully to nuclear physics many years ago. It should be clear that this scheme is constructed to work around nuclear saturation density but will fail at small density. We note that also conventional approaches like the Walecka mean field or the Brueckner scheme are known to be incorrect at small density.

In terms of diagrams one may arrive at the proposed expansion (28) by a perturbative evaluation of the pionic nuclear many-body effects on top of the the high density limit of the parquet resummation for the local two-nucleon interaction. The underlying assumption is of course that the typical small scale in the parquet, Λ_S, is such that the high density limit is reached already at densities much smaller than the nuclear saturation density $k_{F,0}$. Technically our scheme can be generated by the effective Lagrangian density

$$\begin{aligned}\mathcal{L}_{int}(k_F) &= \frac{g_A}{2 f_\pi}\, \bar{N}\gamma_5\, \gamma^\mu \cdot (\partial_\mu \vec{\pi}) \cdot \vec{\tau}\, N \\ &+ \frac{1}{8 f_\pi^2}\left(g_0(k_F) + \frac{1}{4}\, g_A^2\right) \left(\bar{N}\, \gamma_5\, \tau_2\, C^{-1}\, \bar{N}^t\right) \left(N^t\, C\, \tau_2\, \gamma_5\, N\right) \\ &+ \frac{1}{8 f_\pi^2}\left(g_1(k_F) + \frac{1}{4}\, g_A^2\right) \left(\bar{N}\, \gamma_\mu\, \vec{\tau}\, \tau_2\, C^{-1}\, \bar{N}^t\right) \left(N^t\, C\, \tau_2\, \vec{\tau}\, \gamma^\mu\, N\right) \quad (29)\end{aligned}$$

where the couplings $g_0 = g_0(k_F)$ and $g_1 = g_1(k_F)$ are density dependent. A more systematic derivation of the expansion (27) and (28) applying suitable resummation techniques will be presented elsewhere [21].

The chiral power counting rules are simplified significantly as compared to a fully microscopic scheme. The presence of a further small scale $k_F \sim Q \sim m_\pi$ does not anymore generate an infinite tower of diagrams, to be considered at a given chiral order, since by construction the troublesome local 2-nucleon vertex need not be iterated, i.e. terms proportional to $g_0^n(k_F)$ and $g_1^n(k_F)$ with $n > 1$ are already included in $g_0(k_F)$ and $g_1(k_F)$ and therefore must not be considered. Note that here we count $g_{0,1} \sim Q^0$ since the non-perturbative structures like the deuteron, which give rise to the anomalous power $g_{0,1} \sim Q^{-1}$ in the vacuum, are long dissolved at densities close to nuclear saturation. The pion dynamics, if properly renormalized, remains perturbative like in the

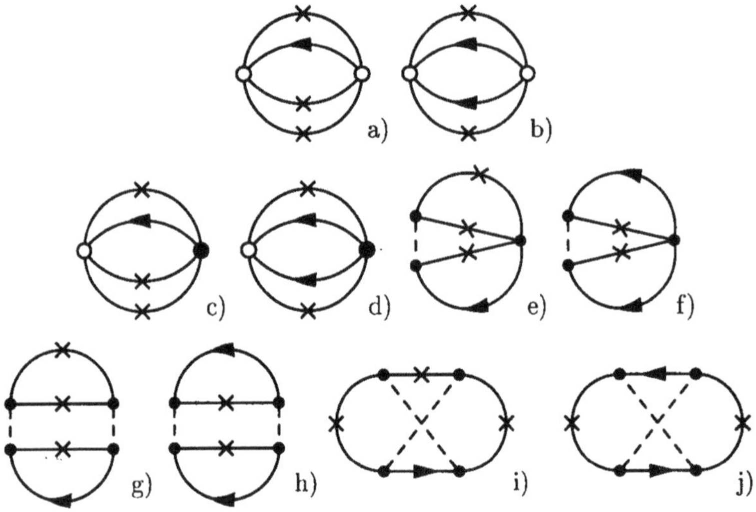

Figure 2: Subleading contribution of chiral order Q^7.

vacuum case. Consider for example the two loop diagrams depicted in Fig. 1 where the nucleon line with a 'cross' represents the projector onto the Fermi sphere

$$\Delta S_N(p) = (\gamma \cdot p + m_N)\, 2\,\pi\, i\, \Theta(p_0)\, \delta(p^2 - m_N^2)\, \Theta\left(k_F^2 - \vec{p}^{\,2}\right) \,. \tag{30}$$

The first diagram in Fig. 1 is proportional to $g_{0,1}(k_F)\, k_F^6$ and is therefore ascribed the chiral order Q^6 since the effective vertex $g_{0,1}(k_F) \sim Q^0$ carries chiral power zero. The second diagram, the one pion exchange contribution, is also of chiral order Q^6 since it is proportional to k_F^6 multiplied with some dimensionless function $f(m_\pi/k_F)$. In Fig. 2 we collected all diagrams of chiral order Q^7. Here we introduced two types of 2-nucleon vertices. The filled circle represents the full vertex of (29) proportional to $g_{0,1} + g_A^2/4$ and the open circle the counter term proportional to $g_A^2/4$. The dashed line is the pion propagator and the directed solid line the free space nucleon propagator. We point out that the diagrams b), d), f), h) and j) in Fig. 2 are divergent. The leading chiral contribution of the sum of all diagrams, however, is finite after including the appropriate counter term in the spin triplet channel.

It is instructive also to display the non relativistic form of the Lagrangian

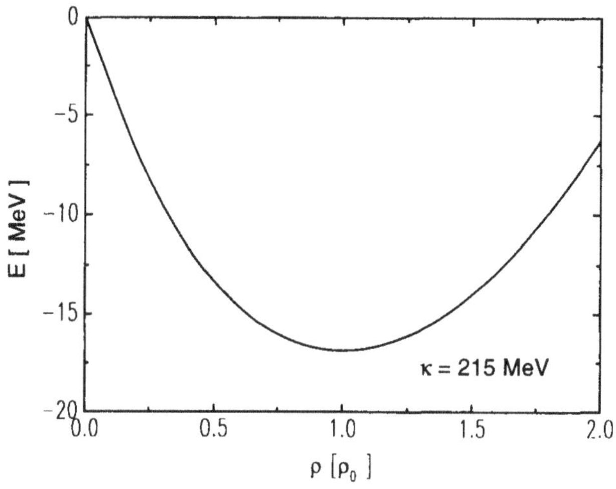

Figure 3: The equation of state for isospin symmetric nuclear matter.

density (29) as suggested by Weinberg [1] in the context of the nucleon-nucleon scattering problem

$$\begin{aligned}\mathcal{L}_{int}(k_F) &= \frac{g_A}{2\,f_\pi}\,\bar{N}\left(\vec{\sigma}\cdot\vec{\nabla}\right)\left(\vec{\pi}\cdot\vec{\tau}\right)N \\ &+ \frac{1}{8\,f_\pi^2}\left(g_0(k_F)+\frac{1}{4}\,g_A^2\right)\left(\bar{N}\,\vec{\sigma}\,\sigma_2\,\tau_2\,\bar{N}^t\right)\left(N^t\,\tau_2\,\sigma_2\,\vec{\sigma}\,N\right) \\ &+ \frac{1}{8\,f_\pi^2}\left(g_1(k_F)+\frac{1}{4}\,g_A^2\right)\left(\bar{N}\,\sigma_2\,\vec{\tau}\,\tau_2\,\bar{N}^t\right)\left(N^t\,\tau_2\,\vec{\tau}\,\sigma_2\,N\right) \qquad (31)\end{aligned}$$

with N now a two component spinor field and $\bar{N} = N^\dagger$ for notational convenience. According to Weinberg the two-particle reducible diagrams are to be evaluated with static pions.

We therefore evaluate all diagrams of Fig. 2 with the appropriate non relativistic interaction vertices and static pion propagators. The solid line with a 'cross' now represents the non relativistic limit of (30) and the directed solid line the free non relativistic nucleon propagator. For technical details we refer to [23]. Both schemes indeed lead to identical results for all diagrams[e].

In Fig. 3 our result for the isospin symmetric nuclear equation of state is

[e]Note that here we correct an error in [10].

shown. The relevant coupling $g_0 + g_1$ is adjusted to obtain nuclear saturation at $k_{F,0} = 265$ MeV. We emphasize that the coupling functions $g_{0,1}(k_F)$ are to be determined from the nuclear equation of state. However there is a strong consistency constraint: according to our scale argument (28) the density dependence of the couplings $g_{0,1}(k_F)$ must be weak for k_F larger than typical small scales integrated out. If the nuclear saturation required a strong density dependence our scheme had to be rejected. The density independent set of parameters $g_0 + g_1 \simeq 3.2$, $g_A \simeq 1.26$, $m_\pi \simeq 140$ MeV and $f_\pi \simeq 93$ MeV give an excellent result for the equation of state. The empirical saturation density $k_{F,0} \simeq 265$ MeV and the empirical binding energy of 16 MeV are reproduced. The incompressibility with $\kappa \simeq 215$ MeV is also compatible with the empirical value (210 ± 30) MeV of [24].

4 The chiral order parameter $\langle \bar{q}q \rangle$

The quark condensate, $\langle \bar{q}q \rangle(\rho)$, is an object of utmost interest. It measures the degree of chiral symmetry restoration in nuclear matter. Furthermore it is an important input for QCD sum rules [25] or the Brown-Rho scaling hypothesis [26]. According to the Feynman-Hellman theorem the quark condensate can be extracted unambiguously from the total energy, $E(\rho)$, of nuclear matter once the current quark mass dependence of $E(\rho, m_Q)$ is known

$$\langle \bar{q}q \rangle(\rho) = \frac{1}{V} \frac{\partial}{\partial m_Q} E(\rho, m_Q) \tag{32}$$

with $m_Q = m_u = m_d$. Recall that $E(\rho)/V = (m_N + \bar{E}(\rho))\, \rho$ is determined by the nuclear equation of state $\bar{E}(\rho)$. Since our chiral approach was set up to treat the pion dynamics and therewith the current quark mass dependence of the equation of state properly it is very much tailored to be applied to the quark condensate.

It is convenient to consider the relative change of the quark condensate since it is renormalization group invariant:

$$\frac{\langle \bar{q}\, q \rangle(\rho)}{\langle \bar{q}\, q \rangle(0)} = 1 - \frac{\Sigma_{\pi N}\, \rho}{m_\pi^2\, f_\pi^2} - \frac{\alpha_\pi(\rho)\, \rho}{2\, m_\pi\, f_\pi^2}\,. \tag{33}$$

The second term in (33) follows from the nucleon rest mass contribution to the total energy of nuclear matter together with the definition of the pion nucleon sigma term, $\Sigma_{\pi N}$:

$$\Sigma_{\pi N} = m_Q\, \langle N | \bar{q}\, q | N \rangle = m_Q\, \frac{d\, m_N}{d\, m_Q}\,. \tag{34}$$

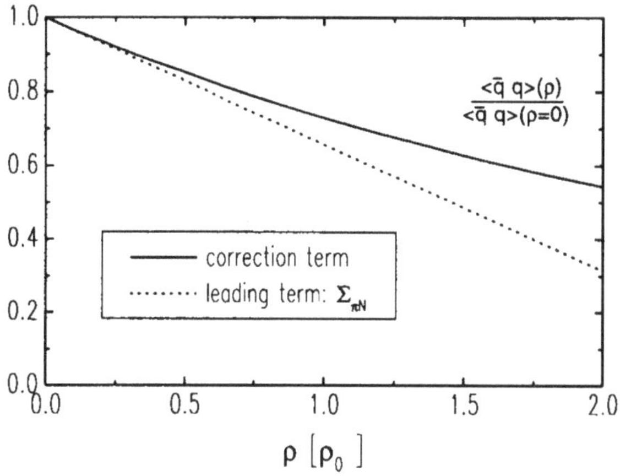

Figure 4: The quark condensate in isospin symmetric nuclear matter.

The last term in (33) measures the sensitivity of the nuclear equation of state, $\bar{E}(\rho, m_\pi)$, on the pion mass

$$\alpha_\pi(\rho) = -\frac{2\,m_\pi\, f_\pi^2}{\langle \bar{q}\, q\rangle(0)}\, \frac{\partial}{\partial\, m_Q}\, \bar{E}(\rho, m_\pi) = \left(1 + \mathcal{O}\left(m_\pi^2\right)\right) \frac{\partial}{\partial\, m_\pi}\, \bar{E}(\rho, m_\pi) \quad (35)$$

where we consider now the current quark mass $m_Q = m_Q(m_\pi)$ as a function of the physical pion mass. We emphasize that the second term in (33) written down first in [27,28,29] does not probe the nuclear many body problem and therefore should be considered with great caution. It is far from obvious that this term is the most important one at nuclear saturation density. The only term susceptible to nuclear dynamics is the last term in (33).

In Fig. 4 we present our result for the quark condensate in nuclear matter. We confront the 'leading' term driven by $\Sigma_{\pi N} \simeq 45$ MeV with our result. The inclusion of pionic many body effects leads to a significantly less reduced quark condensate in nuclear matter. Note that here we do not include an explicit m_π dependence of the coupling $g_{0,1}(k_F)$ simply because there is not yet any reliable estimate available. Our results confirm calculations performed within the Brueckner [30] and Dirac-Brueckner [31] approach qualitatively insofar that the nuclear many body system appears to react against chiral symmetry restoration. They have strong implications for the QCD sum rule approach of hadron properties in nuclear matter.

References

1. S. Weinberg, *Phys. Lett.* B **251**, 288 (1990); *Nucl. Phys.* B **363**, 3 (1991); *Phys. Lett.* B **295**, 114 (1992);
2. C. Ordóñez, L. Ray, and U. van Kolck, *Phys. Rev. Lett.* **72**, 1982 (1994); *Phys. Rev.* C **53**, 2086 (1996).
3. M. Lutz, in *The standard Model at Low Energies*, hep-ph/9606301.
4. M. Lutz, nucl-th/9906028.
5. D.B. Kaplan, M.J. Savage and M.B. Wise, *Nucl. Phys.* **B 534** (1998) 329; T. Mehen and I. Stewart, nucl-th/9809095; nucl-th/9906010.
6. T.D. Cohen and J. M. Hansen, nucl-th/9808038.
7. K.-G. Fogel, *Arkiv f. Fysik* **42**, 4 (1952).
8. A. Friman, *Acta Acad. Aboensis Math. et Phys.* **41**, 21 (1957).
9. J.V. Steele and R.J. Furnstahl, *Nucl. Phys.* A **645**, 439 (1999).
10. M. Lutz, *Nucl. Phys.* A **642**, 171c (1998).
11. S. Weinberg, *Phys. Rev.* **166**, 1568 (1968); *Physica* A **96**, 327 (1979).
12. A. Krause, *Helv. Phys. Acta* **63**, 3 (1990).
13. J. Gasser, M.E. Sainio and A. Svarc, *Nucl. Phys.* B **307**, 779 (1988).
14. E. Jenkins and A.V. Manohar, *Phys. Lett.* B **255**, 558 (1991).
15. V. Bernard, N. Kaiser and U.G. Meißner, *Int. J. Mod. Phys.* E **4**, 193 (1995); T.-S. Park, D.-P. Min and M. Rho, *Phys. Rep.* **233**, 341 (1993).
16. P.J. Ellis and H-B. Tang, *Phys. Rev.* C **57**, 3356 (1998).
17. S.R. Beane, D.R. Phillips and T.D. Cohen, *Nucl. Phys.* A **632**, 445 (1998).
18. M. Mojzis, *Eur. Phys. J.* C **2**, 181 (1998).
19. R.E. Cutkosky, *J. Math. Phys.* **1**, 429 (1960).
20. T. Mehen and I. Stewart, nucl-th/9901064.
21. M. Lutz, in preparation.
22. M. Beiner et al, *Nucl. Phys.* A **238**, 29 (1975).
23. M. Lutz, B. Friman and Ch. Appel, in preparation.
24. J.P. Blaizot, *Phys. Rep.* **64**, 172 (1980).
25. T. Hatsuda and Su H. Lee, *Phys. Rev.* C **46**, R34 (1992).
26. G.E. Brown and M. Rho, *Phys. Rev. Lett.* **66**, 2720 (1991).
27. E.G. Drukarev and E.M. Levin, *Nucl. Phys.* A **511**, 679 (1990).
28. M. Lutz, S. Klimt and W. Weise, *Nucl. Phys.* A **453**, 521 (1992).
29. T.D. Cohen, R.J. Furnstahl and D.K. Griegel, *Phys. Rev.* C **45**, 1881 (1992).
30. G.Q. Li and C.M. Ko, *Phys. Lett.* B **338**, 118 (1994).
31. R. Brockmann and W. Weise, *Phys. Lett.* B **367**, 40 (1996).

Discussion during M. Lutz's presentation

Chair: *M. Birse (University of Manchester)*
Notes taken by: *M. N. Butler (Saint Mary's University)*

The discussion addressed, primarily, the treatment of pions. David Kaplan asked why Lutz did not simply put in relativistic pions and perform a $1/m$ expansion. Harald Grießhammer noted that it has been shown that both static and radiation pions are needed. Lutz agreed, but made it clear that what he was saying is that static pions are not enough by themselves. Tom Cohen asked whether p/m corrections from the nucleon sector might fix this, but Lutz argued that p/m_π corrections were needed. Kaplan asked why one could not simply calculate the box diagram exactly, and then perform an expansion on the result. Lutz noted that the dispersion relation approach simplified the analysis, as one needed to be careful about modes. Grießhammer noted that it had been known for a year that the separation into radiation and static pions was well defined. Martin Savage pondered whether the issue was trivial. The physics is not new and one can set up a Lagrangian with the right symmetry, and get the right answers with static pions. Kaplan noted, for example, that the hydrogen atom can be described perfectly well with static photons. Lutz responded that finite pion mass is a subtle issue with respect to expanding dispersion relations, and one needs to be careful.

Jerry Miller raised the question of k_F setting the scale. Is it the only scale entering when diagrams are summed? Cohen extended that to ask if correlation scales entered the problem. Lutz argued that if there were more scales than he proposed, then there is no sense to the high-density limit. Aurel Bulgac noted later that this could not work in a low-density limit because of cross-channels, and the resulting lack of stable modes. Lutz noted that this is why one starts at higher density first.

Paulo Bedaque wondered whether it was obvious that the leading order calculation can't have more bubbles. Lutz responded that one already has an effective vertex from previous, constructive, iterations. Cohen asked whether pions would mess this up, but Lutz responded that they are higher order. Cohen then asked whether one would get effective vertices between pions, but Lutz responded that they are suppressed through the vertex being Q^0 and the added loop. Kaplan queried loops counting as Q^1, and Lutz noted that one had to be careful of vacuum loops versus matter loops. Bira van Kolck reiterated then that at leading order (LO) there are no pions in this calculation, so saturation is coming from balancing LO against next-to-leading order (NLO). Lutz argued that both LO and NLO are needed to get the physics. van Kolck asked then about next-to-next-to-leading order, but Lutz noted that the tensor effects were picked up at NLO. Savage pointed out that the static versus non-static discussion was arising primarily because of the large momentum flowing through the pion lines. When asked by Cohen where the attraction and repulsion required for saturation are coming from in the calculation, Lutz noted that the counterterm is attractive and Pauli blocking gives the repulsion. Ryoichi Seki noted that having an effective counterterm is okay, but that counterterm depends on density and is not constant. Lutz replied that he was not ignoring the density dependence, but was using scale arguments to say the dependence is weak at intermediate density.

After the talk, the discussion of static vs. non-static pions continued. Miller noted that in treatments of nucleon-nucleon scattering, the crossed box diagram cancels the non-static effects of the box diagram, so that the sum is effectively just static pions. Lutz replied that the crossed box is Q^2 in his counting, while the box diagram is Q^1. He stated that it might be a good numerical approximation to use static pions, but argued that it is not good in a systematic expansion.

THE CANONICAL NUCLEAR MANY-BODY PROBLEM AS A RIGOROUS EFFECTIVE THEORY

W. C. HAXTON [a]
Institute for Nuclear Theory, Box 351550, and Department of Physics, Box 351560
University of Washington, Seattle, WA 98195, USA
E-mail: haxton@phys.washington.edu

The shell model is the standard tool for addressing the canonical nuclear many-body problem of nonrelativistic nucleons interacting through a static potential. We discuss several of the uncontrolled approximations that are made in this model to motivate a different approach, one based on an exact solution of the Bloch-Horowitz equation. We argue that the necessary self-consistent solutions of this equation can be obtained efficiently by a Green's function expansion based on the Lanczos algorithm. The resulting effective theory is carried out for the simplest nuclei, d and ^{3}He, using realistic NN interactions such as the Argonne $v18$ and Reid93 potentials, in order to contrast the results with the shell model. We discuss the wave function normalization, the evolution of the wave function as the "shell model" space is varied, and the magnetic elastic effective operator. The numerical results show a simple renormalization group behavior that differs from effective field theory treatments of the two- and three-body problems. The likely origin of this scaling is discussed.

1 Introduction

In many text books the shell model (SM) is motivated by the analogy with Brueckner's treatment of nuclear matter. While the exact many-body Hamiltonian operates in an infinite Hilbert space

$$H = \frac{1}{2} \sum_{i,j=1}^{A} (T_{ij} + V_{ij}), \tag{1}$$

where T_{ij} is the relative nonrelativistic kinetic energy operator and V_{ij} the nucleon-nucleon potential, the SM Hamiltonian acts in a restricted space and employs a softer "effective" potential,

$$H_{SM} = \frac{1}{2} \sum_{i,j=1}^{A} (T_{ij} + V_{ij}^{eff}). \tag{2}$$

Motivating H_{SM} is the notion that the determination of V^{eff} might be simpler than solving the original A-body problem: the foundation of Brueckner theory

[a] Work presented here has been done in collaboration with C.-L. Song

is that high-momentum contributions to the wave function might be integrated out in a rapidly converging series in $\rho_{nuclear}$ or, equivalently, in the number of nucleons in high-momentum states interacting at one time outside the SM space.

The SM thus represents explicitly $\sim 60\%$ of the wave function that resides at long-wavelengths, treating the A-body correlations important to collective modes by direct diagonalization. Implicitly the high-momentum components are swept into a rather poorly defined "effective interaction," often determined empirically. The strength of the SM resides in the first of these two aspects: the technology developed for direct diagonalizations is quite remarkable, including recent progress in Lanczos-based methods [1], in treatments of light nuclei involving many shells [2], and in Monte Carlo sampling [3,4]. Its weakness is the numerous uncontrolled approximations that become apparent when one tries to view the shell model as a faithful effective theory (ET). The thesis of this talk is that the same numerical strides that have advanced shell model diagonalizations now allow us to remove these uncontrolled approximations. The resulting ET has many differences with the shell model and many similarities to the effective field theories under discussion at this workshop.

Among the SM uncontrolled approximations are the following:
1) Even in lowest order, where only the pairwise interaction of high-momentum nucleons is included in H^{eff}, the functional form of the resulting effective interaction is not as simple as assumed in the SM,

$$\langle|H^{eff}|\rangle_{SM} \equiv \langle\alpha\beta|H^{eff}|\gamma\delta\rangle \tag{3}$$

where the Greek symbols label single-particle shell-model states. For example, if the Slater determinants are formed from harmonic oscillator states, the two-body matrix elements must carry an additional index labelling the total number of quanta in the configuration on which H^{eff} operates [5]. Thus H^{eff} reduces to the shell-model form only when that index is unnecessary, e.g., when a lowest-order calculation is restricted to a single shell. Beyond lowest order, three-, four-, and higher-body operators are successively added to H^{eff}.
2) Typically H^{eff} lacks the symmetries of the original bare H, e.g., translational invariance and Hermiticity (though the latter is often enforced by hand).
3) SM wave functions are orthogonal and normed to unity. In ET the effective wave functions are naturally defined as the restrictions of the true wave functions $|\Psi_i\rangle$ to the model space

$$|\Psi_i\rangle \underset{ET}{\longrightarrow} \langle SM|\Psi_i\rangle|SM\rangle \equiv |\Psi_i^{eff}\rangle. \tag{4}$$

Thus the norms are less than unity and orthogonality, which holds for the true

wave functions, is lost when these wave functions are restricted to the model space.
4) Shell model interactions frequently depend on fictitious parameters such as "starting energies," introduced to adjust the energy denominator in the two-body G-matrix or to account for intermediate-state average energies when the two-body G-matrix is iterated to produce some approximation to a higher order H^{eff}.
5) Perhaps most serious, the important issue of effective operators is almost never addressed in a meaningful way. In many cases practitioners adopted a phenomenological H^{eff} which, while successful in producing spectra, provides no diagrammatic basis for calculating effective operators or wave function normalizations. Even in cases where H^{eff} is derived from some underlying NN interaction, the practice is generally to then employ bare operators. In some well-studied cases, such as allowed β decay in the $1p$ and $2s1d$ shells, it is then recognized that a phenomenological renormalization (e.g., $g_A \rightarrow 1$) of operators greatly improves agreement with experiment. But the origin of this renormalization and its evolution with momentum transfer q are left unclear. The situation is very unsatisfactory and undercuts the shell model as a predictive tool.

2 Self-consistent Bloch-Horowitz Solutions

We consider the cononical nuclear structure problem of nonrelativistic point nucleons interacting through a realistic NN interaction, such as the Argonne $v18$ [6] and Reid93 [7] potentials. The question is whether the uncontrolled approximations in the shell model can be removed, leaving a more complicated but still tractable effective theory. The approach involves three major steps:

• Formulating a treatment of effective interactions and operators that exploits the basic assumption in Brueckner theory — that interactions at high momenta can be integrated out in a cluster expansion (essentially an expansion in ρa^3, where ρ is the nuclear density and a an interaction range) — but is otherwise exact. The convergence of the expansion could then be tested numerically and should depend on the operator under study and the momentum transfer. The goal would be to distinguish fully converged results from those which require higher order calculations.

• To find numerical tricks for implementing this formulation, demonstrating their validity in cases (e.g, A=2,3,4) where the expansion can be carried to all orders, so that the answers should then agree with Faddeev and other exact methods.

• To imbed the formulation in a heavier nucleus, where the cluster expansion can be carried out only partially.

There is some reason for optimism that if the first two goals can be achieved, the third might yield very accurate results: the Argonne group cluster variational Monte Carlo effort on ^{16}O appeared to yield nearly exact results when clusters up to A = 5 were included. In this talk our efforts on the first two points will be described. In particular, we will be able to contrast an exact ET of the deuteron and ^{3}He with the shell model to illustrate the shortcomings of the later: we think the differences are surprising.

The approach is sketched in Fig. 1. The Hilbert space is divided into a long-wavelength "shell model" space, defined by some energy scale Λ_{SM}, and a high-momentum space. One can truncate the latter at some scale $\Lambda_\infty \sim 3$ GeV, characteristic of the cores of realistic potentials, as above this energy, excitations make a negligible contribution. That is,

$$H^{eff}_{(i)}(\Lambda_{SM},\Lambda_\infty) \underset{\Lambda_\infty \text{ large}}{\longrightarrow} H^{eff}_{(i)}(\Lambda_{SM}). \tag{5}$$

All correlations within the "SM" space are included, but the high-momentum correlations in the excluded space are limited to n-body, where n is the cluster size. Thus the lowest order effective interaction is

$$H^{eff}_{(n=2)} \equiv H^{eff}_{(n=2,0)} \tag{6}$$

It corresponds to embedding the A-body ladder diagram of Fig. 1b between SM states: A-2 of the nucleons are spectators, with the remaining pair scattering via a two-body ladder. The notation (n=2,0) states that the two-body cluster has no explicit dependence on the nuclear density, varying as ρ^0. The n=3 A-body ladder of Fig. 1c is similarly

$$H^{eff}_{(n=3)} \equiv H^{eff}_{(n=2,1)} + H^{eff}_{(n=3,0)} \tag{7}$$

where $H^{eff}(n=2,1)$ is the two-body part of the three-body ladder

$$\langle\alpha\beta|H^{eff}_{(n=2,1)}|\alpha'\beta'\rangle = \sum_{\gamma\leq k_F}\langle\alpha\beta\gamma|H^{eff}_{(n=3)}|\alpha'\beta'\gamma - \alpha'\gamma\beta' + ...\rangle, \tag{8}$$

where k_F denotes the Fermi level. This decomposition – which is done only to emphasize the content of the cluster expansion – illustrates that $H^{eff}_{(n=3)}$

contains $H^{eff}_{(n=2,0)}$ as well as a correction to the two-body interaction that depends linearly on the density and is obtained by identifying one ingoing SM single-particle leg of the three-body ladder with one outgoing leg, summed over all occupied states. It also contains a true three-body piece $H^{eff}_{(n=3,0)}$, where three SM ingoing single-particle states connect to three distinct outgoing states after undergoing a series of scatterings outside the SM space. The point is simple pedagogy: treatments of successively larger clusters in the high-momentum space correct the lowest-order two-body $H^{eff}_{(n=2,0)}$ by adding terms proportional to the ρ, ρ^2, etc., in the spirit of Brueckner theory. It also adds true three-body terms, true four-body terms, etc. Thus an expansion through four-body clusters yields $H^{eff}_{(n=2,2)}$, the two-body interaction corrected through order ρ^2, $H^{eff}_{(n=3,1)}$, the three-body interaction through order ρ, and $H^{eff}_{(n=4,0)}$, a density-independent true four-body interaction.

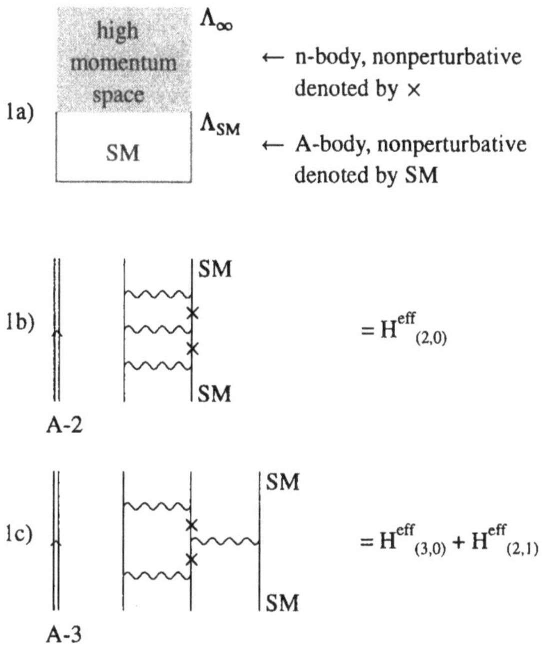

Figure 1: Cluster expansion of the effective interaction.

The calculation begins with a definition of the "SM" space. The goals of handling bound states and of generating an effective interaction that is translationally invariant leaves one sensible choice, many-body states constructed from harmonic oscillator Slater determinants. To exploit the relative/center-of-mass separability of harmonic oscillator Slater determinants, one must separate the SM and high-momentum spaces so that all configurations satisfying

$$E \leq \Lambda_{SM} \hbar\omega \tag{9}$$

are retained in the former. For example, a SM calculation of ^{16}O with $\Lambda_{SM} = 4 + \Lambda_0$, where Λ_0 is the number of quanta in the ^{16}O closed shell, would include all $4\hbar\omega$ configurations, e.g., $0p0h$, $2p2h$, and $4p4h$ excitations of nucleons from the $1p$ shell into the $2s1d$ shell, $1p1h$ excitations of a $1s$ shell nucleon into the $3s2d1g$ shell, etc. One can define the projection operator onto the high-momentum space by

$$Q_{SM} = Q(\Lambda_{SM}, b). \tag{10}$$

where b is the oscillator parameter. Thus the included or "SM" space is defined by two parameters, Λ_{SM} and b. The preservation of translational invariance is also important numerically, as it reduces the two-body ladder to an effective one-body problem, etc.

The resulting Bloch-Horowitz equation [8] is then

$$\begin{aligned} H^{eff} &= H + H \frac{1}{E - Q_{SM} H} Q_{SM} H \\ H^{eff} |\Psi_{SM}\rangle &= E|\Psi_{SM}\rangle \quad |\Psi_{SM}\rangle = (1 - Q_{SM})|\Psi\rangle \end{aligned} \tag{11}$$

where $|\Psi\rangle$ is the exact wave function and $H|\Psi\rangle = E|\Psi\rangle$. The difficulty posed by this equation is the appearance of the unknown energy eigenvalue in the equation for H^{eff}. Thus this system must be solved self-consistently. Note that there is no explicit reference to the harmonic oscillator in this equation: it enters only implicitly through Q_{SM} in distinguishing the long-wavelength "SM" space from the remainder of the Hilbert space.

There is an extensive literature on this and similar equations, often involving a division of H into an unperturbed H_0 and a perturbation $H_1 = H - H_0$ [9,10]. There are well-known pathologies with this division involving the effects of near-by intruder states on the perturbation expansion [11]. Here we explore another approach that is nonperturbative and involves, in effect, a computer summation of diagrams. The method is based on the Lanczos algorithm and offers a remarkably simple solution to the issue of self-consistency.

In the Lanczos algorithm a basis for representing a Hamiltonian is formed recursively in such a way that the resulting Hamiltonian is tridiagonal. Given a Hermitian operator H and and an initial normalized vector $|v_1\rangle$, the successive steps are

$$
\begin{aligned}
H|v_1\rangle &= \alpha_1|v_1\rangle + \beta_1|v_2\rangle \\
H|v_2\rangle &= \beta_1|v_1\rangle + \alpha_2|v_2\rangle + \beta_2|v_3\rangle \\
H|v_3\rangle &= \qquad\qquad \beta_2|v_2\rangle + \alpha_3|v_3\rangle + \beta_3|v_4\rangle \quad etc.
\end{aligned}
\tag{12}
$$

so that the H takes the form

$$
H \rightarrow \begin{pmatrix} \alpha_1 & \beta_1 & 0 & \cdots \\ \beta_1 & \alpha_2 & \beta_2 & \cdots \\ 0 & \beta_2 & \alpha_3 & \cdots \\ \vdots & \vdots & \vdots & \end{pmatrix} \begin{pmatrix} |v_1\rangle \\ |v_2\rangle \\ |v_3\rangle \\ \vdots \end{pmatrix} \tag{13}
$$

The remarkable property of this algorithm has to do with truncating the process in Eq. (12) after n steps, where n can be much smaller than the dimension of the Hilbert space. The resulting truncated matrix in Eq. (13) then contains the information needed to reconstruct the exact $2n$-1 lowest moments of H over the eigenspectrum. As extremum eigenvalues are crucial to higher moments, one common application of the Lanczos algorithm is in determining such eigenvalues and their associated eigenfunctions. Another is to begin with the vector $|v_1\rangle = \hat{O}|g.s.\rangle$ and then use the algorithm to calculate the moments of the response of the ground state $|g.s.\rangle$ to the operator $\hat{O}$. A small number of moments, e.g., ~ 100, often is sufficient to construct a response function with a numerical resolution comparable to that achieved experimentally.

A third application[12] is in constructing fully interacting Green's functions. One finds

$$
\frac{1}{E-H}|v_1\rangle = g_1(E)|v_1\rangle + g_2(E)|v_2\rangle + \cdots \tag{14}
$$

where the $g_i(E)$ are continued fractions that depend on α_i, β_i and where E appears only as a parameter. For example,

$$
g_1(E) = \cfrac{1}{E - \alpha_1 - \cfrac{\beta_1^2}{E - \alpha_2 - \cfrac{\beta_2^2}{E - \alpha_3 - \beta_3^2 \cdots}}} \tag{15}
$$

It follows that the Bloch-Horowitz equation can be solved self-consistently with only a single solution of the effective interactions problem, even in cases where multiple bound states are needed. The procedure is:

- For each relative-coordinate vector in the SM space $|\gamma\rangle$, form the excluded-space vector $|v_1\rangle \equiv Q_{SM}H|\gamma\rangle$ and the corresponding Lanczos matrix for the operator $Q_{SM}H$. Retaining the resulting coefficients α_i, β_i for later use, construct the Green's function for some initial guess for E and then the dot product with $\langle\gamma'|H$ to find $\langle\gamma'|H^{eff}(E)|\gamma\rangle$.

• Perform the "SM" calculation to find the desired eigenvalue E' which, in general, will be different from the guess E. Using the stored α_i, β_i, recalculate the Green's function for E' and $H^{eff}(E')$ then redo the "SM" calculation. Repeat until convergence, i.e., until the input E' in the Green's function equals the desired output "SM" eigenvalue.

• Then proceed to the next desired bound state, e.g., the first excited state, and repeat the above step. Note that it is not necessary to repeat the H^{eff} calculation. The eigenvalue taken from the "SM" calculation is, of course, that of the first excited state. The procedure then generates distinct $H^{eff}(E')$s for each desired state.

The attractiveness of this approach is that the effective interactions part of the procedure, which is relatively time consuming as it requires one to perform a large-basis Lanczos calculation for each relative-coordinate starting vector in the "SM" space, is performed only once. The diagonalization in the model space is generally much faster: modern workstations can handle even large-dimension shell model calculations (sparse matrices of $d \sim 10^6$) quickly ($\sim$ 30 minutes). In practice we found that self-consistency is achieved easily: six to eight cycles is typical. (More cycles are required for states with small binding energies.) Thus it is quite practical to derive the exact $H^{eff}(E)$s for a series of bound states.

Now we discuss the results of applying this procedure to the simplest nuclei, d and ^{3}He, carrying the above process to completion (two-body and three-body ladders, respectively). The motivation is two-fold: demonstrate the numerical procedures we described above, and provide for the first time exact effective theory results that can be compared to those of the shell model.

The harmonic oscillator mode expansion must be sufficient to represent both the long-distance tails of bound states and the short-distance "hard core" scattering predicted by realistic NN potentials. (The Argonne v18 $^1S_0$ potentials are shown in Fig. 2.) Inclusion of high-momentum states through $\Lambda_\infty \sim 50$ yields a deuteron binding energy accurate to $\sim$ 60 keV; extending this to $\sim$ 140 produces a result accurate to one keV, provided one chooses an oscillator parameter that optimizes the convergence. (However, as we will discuss in Section 3, there appears to be a simple scaling with Λ_∞ that allows one to extrapolate results to $\Lambda_\infty \to \infty$. When this is done, the small binding energy differences that result at $\Lambda_\infty = 140$ for a reasonable choices of b all but disappear. We will return to this point in Section 2.) Fig. 3 shows the rate of convergence as a function of b and Λ_∞. The rate of convergence for ^{3}He is similar to that for the deuteron: a $\sim$ 60 keV energy accuracy is achieved at a Λ_∞ of $\sim$ 50.

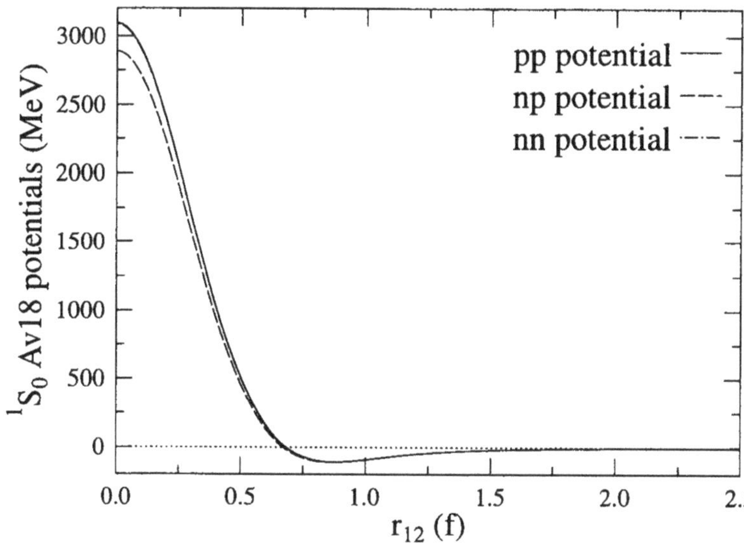

Figure 2: The Argonne $v18$ $^1S_0$ potentials.

The binding energies and operator matrix elements for simple systems like ^{3}He can, of course, be calculated by other methods, e.g., Faddeev techniques or Green's function Monte Carlo. We thus want to stress that the point of the following discussion is to do analogous calculations in the context of an effective theory, so that we begin to see the shortcomings of conventional techniques like the shell model as well as possibilities for overcoming those shortcomings. A first test of the techniques outlined above is to solve the Bloch-Horowitz equation for some SM-like space to then see if the resulting self-consistent energy is, indeed, the correct value. For model spaces of 2, 4, 6, and $8\hbar\omega$ in the case of the deuteron we obtained a binding energy of -2.224 MeV (using $\sqrt{2}b = 1.6f$ and $\Lambda_\infty = 140$). The exact result is -2.2246 MeV. Similar agreement was obtained for ^{3}He.

More interesting is the evolution of the wave functions, shown in Tables 1

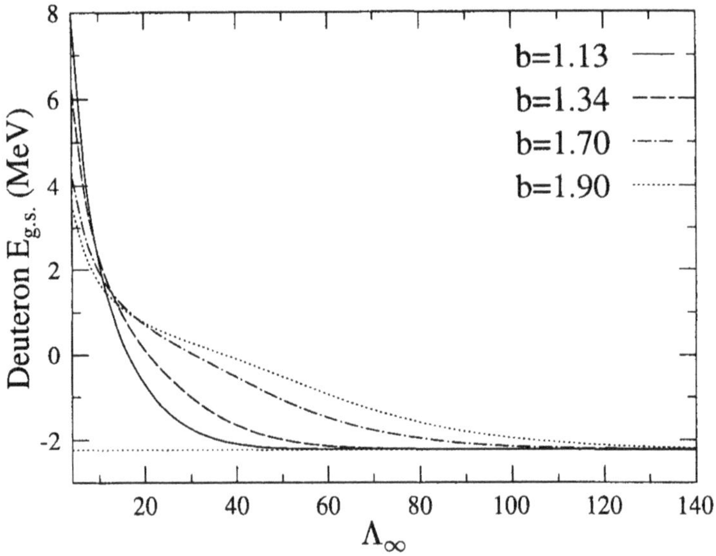

Figure 3: Deuteron ground state energy convergence as a function of Λ_∞ for several choices of the oscillator parameter b. This b is defined as in the independent-particle SM: the corresponding b for relative motion is $\sqrt{2}$ times the values shown. It is clear that if an inappropriate size is chosen for the basis states (e.g., b=1.9), the rate of convergence can be greatly slowed.

and 2. Various ET calculations were done in small model spaces, analogous to shell model spaces, consisting of all $2\hbar\omega$ configurations, all $4\hbar\omega$ configurations, etc. For each such space we then solved the Bloch-Horowitz equation via the Lanczos Green's function method described above, iterating the shell model calculation until the self-consistent energy is fully converged. The deuteron and ^{3}He calculations involve two- and three-body ladder sums in the excluded space, yielding sets of two- and three-body "SM" matrix elements of H^{eff} for the model spaces. The deuteron calculation is rather trivial; for $\Lambda_\infty \sim 50$ the ^{3}He BH calculation involves a dense matrix of dimension $\sim 10^4$, still rather modest by current SM standards. The matrix is dense because we work in relative Jacobi coordinates, rather than an m-scheme, utilizing standard Talmi-Brody-Moshinshy methods [13]. (See Ref. [14] for details.) The results in Table 2 were obtained with approximately 100 Lanczos iterations: it is apparent that the convergence is then quite good. The wave functions must be normalized according to Eq. (11): this involves calculating unity as an effective operator.

Table 1: ET results for the deuteron ground state wave function calculated with the Argonne $v18$ potential. The columns on the right correspond to different choices of the ET model space, the analog of a SM space. The rows correspond to the resulting amplitudes for the designated, selected configurations $\mid n, {}^SL_J\rangle$. The quantities within the parentheses are the square of the norm of the effective wave function, e.g., the probability that the deuteron resides in the corresponding "SM" space.

basis state	amplitude					
	$0\hbar\omega$	$2\hbar\omega$	$4\hbar\omega$	$6\hbar\omega$	$8\ \hbar\omega$	exact
	(65.9%)	(79.5%)	(86.1%)	(91.3%)	(93.0%)	(100%)
$\mid 1, {}^3S_1\rangle$	0.81155	0.81154	0.81155	0.81155	0.81152	0.81155
$\mid 2, {}^3S_1\rangle$	0.00000	-0.31483	-0.31483	-0.31483	-0.31482	-0.31483
$\mid 1, {}^3D_1\rangle$	0.00000	0.19524	0.19524	0.19524	0.19523	0.19524
$\mid 3, {}^3S_1\rangle$	0.00000	0.00000	0.24945	0.24945	0.24944	0.24945
$\mid 4, {}^3S_1\rangle$	0.00000	0.00000	0.00000	-0.20851	-0.20850	-0.20851
$\mid 5, {}^3S_1\rangle$	0.00000	0.00000	0.00000	0.00000	0.12596	0.12596

We will return to this point below.

The tables show the lovely evolution of the wave function in ET, an evolution quite unlike that of typical shell model calculations. The wave functions obtained in different model spaces agree over overlapping parts of their Hilbert spaces. Thus as one proceeds through $2\hbar\omega$, $4\hbar\omega$, $6\hbar\omega$, ... calculations, the ET wave function evolves only by adding new components in the expanded space. The normalization of the wave function grows accordingly. Thus, for ^{3}He, the $0\hbar\omega$ ET calculation contains 0.311 of the full wave function in the effective space; the $0+2+4\hbar\omega$ result is 0.700.

This evolution will not arise in the standard SM because the wave function normalization is set to unity regardless of the model space. It will also not arise for a second reason, illustrated in Table 3. The matrix elements of H^{eff} are crucially dependent on the model space: the listed results for ^{3}He show that a typical matrix $\langle\alpha|H^{eff}|\beta\rangle$ changes very rapidly under modest expansions of the model space, e.g., from $2\hbar\omega$ to $4\hbar\omega$. Yet it is common practice in the shell model to expand calculations by simply adding to an existing SM Hamiltonian new interactions that will mix in additional shells. We suspect the behavior found for ^{3}He is generic in ET calculations: it arises because a substantial fraction of the wave function lies near but outside the model space (e.g., see Table 2). An expansion of the model space changes the energy denominators for coupling to some of these configurations, and moves other nearby configurations from the excluded space to the model space. Naively, relative changes in effective interaction matrix elements of unity are expected.

Now we turn to the question of operators. The standard procedure in the SM is to calculate nuclear form factors with bare operators, or perhaps with bare operators renormalized according to effective charges determined phenomenologically at $q^2 = 0$, using SM wave functions normed to 1. As we now have a series of exact effective interactions corresponding to different model spaces, we can test the validity of this approach. The results for the elastic magnetic form factors for the deuteron and ^{3}He are shown in Figs. 4 and 5. One sees in each case that by the time one reaches a momentum transfer $q \sim 2.5/f$, random numbers are being generated: bare operators used in conjunction with exact effective wave functions generate results that differ by an order of magnitude, depending on the choice of the model space. This is not surprising, of course. If one considers the operation

$$e^{i\vec{q}\cdot\vec{r}}\sigma\tau_3|g.s.\rangle \tag{16}$$

at momentum transfers $\gtrsim 2k_F$, where k_F is the Fermi momentum, most of the resulting amplitude should reside outside the long-wavelength model space, in any simple view of the nucleus. That is, the strength resides entirely in the effective contributions to the operator. If these components are ignored, the results have to be in error.

Clearly the effective interaction and effective operator have to be treated consistently and on the same footing. If $\hat{O}$ is the bare operator, one finds

$$\langle\Psi_f|\hat{O}|\Psi_i\rangle \equiv \langle\Psi_f^{eff}|\hat{O}^{eff}|\Psi_i^{eff}\rangle \tag{17}$$

where

$$\hat{O}^{eff} = (1 + HQ_{SM}\frac{1}{E_f - HQ_{SM}})\hat{O}(1 + \frac{1}{E_i - Q_{SM}H}Q_{SM}H) \tag{18}$$

and where the effective wave function normalization of $|\Psi_i^{eff}\rangle$ and $|\Psi_f^{eff}\rangle$, mentioned earlier, must be determined using the effective operator $\hat{1}$, e.g.,

$$1 = \langle\Psi_i|\Psi_i\rangle = \langle\Psi_i^{eff}|(1+HQ_{SM}\frac{1}{E_i - HQ_{SM}})(1+\frac{1}{E_i - Q_{SM}H}Q_{SM}H)|\Psi_i^{eff}\rangle \tag{19}$$

These expressions can be evaluated with the Lanczos Green's function methods described earlier. When this is done, all of the effective calculations, regardless of the choice of the model space, yield the same result, given by the solid lines in Figs. 4 and 5.

We would argue, based on this example, that many persistent problems in nuclear physics — ranging from the renormalization of g_A in β decay [15] to

Table 2: As in Table 1, only for ^{3}He. The basis states are now designated somewhat schematically as $| N, \alpha\rangle$, where N is the total number of oscillator quanta and α is an index representing all other quantum numbers.

state	amplitude					
	$0\hbar\omega$ (31.1%)	$2\hbar\omega$ (57.4%)	$4\hbar\omega$ (70.0%)	$6\hbar\omega$ (79.8%)	$8\ \hbar\omega$ (85.5%)	exact (100%)
$\|0,1\rangle$	0.55791	0.55791	0.55791	0.55795	0.55791	0.55793
$\|2,1\rangle$	0.00000	0.04631	0.04613	0.04618	0.04622	0.04631
$\|2,2\rangle$	0.00000	-0.48255	-0.48237	-0.48243	-0.48243	-0.48257
$\|2,3\rangle$	0.00000	0.00729	0.00731	0.00730	0.00729	0.00729
$\|2,4\rangle$	0.00000	0.16707	0.16698	0.16706	0.16706	0.16708
$\|2,5\rangle$	0.00000	0.00566	0.00564	0.00565	0.00565	0.00566
$\|2,6\rangle$	0.00000	-0.00017	-0.00017	-0.00017	-0.00017	-0.00017
$\|4,1\rangle$	0.00000	0.00000	-0.02040	-0.02042	-0.02043	-0.02047
$\|4,2\rangle$	0.00000	0.00000	0.11267	0.11274	0.11275	0.11289
$\|4,3\rangle$	0.00000	0.00000	-0.04191	-0.04199	-0.04208	-0.04228
$\|4,4\rangle$	0.00000	0.00000	0.28967	0.28978	0.28978	0.29001
$\|4,5\rangle$	0.00000	0.00000	0.01059	0.01059	0.01059	0.01059
$\|4,6\rangle$	0.00000	0.00000	-0.00213	-0.00212	-0.00211	-0.00210
$\|4,7\rangle$	0.00000	0.00000	0.00998	0.01000	0.01000	0.01000
$\|4,8\rangle$	0.00000	0.00000	-0.11319	-0.11327	-0.11330	-0.11335
$\|4,9\rangle$	0.00000	0.00000	0.08446	0.08447	0.08446	0.08448
$\|4,10\rangle$	0.00000	0.00000	-0.08613	-0.08626	-0.08632	-0.08638
$\|4,11\rangle$	0.00000	0.00000	-0.00210	-0.00211	-0.00211	-0.00211
$\|4,12\rangle$	0.00000	0.00000	-0.00252	-0.00254	-0.00256	-0.00257
$\|4,13\rangle$	0.00000	0.00000	-0.00020	-0.00020	-0.00020	-0.00020
$\|4,14\rangle$	0.00000	0.00000	-0.00010	-0.00010	-0.00010	-0.00010
$\|4,15\rangle$	0.00000	0.00000	-0.00012	-0.00013	-0.00012	-0.00012

Table 3: Selected BH 3-body effective interaction matrix elements for ^{3}He, in MeV, illustrating the strong dependence on the "SM" space. The Argonne v18 potential was used.

	$2\hbar\omega$	$4\hbar\omega$	$6\hbar\omega$	$8\ \hbar\omega$
$\langle 0,1 \| H^{eff} \| 2,1\rangle$	-4.874	-3.165	-0.449	1.279
$\langle 0,1 \| H^{eff} \| 2,5\rangle$	-0.897	-1.590	-1.893	-2.208
$\langle 2,1 \| H^{eff} \| 2,2\rangle$	6.548	-2.534	-4.144	-5.060

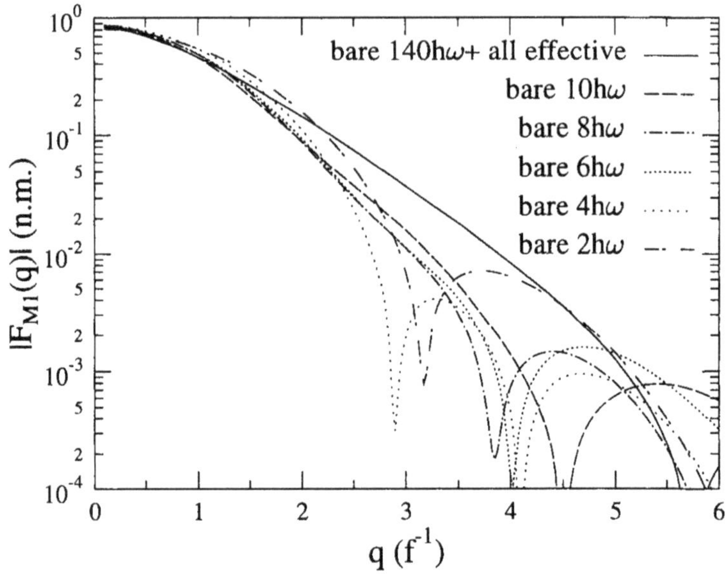

Figure 4: The magnetic elastic form factor for the deuteron calculated with the exact H^{eff}, SM wave functions normalized to unity, and a bare operator are compared to the exact result (solid line). When effective operators and the proper wave function normalizations are used, all results become identical to the solid line.

the systematic differences between measured and calculated M1 form factors at Bates [16] — very likely are due to our naive treatments of operators. It should be apparent from the above example that no amount of work on H^{eff} will help with this problem. What is necessary is a diagrammatic basis for generating H^{eff} that can be applied in exactly the same way to evolving $\hat{O}^{eff}$. From this perspective, phenomenological derivations of H^{eff} by fitting binding energies and other static properties of nuclei are not terribly helpful, unless one intends to simultaneously find phenomenological renormalizations for each desired operator in each q^2 range of interest.

3 Numerical Renormalization Group Speculations

While the Lanczos Green's function method is reasonably elegant, the calculations described above have a "brute force" aspect in that the high-momentum ladders must be summed to very large Λ_∞. In the of the deuteron we carried the sums to $\Lambda_\infty = 140$ to assure one keV accuracy. The calculation for ^{3}He was stopped at $\Lambda_\infty = 52$, with a resulting binding energy of -6.842 MeV. As our

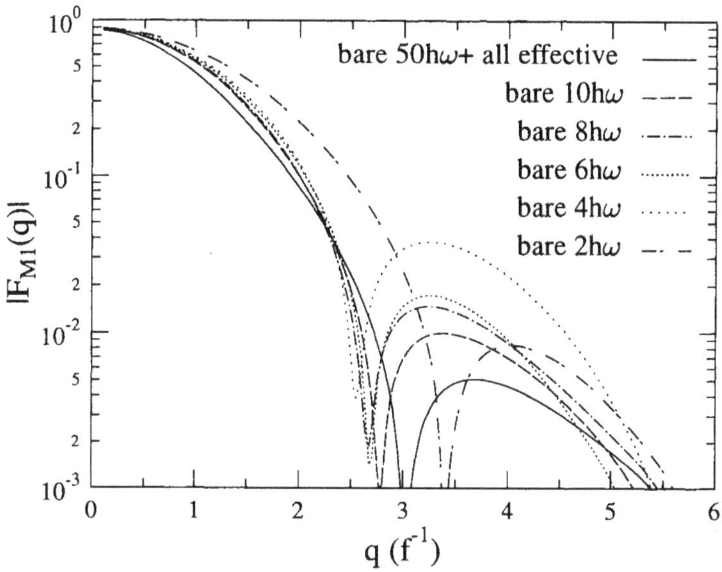

Figure 5: As in Fig. 4, only for ^3He.

calculations are otherwise exact, this is an upper bound: a variational principle operates as Λ_∞ is increased. The corresponding Green's function Monte Carlo result is -6.87 ± 0.03 MeV; Faddeev results for the v18 potential give -6.895 ± 0.001 MeV.

The ^{3}He calculations can (and will) be carried further: the existing matrices are only of dimension 12000. However, there are alternatives to the brute-force approach. One rather simple idea is to evaluate, instead of the three-body ladder, the difference of the three-body ladder and the appropriate sum over the three corresponding two-body ladders, evolving this difference with Λ_∞. The resulting H^{eff} then would be obtained by adding to this difference the two-body ladder. While this is a tautology, our expectation is that the difference will converge more quickly than the full 3-body ladder as a function of Λ_∞, since the difference would start with all pair wave functions being properly correlated. If one were able to truncate the 3-body difference calculation at a lower Λ_∞ than the corresponding 2-body ladders, considerable efficiency would be gained. This possibility will soon be studied.

Another possibility has to do with an observation about the way both ground-state energies and matrix elements of H^{eff} evolve with Λ_∞. Our calculations are based on the division of the Hilbert space into two sectors

$$0 \leq \Lambda \leq \Lambda_{SM} \qquad \Lambda_{SM} < \Lambda \leq \Lambda_{\infty} \tag{20}$$

While the deuterium calculation was done with a large value of Λ_∞, we can bring it to a much lower value in order to study the evolution with Λ_∞ (see Fig. 3). When the numerical results were examined, it was found that they scaled with Λ_∞ simply, approximately as an exponential in Λ_∞^2. For example, calculations were done for the Argonne $v18$ potential and b = 1.7f, with Λ_∞ = 46, 50, and 54, to which we fit the three-parameter functional form

$$E_{g.s.}(\Lambda_\infty) = -2.175\text{MeV} + 3.178\text{MeV}e^{-1.055(\Lambda_\infty/50)^2} \tag{21}$$

Thus -2.175 MeV is identified as the binding energy for $\Lambda_\infty \rightarrow \infty$, if this numerical renormalization group equation were exact. The deuteron binding energy at $\Lambda_\infty \sim 50$ is off by more than an MeV. Table 4 shows that 95% of this difference can indeed be anticipated by examining the running of the results for $\Lambda_\infty \sim 50$: the projected valued at $\Lambda_\infty = 140$ is correct to $\sim$ 30 keV. For $b = 1.13f$, the case converging most quickly in Fig. 3, a similar 30 keV accuracy is achieved for an extrapolation from still lower Λ_∞, 34-40. Numerically, such an extrapolation can be done quite efficiently: with proper coding, it costs relatively little "overhead" to obtain additional results in the neighborhood of some starting Λ_∞.

The corresponding exercise was done for ^{3}He using results for $\Lambda_\infty = 44$, 48, and 52. A very similar functional form is found

$$E_{g.s.}(\Lambda_\infty) = -6.906\text{MeV} + 0.403\text{MeV}e^{-1.57(\Lambda_\infty/48)^2} \tag{22}$$

While we do not have exact results similar to those in Table 4, the resulting asymptotic value of -6.906 MeV is in good agreement with the Faddeev value of 6.895 $\pm$ 0.001 MeV. Such 10 keV accuracy compares well with the Green's function Monte Carlo result, with its 30 keV accuracy.

Now all of this poses an interesting question: our "renomalization group" evolution is a nontrivial one, as it involves a truncation in harmonic oscillator energies for few-body Slater determinants. Yet it appears this truncation is leading to a very simple exponential convergence in results. This can be compared to effective field theory, where the truncation is made on the range of the potential and the convergence is a weaker $1/\Lambda_\infty$. Can one derive and therefore understand this attractive aspect of a harmonic oscillator mode expansion for bound states?

To attack this problem, we envision setting Λ_∞ to some intermediate scale $\sim$ 50 (around 750 MeV) — similar to the numerical examples above — and

Table 4: A "numerical renormalization group" study of the running of the deuteron binding energy with Λ_∞. The results in the middle column are exact. Those in the right column were obtained from Eq. (21), determined from the variation of the $E_{g.s.}(\Lambda_\infty)$ near $\Lambda_\infty = 50$. An oscillator parameter is $b = 1.7f$, chosen so that this is a slowly converging case requiring significant extrapolation. The potential used is Argonne $v18$.

Λ_∞	$E_{g.s.}^{calculated}$ (MeV)	$E_{g.s.}^{projected}$ (MeV)
46	-0.874	fit
50	-1.069	fit
54	-1.247	fit
60	-1.476	-1.480
70	-1.771	-1.773
80	-1.961	-1.962
90	-2.077	-2.071
100	-2.143	-2.128
110	-2.179	-2.156
120	-2.196	-2.167
130	-2.204	-2.172
140	-2.207	-2.174

performing, prior to our SM effective theory, a preliminary integration over modes above this scale. If we write

$$H \equiv H_0 + H' \qquad \text{and} \qquad \mathrm{H}^{\mathrm{eff}} = \mathrm{H}_0 + \mathrm{H}'^{\mathrm{eff}} \tag{23}$$

where H_0 is the harmonic oscillator Hamiltonian and H^{eff} is the appropriate effective interaction to use below Λ_∞, then we find

$$H'^{eff} = H' + H' G^0_{\Lambda_\infty} H'^{eff} \tag{24}$$

where $G^0_{\Lambda_\infty}$ is the harmonic oscillator Green's function corresponding to high-momentum scattering, e.g.,

$$\begin{aligned} G^0_{\Lambda_\infty}(\vec{r}_1, \vec{r}_2) &= \sum_{\Lambda > \Lambda_\infty} \frac{|\Psi^{HO}_\Lambda(\vec{r}_1)\rangle\langle\Psi^{HO}_\Lambda(\vec{r}_2)|}{E - E_\Lambda} \\ &= G^0(\vec{r}_1, \vec{r}_2) - \sum_{\Lambda \le \Lambda_\infty} \frac{|\Psi^{HO}_\Lambda(\vec{r}_1)\rangle\langle\Psi^{HO}_\Lambda(\vec{r}_2)|}{E - E_\Lambda} \end{aligned} \tag{25}$$

In Eq. (24) the high-momentum ladder sum involves terms of the form

$$\Psi^{f*}_{SM}(\vec{r}_1) H'(\vec{r}_1) G^0_{\Lambda_\infty}(\vec{r}_1, \vec{r}_2) H'(\vec{r}_2) \Psi^i_{SM}(\vec{r}_2) \tag{26}$$

The harmonic oscillator Green's function has the attractive property that it can be decomposed as

$$G^0_{\Lambda_\infty} = G^0_{\Lambda_\infty}(r_+, r_-) \tag{27}$$

where $r_+ = |\vec{r}_1 + \vec{r}_2|$ and $r_- = |\vec{r}_1 - \vec{r}_2|$. Now the restriction to large Λ corresponds to short times and thus to short propagation distances. Thus propagation lengths should shorten as Λ_∞ is increased. Furthermore the overall behavior of the Green's function is governed by the factor $e^{-r_+ r_-}$. Thus the propagator is most contracted along the *smaller* of these two coordinates, $r_< = min(r_+, r_-)$. As the ladder will be evaluated between long-wavelength SM states, these observations suggest expanding $\Psi^{f*}_{SM}(\vec{r}_1)H'(\vec{r}_1)$ and $H'(\vec{r}_2)\Psi^{i}_{SM}(\vec{r}_2)$ in Eq. (26) around $r_>$ in a Taylor series in $r_<$. (There is a more pedagogical discussion of this technique and its relevance to effective field theory in Ref. [17].)

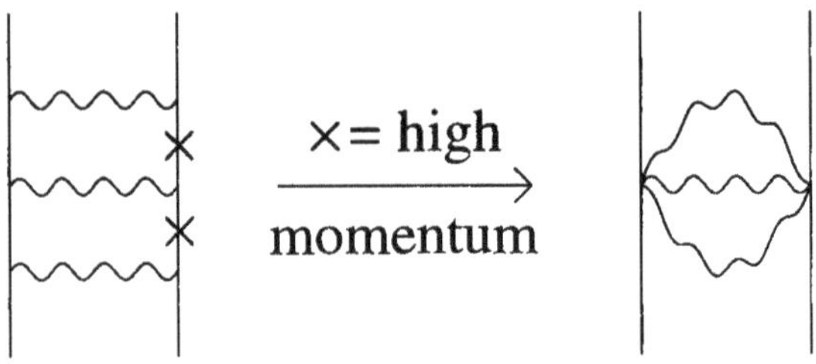

Figure 6: Contraction of a ladder diagram involving only very high momentum excluded states to a local operator.

After a bit of algebra, the net result is the contraction of the ladder sum to a local "rainbow" diagram, as illustrated in Fig. 6. The contracted Green's function has a form familiar to most effective field theorists,

$$G_{\Lambda_\infty} \to g^0(r_>) + \vec{\nabla}_{r_>} g^2(r_>) \vec{\nabla}_{r_>} + \ldots \tag{28}$$

where this is to be inserted between wave functions evaluated at $r_>$. The function $g^0(r_>)$ is obtained by integrating $G_{\Lambda_\infty}(r_+, r_-)$ over $r_<$, while $g^2(r_>)$ involves a second moment in the small coordinate $r_<$, etc. Effectively this sums the ladder according to a local density approximation (g^0), in an approximation that also takes into account the gradient of the local density (g^2), etc.

One of the most attractive features of the harmonic oscillator is that the full Green's function was recently derived in closed form [18], so that $g^0(r_>)$ can be calculated by integrating the last expression in Eq. (25): an infinite sum over high-momentum modes can be replaced by a finite sum over low-momentum modes. The full Green's function is shown in Fig. 7, and the integration that produces g^0 is depicted in Fig. 8. The resulting g^0 for $\Lambda_\infty = 30$ and 50 are shown in Figs. 9 and 10 (dashed lines) along with a similar contraction of the full Green's function G^0. One finds that the dashed lines – the residual contribution of very high momentum excitations – rapidly oscillates, with a frequency very close to Λ_∞. Thus a harmonic oscillator expansion converges not because every point in coordinate space is well represented by the included states, but because integration of a product of such oscillations with a smooth Hamiltonian yields a small remainder. Note that increasing Λ_∞ also extends the range in $r_>$ spanned by the harmonic oscillator basis functions: extended wave functions require more work.

We are now in the process of implementing this local approximation to see how far the integration scale Λ_∞ can be lowered without loss of accuracy. Unlike effective field theory treatments, our Taylor series expansion involves a known potential and can be carried out term by term until exhaustion sets in. The results may provide some estimate of the scales that effective field theories need to reach to become very accurate.

We will close with two speculations. If one folds the dashed-line functions of Figs. 9 and 10 with the potential of Fig. 2, the result is not dissimilar to a $j_0(\Lambda r_>)$. Such a function, integrated with a product of harmonic oscillator wave functions, yields a factor $e^{-a\Lambda^2}$. Thus we are hopeful that this line of work may indeed explain the numerical scaling we have previously discussed. Our goal is to add to Table 4 a fourth column in which our phenomenological scaling function is replaced by a derived one that yields improved results: the "break" seen in Table 4 at $\Lambda_\infty \sim 100$ (a scale characteristic of the Argonne $v18$ hard core) suggests there is a bit of physics missing from our phenomenological function. Second, this suggests that the rapid convergence of the harmonic oscillator mode expansion, as compared to effective field theory scaling of $1/\Lambda_\infty$ which results from absorbing short-range contributions to the NN potential into contact terms, may be because it offers a better compromise between coordinate and momentum space. Hopefully we will be able to be more specific soon.

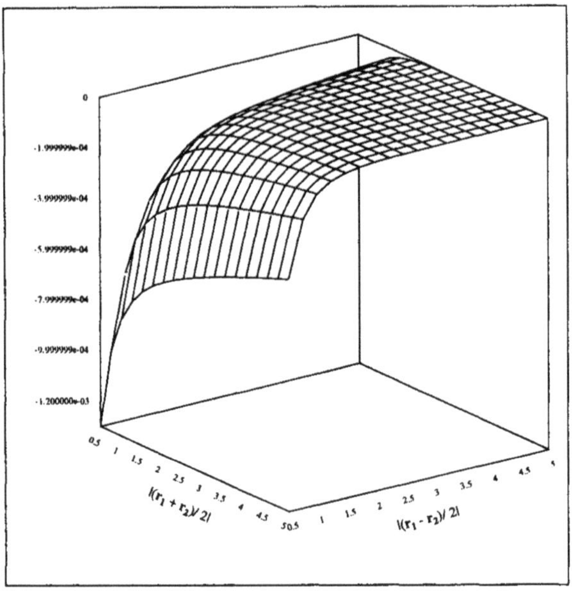

Figure 7: The full (summed over all modes) harmonic oscillator Green's function. For parity conserving interactions, only the even or odd projections are needed. The even projection is shown.

4 Outlook and Future Plans

The ability to efficiently solve ^{3}He as an effective theory in a SM-like space is, in itself, quite significant: this means it is relatively straightforward to execute a faithful BH treatment of heavier nuclei through order ρ in both the effective interaction and effective operators. The numerical effort is comparable.

The example of ^{3}He also suggested that many of the uncontrolled approximations made in the SM cannot be justified. Recently sophisticated numerical advances have been made in SM calculations. We have argued that it may be equally important to train this numerical power on the shakey foundations of that model. A specific example is our use of the Lanczos algorithm – a favorite among SM practitioners – to generate the Green's function, leading to a tractable strategy for solving the BH equation self-consistently.

We believe that, if the program in Section 3 has modest success, we should

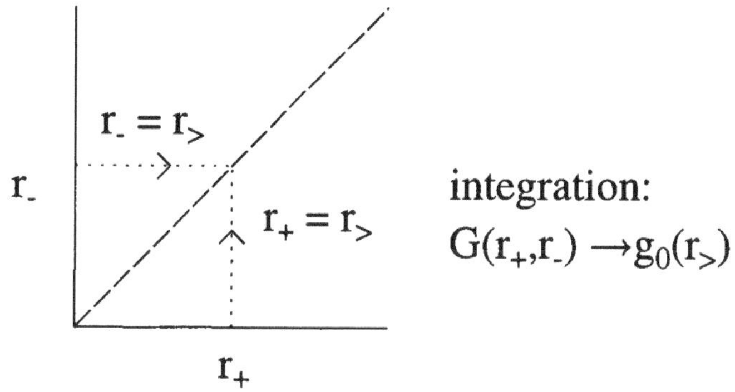

Figure 8: The integration path in the (r_+, r_-) plane that is used to contract $G(r_+, r_-)$ to $g_0(r_>)$.

be able to handle ^{4}He with an accuracy similar to that achieved already for ^{3}He. That would imply that it is now possible to handle effective interactions and operators in finite nuclei through order ρ^2. At that point the exciting challenge will be to determine where convergence in the cluster expansion is being approached.

We would like to think that the work reported here will form a bridge between the ideas growing out of effective field theories, and the successful phenomenology we have achieved in traditional nuclear physics (particularly in modeling the NN potential and in solving few-body systems with nearly exact methods). It may also stimulate effective field theory, which has not yet had an impact on nuclear problems beyond NN and three-body systems. It is not inconceivable that a marriage of EFT – which strives to handle few-body problems rigorously, including relativity, fundamental few-body forces, etc. — with some kind of cluster expansion of the Brueckner type might someday yield a controlled theory of finite nuclei. Furthermore, the attractive exponential convergence of our harmonic oscillator mode expansion is also EFT "food for thought."

Acknowledgments

This work was supported in part by the Division of Nuclear Physics, US Department of Energy.

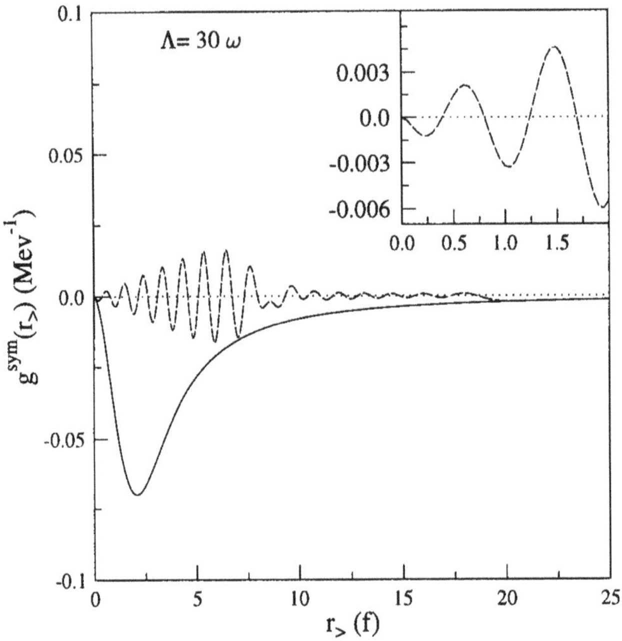

Figure 9: The dashed line is the $g_0(r_>)$ that results when excitations $\Lambda > 30$ are summed. The solid line is the corresponding result for the full Green's function. It is the rapid oscillation of the dashed residual – rather than the accuracy of the harmonic oscillator expansion at all values of $r_>$ – that is responsible for the convergence. The oscillations are very regular and quite close in frequency to $\Lambda \sim 30$.

References

1. E. Caurier, G. Martinez-Pinedo, F. Nowacki, A. Poves, J. Retamosa, and A. P. Zuker, *Phys. Rev.* C **59**, 2033 (1999) and references therein.
2. P. Navratil and B. R. Barrett, *Phys. Rev.* C **57**, 562 (1998) and *Phys. Rev.* C **59**, 1906 (1999).
3. S. E. Koonin, D. J. Dean, and K. Langanke, *Phys. Rep.* **278**, 1 (1997) and references therein.
4. T. Otsuka and T. Mizusaki, *J. Phys.* G **25**, 699 (1999).
5. D. C. Zheng, B. R. Barrett, J. P. Vary, W. C. Haxton, and C.-L. Song, *Phys. Rev.* C **52**, 2488 (1995).
6. R. B. Wiringa, V. G. J. Stoks, and R. Schiavilla, *Phys. Rev.* C **51**, 38 (1995).
7. R. V. Reid, Jr., *Ann. Phys. (N.Y.)* **50**, 411 (1968).

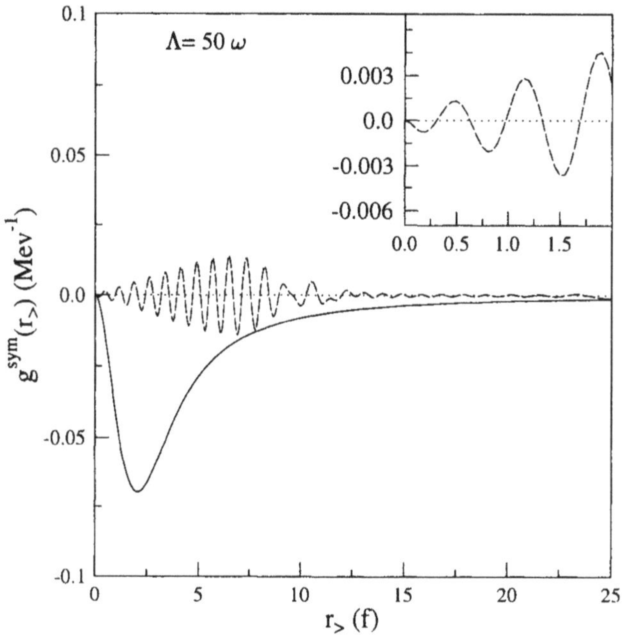

Figure 10: As in Fig. 9, only for $\Lambda > 50$.

8. C. Bloch and J. Horowitz, *Nucl. Phys.* **8**, 91 (1958).
9. B. R. Barrett and M. W. Kirson, in *Advances in Nuclear Physics*, ed. M. Baranger and E. Vogt, vol. 6 (Plenum, New York, 1973) p. 219.
10. T. T. S. Kuo and E. Osnes, *Lecture Notes in Physics*, ed. H. Araki et al., vol. 364 (Springer-Verlag, Berlin, 1990).
11. T. H. Shucan and H. A. Weidenmuller, *Ann. Phys. (N.Y.)* **73**, 108 (1972) and **76**, 483 (1973).
12. R. Haydock, *J. Phys.* A **7**, 2120 (1974).
13. M. Moshinsky, *The Harmonic Oscillator in Modern Physics: From Atoms to Quarks* (Gordon and Breach, New York, 1969).
14. C.-L. Song, *An Improved Procedure for Calculating Effective Interactions and Operators*, Univ. Washington Ph.D. thesis, 1998.
15. I. S. Towner, *Phys. Rep.* **155**, 263 (1987); B. A. Brown, *Nucl. Phys.* A **522**, 221c (1991); G. Martinez-Pinedo, A. Poves, E. Caurier, and A. P. Zuker, *Phys. Rev.* C **53**, 2602 (1996).
16. R. S. Hicks et al., *Phys. Rev. Lett.* **60**, 905 (1988) and references therein.

17. P. Lepage, "How to Renormalize the Schrodinger Equation", `nucl-th/9706029`.
18. D. B. Khrebtukov and J. H. Macek, *J. Phys.* A **31**, 2853 (1998).

Discussion during W. Haxton's presentation

Chair: *M. Birse (University of Manchester)*
Notes taken by: *M. N. Butler (Saint Mary's University)*

Aurel Bulgac disagreed with Haxton's statement that the range was not well defined in shell-model calculations. Haxton responded that this issue is explicitly linked to high momentum states, so that where the model space was truncated was key to the range. This sensitivity leads to the range being ill-defined. Jerry Miller noted that one could tie the picture to Brueckner theory and use the shell model to handle physics near the Fermi surface.

Ulf Meißner asked whether it was necessary to use the Bloch-Horowitz method, to which Haxton responded yes, in the sense that it is an easier formalism in which to use a real potential. David Kaplan asked whether quenchings were caused by the fact that there was only a small fraction of the wavefunction in the shell model space. The response was a very succint 'yes'.

Jim Steele asked about the nature of ρa^3, a being the scattering length, as a dimensionless parameter in the shell model calculation. Haxton responded that he did not know, but that it was an interesting problem. What one wants to do is test cluster convergence, i.e. look at dynamics that require large clusters versus those that require a small cluster response.

SESSION 6

Session Chair: *R. Seki (CSU Northridge and Caltech)*
Notes taken by: *H. Grießhammer (University of Washington)*

OPEN DISCUSSION

R. SEKI
Department of Physics and Astronomy, California State University, Northridge
Northridge, CA 91330, USA
and
W. K. Kellogg Radiation Laboratory, 106-38, California Institute of Technology,
Pasadena, CA 91125, USA
Email: rseki@krl.caltech.edu

H. GRIEßHAMMER
Nuclear Theory Group, Department of Physics, University of Washington,
Box 351560, Seattle, WA 98195-1560, USA
Email: hgrie@redhook.phys.washington.edu

Summary of the open discussion at the end of the second day of the workshop. (The chronological sequence of discussions was slightly altered for a coherent presentation.)

Seki: Before we start, I would like to extend the cordial thanks of the organizers and of all participants to Wick Haxton and the Institute of Nuclear Theory staff for their generous hospitality, without which it would have been very hard to organize this workshop.

I want to start off this discussion by reviewing last year's final session, so that we know where we stand and where to go from here. Last year, we concentrated on the question whether cutoff or dimensional regularization is the more feasible way to go, and also on the perhaps more important question of power counting. I guess we all agree that dimensional regularization in the way that Kaplan, Savage, and Wise (KSW) proposed is a viable way to formulate nuclear effective field theory. When I look at last year's shopping list, there are only a few (but essential) points which remain unanswered. Tom Cohen's suggestion of analyzing a toy model to see how the power divergence subtraction (PDS) scheme works was taken up by David Kaplan[1] and Gautam Rupak[2]. Paulo Bedaque's remark of last year was well taken on an examination of the observables that are either independent of, or only weakly dependent on the EFT counterterms. And, so was Mark Wise's call for explicit calculations. Let me repeat here the old laundry list of priority calculations and what have been done on them.

- Deuteron form factors (electromagnetic and axial): They were computed to next-to-leading-order (NLO) using the KSW power counting with pions and to next-to-next-to-leading-order (NNLO) in the theory without pions[3].

- Deuteron polarizabilities and Compton scattering: They were also computed to NLO in the KSW approach with pions [3] and in Weinberg's approach[4].

- Inelastic reactions involving photons: $np \to d\gamma$ both parity conserving and parity violating was computed to NLO using KSW [3], and in Weinberg's approach [5]. Pion photo-production off the deuteron was worked out using the Weinberg counting[6].

- Three-body systems: Nucleon-deuteron scattering has been treated using a cutoff regularization[7,8].

- Spin observables: Tensor-polarized Compton scattering from deuteron was computed in the KSW approach[9]. Still, work remains to be done, for example, on the A_y puzzle in the three-nucleon system, which – as Jim Friar said last year – is an intriguing topic.

- Nuclear matter: Apart from an exception[10], nuclear matter has not been addressed at this workshop. A lot remains to be done here.

- Matching to QCD: How to match the effective theory to QCD was also not discussed during this workshop.

I would like to divide our future plans into two categories: The first is open issues that still remain, requiring to be cleaned up (perhaps among us as "in-house"), and the second is new calculations that would make an impact in the community as a whole.

The nucleon-nucleon (NN) phase shifts seem to be basically understood[2,11], apart from the important pion issue. Michael Birse gave an interesting interpretation of the KSW versus the Weinberg approach from the renormalization group point of view in yesterday's discussion. We have to see how useful consequences such arguments lead us. We also have to agree whether to include pions non-perturbatively as Weinberg suggests, or perturbatively as KSW want. A related question is where we expect the KSW treatment to break down: For Λ_{NN}, we heard numbers from 300MeV to 500MeV at this workshop. I don't have to emphasize that the question of how far we can push the EFT is crucial both for pion production processes, in which the average momentum is at least $\sqrt{M_N m_\pi} \sim 360$MeV, and for nuclear matter , in which the Fermi momentum is $p_F \sim 280$MeV. But how can we settle this issue? Error analysis in phase shifts at higher order may be a good way and will also settle the issue of convergence [1,2,11,12,13]. This brings me to the next point: We agreed that more calculations have to be done to NNLO including pions.

I stress that in all these, establishing benchmarks is crucial. As Mark Wise remarked, nuclear physics is an old subject with much data and much time to fine-tune models and predictions. We therefore have to choose the calculations wisely to make an impact. Many expressed the feeling that in general, all existing calculations can be pushed one order higher: NNLO with pions gives about 3% accuracy, and next-to-next-to-next-to-leading order (NNNLO) without pions even 1%. In the two-body system, in addition to the calculations I previously noted, Compton scattering off the deuteron as well as πd processes[3,4,6] are very useful. Mannque Rho[5] challenged us to predict spin observables of polarized targets like $\vec{n}+\vec{p} \rightarrow d+\gamma$. Ulf Meißner's examination[14] of charge independence and charge symmetry breaking in NN interactions should be followed up by investigations of symmetry breaking in other processes.

Calculations are also in progress for three-body systems by Bedaque *et. al.* Three-body forces in EFT depend on the scale or resolution under consideration, and their enhancement in the calculation is explainable by a renormalization-group flow [7,8]. Clearly, three-body forces depend on the two-body forces used, as it was commented by Robert Perry. Could we have a more concrete handle on this elusive, but important issue? And how about four-body forces in multi-nucleon systems [7]? Nuclear matter has begun to be investigated[10] in the EFT approach, and for finite nuclei EFT ideas are shown to be promising[15]. Understanding of these systems in EFT would be our ultimate objective, but we have much to do before achieving it.

What else can we think of? What can we do in the two-body sector? For example, the $\gamma d \rightarrow \pi^0 d$ calculation [6] was a successful prediction by EFT. I believe something in the same spirit can be done with polarized targets.

van Kolck: Yes, that's only a question of manpower. I am not doing it.

Wise: In general, a couple of percent accuracy is possible, especially without pions. On the other hand, some things might not converge too well. And the number of diagrams just becomes too big after a few orders to suggest the series should converge to arbitrary accuracy. After all, we know the series is only asymptotic.

Savage: Especially when more and more counterterms occur at higher orders. At some point, there will be more of them than observables. But let me come back to the question: What can we beat on and hand to the experimentalists? Mannque Rho's bet gives us the chance to compare potential models and effective field theory both in the Weinberg and KSW approach in a clear, unbiased setting: Nobody knows the answer yet.

Rho: Only the largest term is available in potential models, and the result is not of very high accuracy because usually potential modelists perform their calculations after the data is available.
Savage: TUNL-FEL at Duke will be on line soon with a vector-polarized deuterium target for Compton scattering spin observables. What is the status of neutrino deuteron scattering and breakup?
Butler: We are pretty much done with a calculation which should be accurate to 10%. Unfortunately, there is one counterterm which we hope to fix from the parity violating part of the deuteron breakup, $\vec{e}d \rightarrow enp$, once data is available there. The project is very interesting because SNO relies only on theoretical predictions of the deuteron breakup.
Haxton: But potential models have an uncertainty of only a few percent there.
Meißner: Potential models cannot get the correct exchange currents in $\gamma d \rightarrow \pi^0 d$. Our key point is that we have to be consistent to include pion loops in the single as well as in the two nucleon system. Our chiral perturbation theory Lagrangian tells us that and how to do it. We could show that the corrections to the large leading exchange currents in $\gamma d \rightarrow \pi^0 d$ are small, which allows to pin down the neutron electric dipole amplitude from a measurement on deuterium. There are certainly a couple of other places where one can play the same trick. And we have control over the error because we do a systematic calculation. Potential models cannot do that.
Haxton: I believe we must be a bit more careful: Potential model meson-exchange currents are fixed to data - triton beta decay - and then applied to the proton-proton reaction. So, it's some kind of effective theory. Why should that not work?
Savage: What about models of the electric quadrupole moment and form factors? The Argonne V18 potential gets that right only to 6 percent.
Rho: The leading exchange current contribution to *pp* fusion is about 2%, taking the Weinberg power counting. The NLO contribution can also be as much as 2%. It's not clear how to do that because of the counterterms.
Savage: It will require some effort to fix the counterterms, but it would be a valuable contribution. I would like to add the following to the shopping list:

- We need high precision two and three body calculations, as many as possible. Spin observables are very interesting. NNLO with pions and NNNLO without is doable and gives 3 and 1% accuracy.

- I want to understand how to regularize the three body problem in dimensional regularization because then we can easily incorporate pions.

- In the many body sector, the task is to win George Bertsch's challenge [16]: "What are the ground state properties of the many-body system composed of spin-1/2 fermions interacting via a zero-range, infinite scattering-length, contact interaction? You may assume the interaction has no two-body bound states. Also, the zero range is approached with finite ranged forces and finite particle number by first taking the range to zero and then the particle number to infinity."

van Kolck: What do you mean by "high precision"? Suppose we do well with the theory breaking down at 400MeV. I believe we cannot claim victory that easily. One has to show that one does better than the effective range expansion.

Mehen: All deuteron calculations to NNLO will show that effective range expansion just fails.

Lepage: I am looking forward to next summer's workshop on effective field theories here in Seattle. We will meet with the experimentalists to see what should be calculated and what the actual error bars on their experiments are. Usually, this is a highly non-trivial issue. One should maybe focus on calculating cross sections because the error analysis of experimental cross sections is in general less subtle than the error analysis of phase shifts.

Savage: The electric form factor of the deuteron is known very well, the charge radius is measured to 2 or 3 decimal places. One photon and multiphoton exchanges are very different in the $p\gamma$ system and difficult. Now we can do this in the $d\gamma$ system. Is there a chance to relate the two?

Meißner: This only makes sense if the $d\gamma$ calculation is much simpler and much more precise, and if the one photon exchange is determined very precisely from experiment.

Savage: Two photon exchange is the biggest unknown correction to the charge radius.

Miller: That's known to be small. The Russians showed that two photon contributions are less than the experimental accuracy.

Rho: When one looks at potential models, electro-disintegration of the deuteron is very sensitive to the interference of s and d wave parts in the deuteron wave function. It is also very interesting. These guys can predict it up to momenta of 2GeV. One should do this process using KSW.

All: That's well above all ...

Savage: But it's straightforward for us in our range of validity, up to say 300MeV. I also believe the three-body program should be pushed to its limits: Higher order calculations, photon probes, adding external currents, ... We also have to understand the renormalization group behavior of the three body force.

It is very important to use dimensional regularization in order to understand the wild variation of the cutoff dependent three body force in the triton channel.

Wise: What data is there in which potential models report high uncertainties?

Bedaque: As Jim Friar said last year, spin observables are in general the place to look. For example, in *pd* scattering, potential models fail by a factor two, that's the A_y puzzle. They are off whether they include a three body force or not. But what if we just encounter a bunch of counterterms?

van Kolck: People claim that Friar's problem is very sensitive to *p*-wave physics in the two body system. This smells sub-dominant.

Savage: What are the plans on the three body problem?

Bedaque: There is more to be done than we can think of right now. Without pions, one can certainly go to one or two higher orders. The precision will of course depend on the process, but for example, the triton binding energy at NNLO will have only 1% uncertainty. Once we know how to do KSW in the three body system, we can just work the list of two body calculations and do the corresponding three-body stuff within a year. We are at the moment calculating the electromagnetic form factors, but it is again hard to beat potential model computations.

Savage: Still, one has to verify that one gets sensible results. Is there a way to systematically power count three body forces?

van Kolck: We get the triton binding energy to be very close to the actual value already at leading order, therefore effective range corrections seem not to require a new counterterm at NLO. Our conservative estimate is that counterterms are of naïve size.

Savage: What if the C_2 term behaves like the one for C_0, and is out of phase with it when one varies the cutoff?

van Kolck: Only convergence in observables is important. But as Paulo said at the end of his talk, we don't have renormalized equations for the three body problem and for the moment certainly have to live with an explicit cutoff.

Bedaque: The problem with dimensional regularization is that nothing seems to change: In the ultraviolet, the integral equation has a finite but oscillating solution, so that it seems one will end up in dimensional regularization with just the same problem as in cutoff regularization. Furthermore, the oscillatory behavior in the UV is found independent of the number of dimensions.

Gegelia: Have you included C_2 already to all orders?

van Kolck: Not together with the three body force, but all diagrams are finite and even more convergent than at LO.

Steele: What about the four body system and its counterterm?

Miller: And are there contact terms in the five, six, etc. body system?

Bedaque: Because of the Pauli principle, the contact term with the highest number of nucleons is the four body contact term ...
Wise: ... assuming that contact terms involving derivatives are suppressed, this is the end of the story.
Bedaque: We know that the two and three body terms are leading, but we have no idea what happens to the four body contact term.
Perry: I believe it would be valuable to look at a toy model for the three body problem, e.g. to look only at the ground state of three bosons. A variational calculation can determine the infinities and tell one how to define a three body counterterm such that the system is stable, giving a lower bound. And we might be able to build the four body system from just an analysis of the renormalization group flow, seeing whether a four body contact term is necessary.
van Kolck: I don't see how this will help. Our three boson system is already very simple, and we don't just want to bound the three body counterterm from below, we have to fix it. The four body system is more complicated: We have a two-body attraction in the two body system at LO, a three body repulsion to stabilize the three body system, and it's not clear that this is repulsive enough to stabilize the four body system, as well.
Kaplan: The three body force gives a quadratic divergence in the nuclear density and stabilizes nuclear matter.
van Kolck: Yes, but we want more than just understand saturation, we want to predict it.
Furnstahl: The Walecka model, i.e. phenomenology, saturates because there is a balance between velocity dependent and velocity independent terms. Clearly, that is something we do not want. On the other hand, the density expansion goes like $\frac{1}{6}$, which is better than the factor $\frac{1}{2}$ or $\frac{1}{3}$ in effective nuclear theory.
Lutz: I have in my model the tensor piece balancing the leading order piece.
Bedaque: If one balances a LO against a NLO contribution, then you don't do something right.
Lutz: Wrong: I can have the case where the expansion converges badly in the first few terms, which hence will be all over the place.
Mehen: Is it conceivable to have saturation only at some higher order?
Kaplan: In my opinion, the cancellation of two different orders against each other means the whole expansion fails and the power counting is wrong. Especially, as Richard says, when one has to balance velocity dependent terms with velocity independent ones.
Lutz: It could be that the next-to-leading order is anomalously large and only when summing it to all orders does one obtain something small.
Kaplan: Yes, but how do you want to achieve that?

Seki: Well, it seems there is no reason for boredom in the coming year.

References

1. D.B. Kaplan, "Effective Field Theory Expansion for Nonrelativistic Scattering", *this volume.*
2. G. Rupak, "Nucleon-Nucleon Scattering in Effective Field Theory", *this volume.*
3. M.J. Savage, "The Two-Nucleon Sector with Effective Field Theory", *this volume.*
4. S.R. Beane, "Gamma-Deuteron Scattering", *this volume.*
5. M. Rho, "On Making Predictions with Effective Field Theory in Nuclear Physics", *this volume*
6. U. van Kolck, "Near-Threshold Pion (Photo) Production: A Success and an Opportunity", *this volume.*
7. P.F. Bedaque, "A Surprising Renormalization on the Way to N-Body Power Counting", *this volume.*
8. H.-W. Hammer, "Effective Theory for Neutron-Deuteron Scattering and the Triton", *this volume.*
9. J.-W. Chen, `nucl-th/9810021`.
10. M. Lutz, "Effective Field Theory For Nuclear Matter", *this volume.*
11. T. Mehen, "Nucleon-Nucleon Effective Field Theory at NNLO: Radiation Pions and $^1S_0$ Phase Shift", *this volume.*
12. J.V. Steele, "Convergence Issues in Nucleon-Nucleon Effective Field Theory", *this volume.*
13. T.D. Cohen, "Perturbative Pions and the Effective Range Expansion", *this volume.*
14. U.-G. Meißner, "Isospin Violation in the Two-Nucleon System", *this volume.*
15. W.C. Haxton, "The Canonical Nuclear Many-Body Problem as a Rigorous Effective Theory", *this volume.*
16. http://www.phys.washington.edu/ mbx/george.html

CONFERENCE SUMMARY

G. P. LEPAGE
Newman Laboratory of Nuclear Studies, Cornell University
Ithaca, NY 14853, USA
E-mail: gpl@mail.lns.cornell.edu

This summary talk reviews recent developments in the ongoing effort to apply the ideas of effective field theory to low-energy nuclear physics. Various topics from the workshop are highlighted.

1 Introduction

The physics of the nucleon-nucleon force is a very old, well developed subject, having played a central role in nuclear physics for much of this century. The traditional conceptual basis for this subject, however, was completely undermined by the discovery that nucleons are not fundamental particles, but rather complicated bound states of quarks and gluons governed by quantum chromodynamics (QCD). The nuclear force was no longer fundamental, but rather a consequence of quark and gluon exchanges between bound states. This development poses a major challenge to nuclear physics: Experiment leaves little doubt that QCD is the correct theory of subnuclear structure, but, unfortunately, even the simplest nuclei are far too complicated to be analyzed directly in QCD. The traditional theory of the nuclear force is tractable, at least for small nuclei, and, given its many successes, must have some truth in it; but what exactly are its limitations, and what is its relation to the true underlying theory, QCD?

Recent years have seen a renaissance of interest in the fundamental nature of the nucleon-nucleon force, and these questions have now been largely answered[1,2,3]. These new developments are the subject of this workshop and of this short review.

2 An NN Primer

The key insight embodied in recent work on NN forces is that a theory of elementary nucleons interacting either through potentials or through meson exchange is necessarily a low-momentum, long-wavelength approximation to QCD. Previously one tried to design nucleon-nucleon interactions that would account for scattering at arbitrarily high momenta. We now realize that this is impossible; QCD is unavoidable at momenta that are sufficiently large to probe and excite the QCD substructure within nucleons. On the other hand, nucleons with momenta p of order $\Lambda_c \equiv 500$–1000 MeV or less, corresponding to wavelengths $\lambda = 2\pi/p$ of 1–2 fm or more, are unable to accurately resolve their own internal quark substructure — hadrons typically have diameters d of 1–2 fm and these can only be resolved with probes for which $\lambda d \ll 1$. Consequently, when analyzing low-momentum collisions, we can mimic the complicated QCD interactions that actually occur with far simpler effective potentials acting between elementary, point-like nucleons; the long-wavelength nucleons involved in the collisions are incapable of resolving the difference between the real theory (QCD) and the effective theory (elementary nucleons plus potentials). Since much of traditional nuclear physics involves nucleons with momenta smaller than Λ_c, the effective theory is a practical alternative to full QCD. The effective theory, however, is never exact, and becomes less so at higher energies; as we will discuss, the errors typically grow as a power of p/Λ_c when p increases. These errors can be predicted and they can be *systematically* reduced.

There is a systematic procedure for designing low-momentum, effective theories that is based on renormalization techniques originally developed for relativistic quantum field theories and statistical field theories. This procedure requires that we treat long-range interactions differently from short-range interactions. To begin, therefore, we must be precise about the meanings of "long-range" and "short-range." We make this distinction quantitative by introducing an ultraviolet momentum cutoff Λ into the effective theory. There are many different ways to impose such a cutoff but in each case the particle momenta in the theory are effectively restricted to be smaller than Λ. Thus there is little or no structure in the theory at distances smaller than $1/\Lambda$; the cutoff determines the ultimate resolution length scale of the effective theory. "Long-range" refers to distances larger than this scale; "short-range" refers to distances that are smaller. The value of the cutoff is tuneable and ideally should be set equal to Λ_c, the scale at which QCD effects appear. The predictions of the effective theory, however, depend only weakly or not at all on the cutoff provided it is of order Λ_c.[a]

Nucleons in the effective theory are necessarily nonrelativistic since the nucleon mass M is larger than $\Lambda \approx \Lambda_c$. Thus the nucleon-nucleon system can be accurately modeled using a nonrelativistic formalism (with relativistic

[a] A procedure for estimating Λ_c is to start with a very low cutoff Λ ($\ll \Lambda_c$) and gradually increase it. The p/Λ errors in predictions of the effective theory shrink rapidly until $\Lambda \approx \Lambda_c$, at which point the errors become constant or might even increase slightly. Thus the point at which the errors stop decreasing determines Λ_c. In many cases the effective theory becomes badly behaved when $\Lambda \gg \Lambda_c$.

corrections as needed): we use the effective hamiltonian

$$H_{\text{eff}} \equiv \frac{-\nabla^2}{M} + V_L + V_S. \tag{1}$$

Here V_L is the long-range potential, and V_S the short-range potential. We have long known that the long-range part of the nucleon-nucleon interaction is dominated by pion exchange, resulting in Yukawa-like potentials,

$$V_L \propto \frac{m_\pi^2}{\Lambda_c^2} \frac{e^{-m_\pi r}}{r} + \cdots. \tag{2}$$

(For simplicity we omit the spin dependence in this formula and in those that follow, below.) Only pions are light enough to contribute to the long-range potential, and their contribution can be computed using standard chiral perturbation theory.

Contributions from the ρ, ϕ, and other heavier mesons are short-range because these particles all have masses of order 1 GeV. While the full short-range interaction is quite complex in QCD, none of the (necessarily low-momentum) nucleons in the effective theory can resolve this complexity. Just as we can model a complex current by a sum of idealized multipole currents, renormalization theory tells us that we can model the short-range potential by a sum of contact interactions — delta functions and derivatives of delta functions. The delta functions must be regulated, so as to restrict momentum transfers in accordance with the cutoff; for example, we can make the replacement

$$\delta^{(3)}(r) \to \delta_\Lambda^{(3)}(r) \equiv \frac{\Lambda^3}{(2\pi)^{3/2}} \exp(-r^2\Lambda^2/2). \tag{3}$$

The resulting short-range potential has the form

$$\begin{aligned} V_S = {} & \frac{c}{\Lambda^2} \delta_\Lambda^{(3)}(r) \\ & + \frac{d_s}{\Lambda^4} \nabla^2 \delta_\Lambda^{(3)}(r) + \frac{d_p}{\Lambda^4} \overleftarrow{\nabla} \delta_\Lambda^{(3)}(r) \overrightarrow{\nabla} \\ & + \frac{e}{\Lambda^6} \nabla^4 \delta_\Lambda^{(3)}(r) + \cdots. \end{aligned} \tag{4}$$

The dimensionless coupling constants c, d_s, d_p, e... are tuned so that V_S mimics the true short-distance behavior; they are analogous to multipole moments. The coupling constants are all of order $\Lambda/M = \mathcal{O}(1)$.

It is important to realize that V_S is not unique. Nor is it even approximately equal to the true short-distance potential. (In fact there is no true short-distance potential for nucleon-nucleon interactions; one needs field theory.) V_S mimics the true short-distance behavior only in its effects on physical quantities such as low-momentum scattering amplitudes, and the binding energies of bound states near threshold. Other V_S's can be designed by using different ultraviolet cutoffs — for example, by using different definitions of δ_Λ. All give the same long-distance physics, while disagreeing completely at short distances.

In principle there are infinitely many terms in the short-range potential, V_S, but in practice only a few are needed to achieve high precision, again in analogy with a multipole expansion. The terms in Eq. (4) for V_S are listed in order of decreasing importance (for a low-momentum process). The key difference between terms is in the number of factors of ∇/Λ; the more factors there are, the less important the term. Each additional ∇/Λ suppresses a term's contribution, after renormalization, by an additional factor of p/Λ, where p is the typical (external) momentum in the process under analysis. The suppression is strong when $p \ll \Lambda$; in such cases the expansion in powers of ∇/Λ converges rapidly, and only the first few terms in V_S need be retained. When p is of order Λ or larger, all terms are equally important, and the effective theory becomes useless — our effective theory is a low-momentum approximation (high momenta require QCD).

The convergence of the effective theory as new terms are added to V_S is illustrated by Fig 1 where experimental results for the $^1S_0$ phase shift are compared with predictions from the effective theory over a range of center-of-mass energies $E = p^2/M$. The plot shows the errors in predictions with no short-range potential, with only the $1/\Lambda^2$ term, and with all terms through $1/\Lambda^4$. At low energies, the errors decrease steadily as each new correction is added to V_S. Generally we expect some sort of improvement when new parameters are added to an approximate theory, but renormalization theory is very specific about the size of the improvement and the number of parameters needed. When we include the $1/\Lambda^4$ terms in V_S, the errors are reduced by an extra power of p^2/Λ^2 where, from the plot, the effective Λ is approximately 700 MeV (as expected *a priori*). The extra factor of $p^2 \propto E$ increases the slope of the curve. Thus, as expected, the largest improvements are at low energies. At high energies there is no improvement, and the effective theory fails to reproduce the data. Similar analyses can be applied to any NN channel. The errors in phase-shift predictions for $^1S_0$, $^1P_1$, and $^3D_2$ scattering, coming from an effective theory corrected through $\mathcal{O}(1/\Lambda^4)$, are shown in Fig. 2; all are roughly equal and have the same slope, as expected.

Predictions for things other than phase shifts also follow from the effective theory. The deuteron binding energy, for example, is 1.5 MeV in the $1/\Lambda^2$

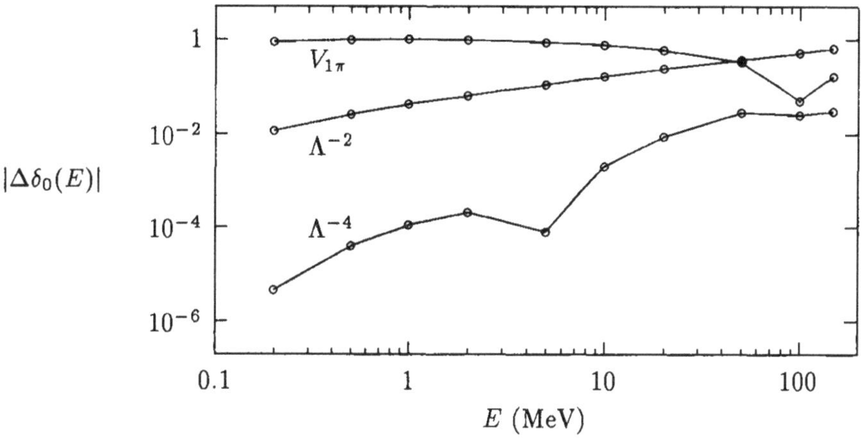

Figure 1: Errors in the nonperturbative $^1S_0$ phase shifts (in radians) versus energy for the effective theory through orders Λ^{-2} and Λ^{-4}. Results from the theory with just pion exchange ($V_{1\pi}$) are also shown.

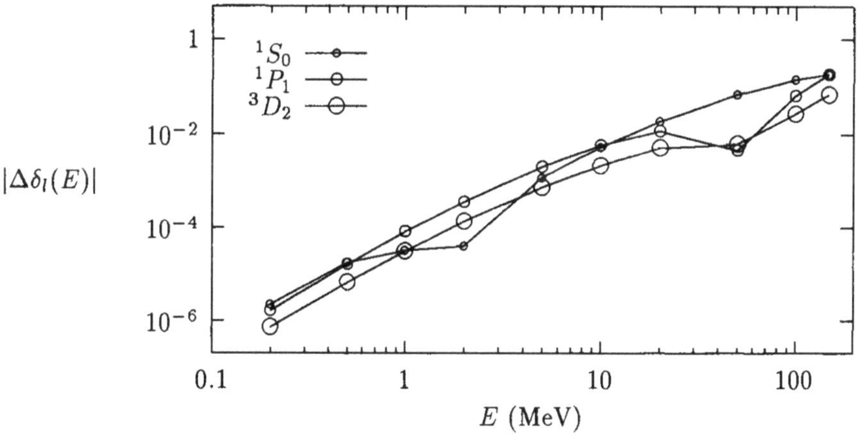

Figure 2: Errors in nonperturbative phase shifts for different channels from the effective theory corrected through order Λ^{-4}.

theory, and 2.17 MeV in the $1/\Lambda^4$ theory (both with $\Lambda = 500\,\text{MeV})^4$; experiment gives 2.22 MeV. In the $1/\Lambda^2$ theory, the deuteron's electric quadrupole moment is .27 fm^2 (experiment gives .29 fm^2), and its charge radius is 1.97 fm (experiment gives 2.11 fm).

Our treatment of the long-range potential, V_L, is very similar to that in traditional nuclear physics. It is our analysis of the short-range potential that is new, and in particular the systematic expansion in p/Λ. The notion that a short-range potential might be expanded in terms of delta functions and derivatives of delta functions seems obvious in Fourier space: The Fourier transform, $\tilde{V}_S(q^2)$, of a short-range potential should be weakly dependent on q since it is short-range, and therefore one is tempted to replace it by the first few terms in its Taylor expansion,

$$\tilde{V}_S(q^2) \approx \tilde{V}_S(0) + q^2 \tilde{V}_S'(0) + \cdots. \tag{5}$$

Transforming back to coordinate space gives a delta-function term and derivatives of the delta function. This approximation would be fine at the level of the (first) Born approximation for low-momentum scatttering amplitudes. But the resulting potential is far too singular for use in full quantum mechanics, and leads inevitably to ultraviolet infinities: Scattering amplitudes and other physical quantities in full quantum mechanics are nonlinear in the potential, and have contributions from virtual intermediate states with arbitrarily high momenta, where the Taylor expansion of $\tilde{V}_S$ is badly behaved.

Our expansion of V_S is different from and more subtle than a Taylor expansion. For example, although predictions of the effective theory converge as higher-order (in $1/\Lambda$) terms are added to V_S, the expectation values of these higher-order terms are typically no smaller than the expectation values of the lower-order terms. This seems contradictory until we realize that the coupling constants of the lower-order terms must be changed, often by 100% or more, when new higher-order terms are added to the potential. The reason is that the nonlinearities in quantum mechanics induce lower-order components in the higher-order terms (and *vice versa*). For example, when we add a $\nabla^2\delta_\Lambda^{(3)}/\Lambda^4$ term to V_S we are also in effect adding some extra $\delta_\Lambda^{(3)}/\Lambda^2$. Consequently the coefficient of the explicit $\delta_\Lambda^{(3)}/\Lambda^2$ term must be adjusted to cancel the extra low-order component implicit in $\nabla^4\delta_\Lambda^{(3)}/\Lambda^4$. The coupling constants in our V_S depend strongly on how many terms are retained, and also on how the cutoff is implemented.

Fortunately it is straightforward to use the effective theory, even with little or no knowledge of renormalization theory. Given the long-range potential, one truncates the expansion of the short-range potential at the desired level of precision. The coupling constants in the potential are then tuned until the effective theory reproduces several pieces of low-momentum data, usually one piece per coupling. For example, the results given above were obtained using an effective theory in which the couplings were tuned to reproduce experimental phase shifts at very low energies (0.05 and 0.1 MeV). The effective Hamiltonian can be analyzed like any other Hamiltonian once the coupling constants are tuned to their correct values. In particular there are no ultraviolet divergences since the cutoff is finite ($\approx \Lambda_c$); and there is no need for further explicit renormalization — tuning the couplings automatically renormalizes the Hamiltonian. The only other complication is that quantum mechanical operators other than the Hamiltonian also have short-distance corrections that are similar in form to those in the Hamiltonian. In particular, currents must be corrected.

3 Issues and Advances

This workshop dealt with a wide range of interesting issues. Here we touch on several of these, with references to the appropriate talks in this volume.

Perturbative versus Nonperturbative: The effective theory is particularly easy to solve numerically since the Hamiltonian is relatively simple and completely analytic at short distances. This puts the theory within easy reach even for novices — undergraduate physics majors, for example, armed only with an introductory course in quantum physics and simple numerical analysis. It also makes for very efficient solutions; most of the earliest work was numerical.

Numerical solutions are usually completely nonperturbative in the potential. Recently analytic solutions have also become possible, thanks in large measure to pioneering work by Kaplan, Savage and Wise (KSW)[2]. These solutions treat the leading contact term in S-wave channels nonperturbatively, which is simple since the interaction is separable; it must be handled nonperturbatively because it forms bound states (the deuteron) or near bound states, which by their nature are nonperturbative (the particles in a bound state interact again and again, forever). All other interactions are included perturbatively. A perturbative approach raises new questions that don't arise in nonperturbative treatments. For example, some quantities ($p \cot \delta$?) might have more convergent expansions than others (δ?); how can we tell which to use? Despite such differences it is clear that both approaches work. Convergence issues are most easily studied by using realistic models to generate synthetic data which is then analyzed using effective theories (see talks by D. Kaplan, J. Steele, and G. Rupak). The results shown in the previous section are all numerical (nonperturbative); comparable analytic results are presented in the talks by T. Mehen and M. Savage.

Power Counting: Several talks, including those by D. Kaplan, M. Wise, and M. Savage, addressed the "power counting" rules for the effective theory, which allow us to predict how much individual terms contribute to a given phase shift, binding energy, etc. These are now understood and codified as the KSW rules. These rules are applied either directly to the effective Hamiltonian, in the case of numerical solutions, or to individual terms in the perturbative expansions used for analytic solutions. The rule is quite simple for most channels in NN scattering: the long-range potential, V_L, dominates, while contributions from the contact terms in V_S are suppressed by powers of p/Λ, just as discussed in the previous section.

The S-wave channel, however, is slightly more complicated. The dominant interaction is the the first contact term in V_S; pion exchange (V_L) is nonleading because S-wave scattering is strongly affected by short-range interactions, having no centrifugal barrier, and pion exchange is substantially weaker at short distances than the contact term. Higher-order contact terms are suppressed by powers of p/Λ relative to the leading order term, just as discussed in the previous section. However they are made relatively more important in this channel because the bulk of the contribution from the leading-order potential cancels that of the kinetic energy. This delicate cancellation, which is purely accidental, is what leads to the tiny 2 MeV binding energy of the deuteron. Because of the cancellation, contributions from higher-order contact terms are all larger by a single power of Λ/p, relative to the leading-order term, than one might have expected otherwise.

Perturbative Pions: Long-range pion exchange, V_L, is a small perturbation in S-wave NN scattering. Furthermore the interaction is indistinguishable from a contact term for momenta $p \ll m_\pi$, where m_π is the mass of the pion. Consequently an effective theory with no long-range interaction accounts very nicely for S-wave scattering; such a theory is essentially the same as traditional effective-range theory for scattering amplitudes. Adding in the long-range potential should result in a theory that is more accurate than effective-range theory.

T. Cohen's talk suggests that the expansion in powers of V_L is slowly convergent, at best, and that pion-exchange does not improve the theory. His analysis is based on the analytic perturbative expansion. A nonperturbative analysis of the theory leads to different conclusions, as illustrated in Fig. 3[3]. This figure shows errors in the $^1S_0$ phase shift for different energies as pre-

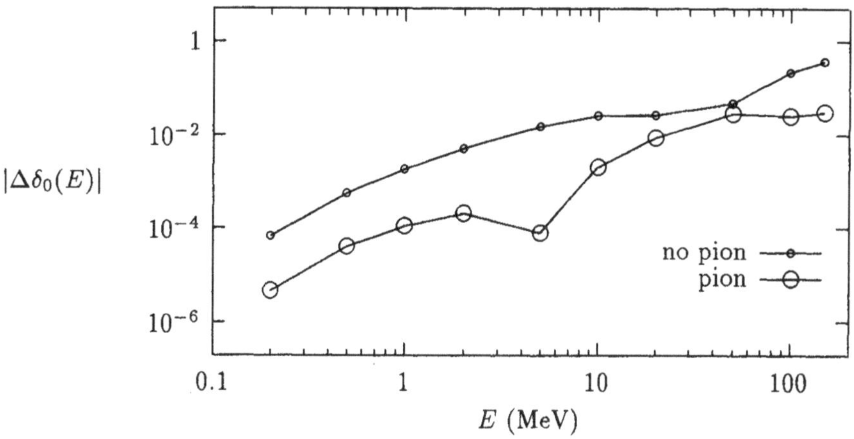

Figure 3: Errors in the nonperturbative $^1S_0$ phase shifts (in radians) versus energy from the effective theory (through $1/\Lambda^4$) with and without pion exchange.

dicted by theories without pion exchange (top curve) and with pion exchange (bottom curve). Pion exchange improves the predictions by about an order of magnitude for energies below about 100 MeV.

Pion Retardation: The exchanged pions in low-momentum exchanges violate energy conservation by $\Delta E \approx m_\pi$. When this is much larger than the kinetic energies of the colliding nucleons, the time required for the pion exchange cannot be resolved by the nucleons. Consequently the interaction is effectively instantaneous, and can be modeled by an instantaneous potential (that is, V_L). This, coupled with the fact that soft pions have a vanishing coupling to nucleons, means that pion retardation is unimportant below the pion-production threshold (at $p = \sqrt{Mm_\pi} \approx 360\,\text{MeV}$). Explict calculations of pion retardation effects were described in T. Mehen's talk; the effects are indeed small.

Beyond Phase Shifts: NN phase shifts in many channels have been thoroughly studied and are quite well understood now. Several talks dealt with the many other quantities that can be measured and also predicted using the effective theory. In M. Savage's talk results are summarized for deuteron static properties, like the binding energy, the electric quadrupole moment, the magnetic moment, and the electric charge radius, as well as for deuteron form factors, polarizabilities, and Compton scattering. Deuteron Compton scattering, pion photoproduction, $np \rightarrow \gamma d$, and pion production in NN scattering are discussed in the talks by S. Beane, B. van Kolck and M. Rho. Most work mentioned above neglects the tiny differences in nuclear properties between neutrons and protons. U. Meissner's talk discusses how such isospin violations are analyzed in the effective theory.

This very wide range of topics is the current forefront in work on the NN system. It demonstrates the versatility of the effective theory. The calculations reported in these talks are all perturbative; comparable nonperturbative studies would be interesting.

Three-Body Problem: Impressive work on three-body systems is discussed in talks by P. Bedaque and W. Hammer. This provides a first look at the role of three-body contact terms in the effective theory. These analyses reduce the problem to an two-body problem by introducing an elementary interpolating field for the deuteron. Again, it would be interesting to compare with a completely numerical study of the full three-body problem using the two-body effective potentials discussed above.

Multi-Nucleon Problem: W. Haxton's talk outlines an ambitious plan to design a more rigorous shell model for multi-nucleon problems. Effective theory techniques enter because the basis sets used to describe multiparticle states must be truncated at short wavelengths to make the analysis tractable. Renormalization theory explains how to modify the Hamiltonian so as to compensate for such truncations. M. Lutz discusses effective theory and nuclear matter in his talk.

4 Conclusions

Many of the basic tools and concepts are now in place for a systematic redesign of low-energy nuclear physics. Much of the current effort involves a recasting old ideas, but much is also new. The new techniques provide a rigorous, systematic framework for computing the effects of the nucleon-nucleon force, and also for relating it to QCD (for example, by tuning the coupling constants in the effective Hamiltonian to reproduce lattice QCD simulations). As is clear from this meeting the effective theory provides an excellent description of NN scattering. It will be interesting to see what impact it has on our understanding of multi-nucleon systems and nuclear matter.

Acknowledgements

This work was supported by a grant from the National Science Foundation.

References

1. For early work see, for example: S. Weinberg, *Phys. Lett.* B **251**, 288 (1990); *Nucl. Phys.* B **363**, 3 (1991); *Phys. Lett.* B **295**, 114 (1992). For a more thorough study see: C. Ordóñez, L. Ray, and U. van Kolck, *Phys. Rev.* C **53**, 2086 (1996).
2. For a modern treatment of the perturbative approach to NN interactions see: D.B. Kaplan, M.J. Savage and M.B. Wise, *Nucl. Phys.* B **534**, 329 (1998); *Phys. Lett.* B **424**, 390 (1998); *Nucl. Phys.* B **478**, 629 (1996).
3. For a pedagogical discussion of renormalization theory, perturbative and nonperturbative, and its application to the NN problems see: G.P. Lepage, "How to renormalize the Schrodinger equation," `nucl-th/9706029`.
4. G.P. Lepage and B. Sabbey, preliminary nonperturbative results.

Discussion during P. Lepage's presentation

Chair: *R. Seki (CSU Northridge and Caltech)*
Notes taken by: *H. Grießhammer (University of Washington)*

Mehen: What are the chances of matching the coefficients in effective nuclear theory to QCD?
Lepage: We can do it. I can give you C_0 from the lattice, but it is hard to get somebody to do it, because it's just a waste of time, and far too boring. Of course, there is the problem that because there are no asymptotic states on the lattice, we cannot compare scattering amplitudes but must compare to EFT computations using standing wave boundary conditions. On the other hand, we know the result for C_0, and we know it has to come out correctly. We will still have learned nothing about why the deuteron binding is so small. All that we will see is that it works: "Gee, what an amazing cancellation!"
Savage: I still would love to see that.
(General agreement in the audience.)
Cohen: The deuteron is a very big object with a very small binding energy and hence quite a challenge on the lattice.
Meißner: The lattice can be used to obtain counterterms which are very hard to determine from experiments, like in the pion sector. There it is extremely useful, and it would be very much worth while. But otherwise ...
Furnstahl: The lattice allows one to dial the pion mass, and this way we can see the fine tuning for the deuteron binding energy.
Lepage: Yes, that is another very useful application.
Kaplan: Because of fine tuning, the fixed point in the two nucleon system is non-trivial. But you suggested we are at a Gaußian one.
Lepage: We are so near the Gaußian one that it does not matter. I believe there is not much value in discussing nuclear theory using fixed points and renormalization group arguments, when we only choose our cutoff right. After all, we know the theory is *not* at its fixed point. What I want to say is that there is another grouping of diagrams besides the KSW one: Counting powers of the cutoff corresponds to the Gaußian fixed point, as M. Birse pointed out in yesterday's discussion. As a rule, it is fairly simple but includes terms which are possibly small. Still, I see it as the best way to proceed. I see renormalization group equations etc. as far less central to the issues addressed here.
Kaplan: What is your value for the cutoff, where does the theory break down?
Lepage: I am hesitant to answer this, because I played only a bit with the cutoff, calibrating my expectations by comparing results with all one pion physics taken into account against results without pions. At the moment, I tend to set breakdown of the EFT at about 500 or 600MeV. With this value,

I see the binding energy of the deuteron off by 5% when I fix the scattering length to the experimental value.
van Kolck: But your analysis has not yet included two-pion exchange in the potential.
Lepage: That is right. One has to calculate the two-pion-exchange potential.
van Kolck: It has already been calculated and published.
Lepage: I know... Your expression is too long, but I will include it.

PARTICIPANTS

M. Alberg, *University of Washington, USA*
S. Beane, *University of Maryland, USA*
P. Bedaque, *National Institute for Nuclear Theory, USA*
M. Birse, *University of Manchester, UK*
A. Bulgac, *University of Washington, USA*
M. Butler, *St. Mary's University, Canada*
J. Chen, *University of Washington, USA*
T. Cohen, *University of Maryland, USA*
H. Fearing, *TRIUMF, Canada*
S. Fleming, *University of Toronto, Canada*
R. Furnstahl, *Ohio State University, USA*
J. Gegelia, *The Flinders University of South Australia, Australia*
H. Griesshammer, *University of Washington, USA*
H. Hammer, *TRIUMF, Canada*
J. Hansen, *University of Maryland, USA*
C. Hanhart, *University of Washington, USA*
W. Haxton, *National Institute for Nuclear Theory, USA*
T. Hemmert, *Institut fur Kernphysik, Juelich, Germany*
E. Henley, *University of Washington, USA*
B. Holstein, *University of Massachusetts, USA*
D. Kaplan, *National Institute for Nuclear Theory, USA*
B. Keister, *National Science Foundation, USA*
X. Kong, *University of Washington, USA*
P. Lepage,*Cornell University, USA*
M. Luke, *University of Toronto, Canada*
M. Lutz, *GSI, Germany*
J. McGovern, *University of Manchester, UK*
T. Mehen, *California Institute of Technology, USA*
U. Meissner, *Institut fur Kernphysik, Juelich, Germany*
G. Miller, *University of Washington, USA*
D. Min, *Seoul University, South Korea*
T. Pappenbrock, *National Institute for Nuclear Theory, USA*
A. Parreno, *National Institute for Nuclear Theory, USA*
T. Park, *Korea Institute of Advanced Study, South Korea*
R. Perry, *Ohio State University, USA*
D. Phillips, *University of Washington, USA*
S. Puglia, *University of Connecticut, USA*
F. Ravndal, *University of Washington, USA*

M. Rho, *Service de Physique Théorique, Saclay, France*
S. Reddy, *National Institute for Nuclear Theory, USA*
G. Rupak, *University of Washington, USA*
M. Savage, *University of Washington, USA*
K. Scaldeferri, *California Institute of Technology, USA*
R. Seki, *CSU Northridge and California Institute of Technology, USA*
N. Shoresh, *University of Washington, USA*
R. Springer, *Duke University, USA*
J. Steele, *Ohio State University, USA*
I. Stewart, *California Institute of Technology, USA*
U. van Kolck, *California Institute of Technology, USA*
M. Wise, *California Institute of Technology, USA*

www.ingramcontent.com/pod-product-compliance
Lightning Source LLC
Chambersburg PA
CBHW042031200926
51615CB00028B/4
* 9 7 8 9 8 1 0 2 4 1 8 1 0 *